Sustainable Regional Planning
http://dx.doi.org/10.5772/intechopen.105225
Edited by Amjad Almusaed and Asaad Almssad

Contributors
Souporni Paul, Suchandra Bardhan, Sankeerthana Ananthula, Haider Jasim Essa Al-Saaidy, Banu Chitra Mookiah, Barakat Tiamiyu, Nancy Odendaal, Kazutoshi Fujihira, Sérgio Lousada, José Manuel Naranjo Gómez, Luís Loures, Constantin Vîlcu, Steve Vanlanduit, Resul Kara, Aurora Ranca, Mohamed Bourouah, Kassahun Gashu, Gobena Feyisa, Vicent Mbonye, Samuel Danjuma Wapwera, Jorge Carro-Suárez, Susana Sarmiento-Paredes, Doroteo Nava, Abdulkarim Hasan Rashed, Mieczysław Adamowicz, Amjad Zaki Almusaed, Asaad Almssad

First published in London, United Kingdom, 2023 by IntechOpen
IntechOpen is the global imprint of INTECHOPEN LIMITED, registered in England and Wales, registration number: 11086078, 5 Princes Gate Court, London, SW7 2QJ, United Kingdom

British Library Cataloguing-in-Publication Data
A catalogue record for this book is available from the British Library

Additional hard and PDF copies can be obtained from orders@intechopen.com

Sustainable Regional Planning
Edited by Amjad Almusaed and Asaad Almssad
p. cm.
Print ISBN 978-1-80356-050-2
Online ISBN 978-1-80356-051-9
eBook (PDF) ISBN 978-1-80356-052-6

Sustainable Regional Planning

Edited by Amjad Almusaed and Asaad Almssad

Published in London, United Kingdom

Meet the editors

Prof. Amjad Almusaed has a Ph.D. in Architecture (Environmental Design) from Ion Mincu University, Bucharest, Romania. He completed postdoctoral research on sustainable and bioclimatic houses at the School of Architecture, Aarhus, Denmark in 2004. His academic focus is on the role of sustainability in building design and city planning. He has done extensive study, research, and technical surveying in these fields. He is an active member of numerous international architectural organizations and has edited several foreign books. His bibliography includes around 185 worldwide academic publications (articles, studies, books, and chapters) written in various languages.

Associate Professor Asaad Almssad has worked in industry, academia, and research for more than 30 years. He has held positions at Umea University, Sweden; Karlstad University, Sweden; and a variety of other European and non-European institutions. His studies concentrate on building materials, structures, environmentally responsible construction, and the energy efficiency of building systems. He has more than sixty academic papers and numerous books to his credit. Currently, he is employed as a docent at Karlstad University, Sweden.

Contents

Preface

There is a broad consensus regarding the definition of "sustainable development" globally, with no significant contentions. Sustainable development simultaneously pursues economic growth and fulfills present and future generations' needs. This multifaceted endeavor entails the coordination of productive capacities, the equitable provision of essential social services, and the safeguarding of the natural environment. However, achieving a consensus on operationalizing sustainable development within specific regions and the necessary actions to be undertaken at the regional level presents a more complex challenge. Consequently, the notion of sustainable development in the global context is primarily theoretical, needing a well-established underlying methodology. A comprehensive set of indicators must accompany sustainable development to be practically meaningful. These indicators serve as valuable tools for establishing long-term targets and tracking progress. Although existing indicators may monitor shifts in sustainable development trajectory, more are needed to precisely identify the specific measures required to realize sustainable development at the regional level. Therefore, it is imperative to develop clearer indicators that local authorities can utilize to effectively monitor sustainability within their respective jurisdictions. Such indicators should be tailored to the specific mandate and context of the area in question.

Regional planning is a type of planning that engages the synthesized physical, political, cultural, and economic interests of a region. Sustainable regional planning calls for an integrated strategy that considers the interdependencies across many sectors, such as land use, transportation, housing, energy, and the management of natural resources. Integrating planning helps to enhance collaboration and cooperation among many stakeholders, which fosters the development of comprehensive solutions to complex regional concerns. Sustainable and regional development has come to the forefront as essential ideas in the later decades of the second century. This is a direct result of increasing awareness of the urgent need to solve various social and environmental issues. These concepts have received significant momentum in the academic and policy debate, with academics and practitioners underlining their crucial role in determining humanity's destiny. Although there may be a gap in time between the original birth of these notions and the subsequent refining of our understanding of them, the relevance of these ideas cannot be emphasized. For both the present generation and the generations to come, it is of the utmost significance to arrive at a precise definition of sustainable regional development and to put that concept into practice effectively. It is generally accepted that to progress toward a sustainable future, there must be an all-encompassing and well-coordinated effort worldwide. Several schools of thought support adopting an integrated and contemporary strategy, emphasizing environmental, social, and economic connectivity. A holistic viewpoint like this one acknowledges that sustainable development cannot be broken down into its parts; instead, it calls for complementary and concordant involvement across several different locations and industries. Despite this, achieving sustainable development on a global scale will not be without its share of difficulties. Due to the many complex interdependencies at play, it is necessary to use targeted, practical solutions firmly rooted in scientific facts. Academics and industry professionals are making concerted efforts to investigate novel approaches and workable solutions that have the potential to bring about good change in specific fields, such as ecological agriculture and ecological tourism. These initiatives at the sectoral level serve as examples of the revolutionary potential inherent in a new paradigm of development that incorporates sustainability principles into the fundamental aspects of economic and social systems. In addition, it is becoming widely acknowledged that regional settings offer fertile ground for implementing sustainable development programs. Because regional structures often demonstrate better degrees of flexibility and a more robust openness to adopt practical solutions, these are frequently vitally essential catalysts for change. The European continent, with its many regions, offers a one-of-a-kind chance to make strides in the direction of regional and sustainable development. When seen in this light, the Scandinavian region is a beautiful location for cultivating sustainable practices. This is due to the region's forward-thinking government, socio-environmental awareness, and well-established infrastructure. Europe, especially the Scandinavian area, has the potential to establish itself as a worldwide leader in sustainable practices if it adopts sustainable development on a regional scale and sees it through to completion. Not only can such a strategic approach boost competitiveness, but it also helps create resilience in the face of global concerns like climate change and the depletion of resources. The efforts at the regional level work as testbeds for new ideas, paving the way for the improvement of sustainable development plans and discovering best practices that can be expanded upon and copied in other locations.

Sustainable and regional development are two fundamental ideas that have come to the fore in recent years due to their growing popularity. The accuracy with which they are defined, the efficiency with which they are implemented, and the degree of strategic integration at which they are included on a global and regional scale all bear enormous consequences for the future of humanity. Societies may pave the way toward a sustainable future by adopting a comprehensive and sector-level approach rooted in practicability and guided by empirical data. This will ensure the health and prosperity of the current and future generations. The fundamental tenets of sustainable regional planning concentrate on integrated planning and decision-making, social fairness and inclusion, environmental conservation and resource efficiency, and resiliency and adaptation. Interdependencies across different industries are considered during integrated planning, which works toward developing comprehensive solutions to complex regional problems. The primary focuses of environmental conservation and resource efficiency are preserving natural resources, protecting ecosystems, and reducing waste output via sustainable methods. Social equity and inclusion guarantee that everyone in an area has equal access to resources, services, and opportunities. The goal is to minimize inequalities and improve general well-being. In conclusion, resilience and adaptation both acknowledge the need to enhance the region's capacity to resist and recover from shocks and stressors, particularly climate change. This capacity includes the ability to endure and recover from natural disasters.

To achieve multidimensional economic growth that is in line with national imperatives, one must have a comprehensive awareness and grasp of the many components that contribute to the achievement of this goal. In this context, adopting a strategy that recognizes and elucidates the presumed role of human capital in preserving environmental integrity becomes an absolute need, especially in notoriously cutthroat competition settings. Sustainable regional planning is a multifaceted and strategic method that aims to integrate economic growth, social fairness, and environmental protection within a particular geographic area. This technique was developed by the Institute of Transportation and Development Policy (ITDP). It acknowledges the complex interdependencies between several industries, including transportation, land use, energy, and the management of natural resources, and it seeks to strike a balance that will maintain the health and happiness of both the present generation and the generations to come. Sustainable regional planning aims to solve urgent environmental concerns while fostering economic expansion and social inclusion. This is accomplished by incorporating sustainability concepts into the procedures that comprise sustainable regional planning. The need for sustainable regional planning has become more evident as the globe confronts tremendous difficulties in recent years, such as climate change, resource depletion, urbanization, and socioeconomic inequities. Traditional development methods often prioritize short-term benefits over long-term sustainability, which may negatively affect society and the natural environment. However, sustainable regional planning provides an innovative alternative by emphasizing integrating environmental issues, social equality, and economic viability. This kind of planning is said to be transformational. The inclusion of sustainable regional development within the planning framework is a fundamental imperative in the pursuit of long-term socioeconomic and environmental objectives. This necessitates striking a delicate equilibrium between fostering economic growth and safeguarding the integrity of the natural environment. A comprehensive approach is deemed essential, wherein the interdependencies among diverse regional sectors and stakeholders are duly considered and harmonized.

The present book provides a comprehensive overview of sustainable regional planning, encompassing its foundational principles and the challenges and prospects associated with its practical implementation. Moreover, it delves into the integration of sustainability concepts within regional planning processes while elucidating the roles played by various actors in advancing sustainable development at the regional scale. Within the context of sustainability, this book explains urban and regional planning principles and practices. Its publication signifies an emerging recognition within planning and environmental studies that the construction of cities, where most of the global population currently resides, must embody sustainability. This acknowledgment is underscored by the significant urbanization trend witnessed worldwide. The work interweaves the concepts of planning, development, and environmental conservation, offering a comprehensive perspective. Students and planning practitioners are equipped with a framework that effectively bridges the gap between environmental considerations and economic aspirations. This book exemplifies sustainable urban planning practices in the outskirts of urban areas, encompassing diverse scales and various planning modalities. Its central objective is to advance the progress of regional planning through a sustainability-focused strategy. The proposed approach embraces a participatory framework, drawing inspiration from and driven by social, cultural, perceptual, sustainability, and economic considerations. Consequently, the plan addresses the conflict between preserving peri-urban areas' inherent natural and cultural attributes and the pressing imperative to allocate land for public utilization. This strategic approach is manifested in the book's organizational structure, which is comprised of three distinct sections: "Principles of Sustainable Urban and Regional Planning", "Case Studies", and "Challenges and Future Directions".

Amjad Almusaed
Department of Construction Engineering and Lighting Science,
Jönköping University,
Jönköping, Sweden

Asaad Almssad
Faculty of Health, Natural Sciences, and Technology,
Department of Engineering and Chemical Sciences Construction Technology,
Karlstad University,
Karlstad, Sweden

Section 1

Principles of Sustainable Urban and Regional Planning

Chapter 1

Introductory Chapter: Sustainable Development and Regional Planning Strategies

Amjad Almusaed and Asaad Almssad

1. Introduction

1.1 Sustainable development challenges

Regional planning entails formulating and enforcing strategies for a region's economic and social growth; it is typically a subset of a larger national plan. As a type of official engagement in the economy, regional planning aims to reduce glaring inequalities and social tensions in certain areas (depression, stagnation of certain regions, excessive concentration of production and population in other areas, regional unemployment, etc.) [1]. In most parts of the world, strategic regional planning is seen as a novel approach to enhance the effectiveness of public administration of regional development. Within these shifts, strategic planning became increasingly important, primarily due to the expansion of the environmental complex and the planning for sustainable growth, but also because of the requirement for increased safety in both markets and states [2]. The exhaustion of the previous state policy of regional development of the country, which was based on the ideas of reducing differences and equalizing the levels of social and economic development of the regions, is the critical factor that determines the need to improve and increase the efficiency of public administration in this area. This factor enhances and increases public administration's efficiency in this area. The notions of sustainable development and regional planning are relatively recent, having emerged in the latter decades of the second millennium. There is consensus that a holistic and integrated approach to economic, social, and environmental processes is necessary for sustainable development, which is why the World Commission on Environment and Development established the idea. Sustainable development emerged from the United Nations Organization on the Human Environment's 1972 Stockholm Conference. Through science diplomacy, this endeavor examines Sweden's participation in the UN in 1967 and 1968, culminating in the landmark UN Conference on the Human Environment in 1972 [3]. In the 60s of the previous centuries, certain industrialized nations of the EU, the USA, Canada, Japan, and other countries began successfully implementing a technique known as regional development [4]. This policy came into existence not because of the imposition of funds, as is currently the case in the case of EU candidate states, but rather because, throughout history, the regions have evolved and developed in different ways and at different rates; this is why we are currently witnessing a very heterogeneous landscape in terms of their respective levels of development [5, 6]. Once counties select the most viable and sustainable economic growth pattern, as indicated in **Figure 1**, regional planning plans offer the framework around which policies and choices should be made.

The issue of utilizing strategic planning to achieve environmentally responsible growth in the region is considered. It has been demonstrated that the outcome of any program intended to ensure socioeconomic security ought to be predicated on the overarching principle of making the transition to sustainable development; conversely, restructuring the economy necessitates enhancing the function of state regulation of the method of achieving sustainable socioeconomic development as a system of strategic planning. European policymakers have prioritized sustainability since the 1992 United Nations Conference on Environment and Development in Rio de Janeiro, where Agenda 21 was ratified [7]. In September 2015, during the Paris Climate Summit, the 193 member nations of the United Nations accepted the new 2030 Agenda, which includes the Sustainable Development Goals (SDGs). Despite this, international cooperation is still needed 5 years forward to fulfill all these promises. New European regulations aim to facilitate this transition from a linear to a circular economic model, and local public and private actors are urged to strive toward these ends. In recent years, the European Union has used the SDGs as part of its Cohesion Policy, which aims to achieve economic, social, and territorial development by reducing regional disparities.

1.2 Regional planning in the context of sustainability

The regional planning and sustainable development plan, also known as the regional planning and development plan (SRADDT), was formerly known as the regional planning and development plan (SRADT) [8]. This document details the fundamental and medium-term guidelines of the sustainable development of a restricted territory as well as its development principles. It is an instructive illustration of the local implementation of sustainable development ideas as they have been developed since the 1980s, most notably by the Brundtland report. European policymakers have prioritized sustainability since the 1992 United Nations Conference on Environment and Development in Rio de Janeiro, where Agenda 21 was ratified. In September 2015, during the Paris Climate Summit, the 193 member nations of the United Nations accepted the new 2030 Agenda, which includes the Sustainable Development Goals (SDGs) [9]. Despite this, 5 years forward, international cooperation is still needed to fulfill all these promises. New European regulations aim to facilitate this transition from a linear to a circular economic model, and local public and private actors are

The basis of strategic sustainable regional planning	Activating land use planning
	Support for a high level of sustainable transport
	Integrates objective strategies for protecting natural resources
	Strong promotion of social equity
	A practical approach to conceiving and managing a region's physical, economic, and social development
	A clear strategy to integrate the different sectors and disciplines
	Efficient uses and application of data and technology
	Promoting resilient communities
	An ongoing process should be revised and updated regularly to reflect the changes
	A global approach to conceive and manage a region's physical, e, economic and social development.

Figure 1.
The bases of strategic regional planning in the modern city.

urged to strive toward these ends. In recent years, the European Union has used the SDGs as part of its Cohesion Policy, which aims to achieve economic, social, and territorial development by reducing regional disparities [8]. Despite taking place against the backdrop of a disagreement between the North, the affluent countries, and the South, the Third Globe countries, this was the first time the governments of the world had examined the pollution problems.

When developing areas' socioeconomic development strategies, current methodologies for producing successful decisions based on strategic planning are utilized to facilitate this process, the information foundation upon which strategic planning is based indicates socioeconomic growth that may be sustained over time. The current climate demands a greater emphasis on sustainable development analysis than ever before. This is due to the increased influence of external and internal variables that contribute to economic instability in the area. Because the quality of the analysis of the sustainable development of regions (considering the factors of the external and internal environment and their assessment) determines the adequacy of the strategic priorities formed during planning, the problem discussed in the article is relevant [10, 11]. The future of humanity depends on progress, and the precision of the definition and the pace with which these concepts are practically implemented are crucial for our generations, even if there is a lag of many decades between their inception and deepening. Many people think a worldwide effort must be made to achieve sustainable development. However, targeted, more realistic methods with higher success rates are required due to the challenges involved. Consequently, current practices regarding the largely positive evolution of fields like ecological agriculture, ecological tourism, demonstrate the viability of a new type of approach at the sectoral level, which can and even must, in our opinion, become a natural perspective on development in the world [12]. The following figure illustrates the connection between the selected fields: Natural resources, discussed in the other guides, are "incorporated" into strategic planning *via* its territorial administrative component, which is another way of saying that planning for sustainable development "incorporates" natural resources. The sustainable utilization of these resources necessitates planning a sectoral strategy (**Figure 2**).

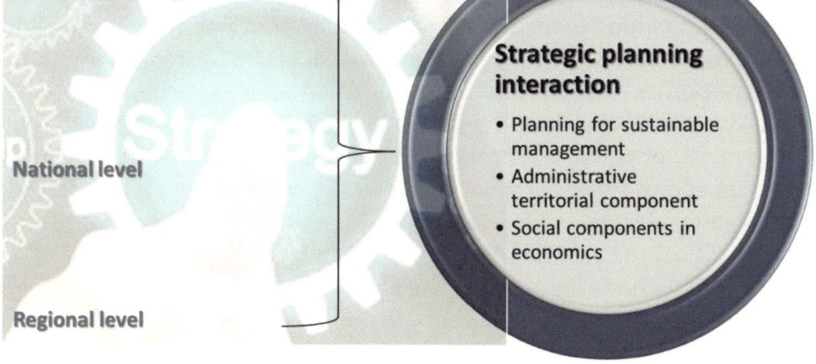

Figure 2.
The planning interaction and components.

1.3 Regional spatial planning

Adapting to a new, today's regional spatial planning is a waste of time. Given that the area's topography is the most significant resource in the region, we need to determine why we should nurture them and how we should do so. In addition, the utilization of a variety of conceptual frameworks helps to bolster an ecological perspective and foster the development of anticipated results, most notably ecosystem services and GI. Co-creation of new solutions for public and private green spaces is encouraged *via* the development of relevant competencies and the use of collaborative working methodologies. However, institutional circumstances such as legal frameworks and landownership may affect the planning process and make it more difficult to improve social and ecological aspects. These institutional conditions impact the planning process [13]. The phenomena responsible for land degradation discussed in the literature on strategic spatial planning are the primary focus of this study. A secondary emphasis is placed on the processes that describe the function of strategic spatial planning in preventing land degradation [14]. Spatial planning has always been a source of contention. It is considered useless, causing chaos and ruin to the landscape, bureaucratic, reliant on painfully slow procedures, even fostering corruption, and so on. It should be noted, however, that nearly all complaints and repair activities involve so-called—local planning, that is, at the commune level [15, 16]. Neither theory nor practice successfully integrated regional and local planning. There was no political will to address the issue.

2. The interaction of regional sustainability planning

The importance of maintaining conditions that are conducive to long-term life is something that must be considered. Unfortunately, the present trend is to implement regulations that prioritize economic benefits over environmental ones, social advantages over environmental ones, or ecological benefits over economic ones [17]. These tendencies are counterproductive, and it will be necessary for societies to find a middle ground in their future growth that considers the interconnectedness of economic, social, and environmental progress. From this vantage point, the previous several decades have been "helpful" in defining new conceptions of global evolution that give rise to optimism for society's sustainable growth and progress. The core of the new is currently found in ideas such as sustainable development and regional development. Most nations' rural and urban areas are transforming, necessitating institutional responses from the relevant authorities. Many governments worldwide have adopted a regional approach to planning, which has aided in reducing spatial inequities in social, physical, and economic development [18]. Plans at the regional level attempt to address issues on a national scale by considering economic, environmental, and spatial objectives. One of the keys to making the necessary shift is doing an integrated critical analysis and developing meaningful links. The political-administrative control structures hardly correlate with the functional interrelationships and requirements of densely inhabited regions, making metropolitan regions challenging for politics, planning, and land-use management. The divergence of administrative borders of areas and their subspaces from their practical settings is one of the primary obstacles [19]. Regional planning focuses on the area's regional policy instead of municipal planning, where land use plans are already developed. Local, regional, and national objectives and criteria are established. When urbanization begins to have a widespread impact, regional plans become obligatory to prevent the waste and underutilization of resources that may otherwise follow from such an occurrence.

Numerous institutions and academics have come to various conclusions about what constitutes effective urban planning and development from the point of view of environmentally responsible regional planning. They have put out several city concepts over time, including garden, ecological, green, and livable cities. Consequently, the growth process incorporates variable degrees of theory and practice [20]. The World Health Organization (WHO, 1992) argued that sustainable urban development should be based on limiting resource usage and that the urban economy should grow in a productive, stable, and creative manner. This proposition was made from a theoretical point of view. Oren Yiftachel and David Hedgcock (1993) proposed that sustainable regional planning should aim for a city that has significantly advanced human interaction, information dissemination, and culture in terms of society. This society is characterized by vitality, stability, and fairness and is free of crime [21]. According to Nijkamp (1994), cities should give full play to their potential, pursue a certain amount of high-quality economic, social, population, and technical output, maintain their stability, and consolidate their status and role in the urban system for a long time. These goals should be accomplished while maintaining their strength [22].

3. Sustainability of urban and regional planning challenges and opportunities

3.1 15-minute city (FMC)

Many academics, urban and regional planners, and policymakers support the benefits of adopting the 15-Minute City (FMC) in major cities worldwide. However, despite the attention generated by the concept and various forms of chrono-urbanism, only a few studies have examined cities from the FMC perspective. This is even though the idea has generated considerable interest. The 15-minute city (FMC) model is a relatively new method of thinking about urban planning and policy-oriented on the city's human size and experience. It is used for urban development and planning. The central idea behind it is that cities ought to be planned so that residents can fulfill all of their day-to-day requirements work, home, food, health, education, culture, sports, and leisure—within a 15-minute walk or cycling distance from where they live [23]. This is the central idea behind it. To this end, it gives prominence to the neighborhood as the essential element of spatial and functional organization. It argues that cities should be organized into areas within which any need should be satisfied within a 15-minute walk or bike ride away. This is accomplished by giving prominence to the neighborhood as the essential spatial and functional organization element. The urban ambiguous planning strategy also hurts the attainment of the 15-Minute City (FMC), a crucial community unit that applies planning and design principles to ensure the safety, sustainability, resilience, and inclusiveness required by the objective of Sustainable Cities and Communities [24].

3.2 Affordable social equity-promoting activities

The environmental, economic, and social concerns reintroduced in the European Union Agendas are again the subject of intense debate. Social housing (SH) policies and initiatives are crucial to urban revitalization [25]. These interventions blend architectural, urban, and environmental excellence with sustainability requirements [26]. Affordable housing, community development, and easy access to services and facilities are just a few examples of the kind of social equity-promoting activities that are common in sustainable regional planning. This is crucial because it ensures that people of all socioeconomic backgrounds can participate and reap the benefits of the region's progress [27, 28].

3.3 Intelligence in urban and regional planning

Today, research needs to concentrate on urban growth management as a strategy for creating sustainable communities. This will allow for the creation of new communities that provide higher living standards while also addressing the challenges that exist within the existing urban mass [29]. This can be accomplished by developing new policies limiting and controlling urban growth. Intelligent urbanism is a concept in urban space planning that seeks solutions to various challenges associated with urban development. It is founded on 10 principles, which are as follows: compatibility with nature, harmony with tradition, appropriate technology, friendliness, the efficiency of infrastructure, human scale, opportunity system, regional integration, balanced movement, and institutional integrity [30].

3.4 Land-use challenges

The process of increasing the amount of land that is used in urban areas is commonly referred to as urbanization. The conventional definition of urbanization considers "land-use change," which refers to the transition from dispersed "exploitation" of a resource to more concentrated land-use practices. Utilizing land-use planning as a compass to direct growth is essential to sustainable regional planning. Creating a vision for the region, determining which regions should experience growth and which should be preserved, and then formulating laws and regulations to direct development in those areas are all part of this process. Plans for land use may, for instance, identify specific areas as ideal for commercial or industrial development while conserving other locations for agricultural or natural usage. Programs may also designate certain areas as suitable for both types of development. This strategy can assist in guaranteeing that growth is concentrated in areas of the region where it will have the least negative influence on the surrounding environment and where it will have the most potential to benefit both the local economy and the people who live there.

3.5 Transportation and mobility in sustainable regional planning

The infrastructure of transportation has a significant influence on the process of sustainable development: to determine various consequences caused by transportation infrastructure and to illustrate developing patterns and difficulties [31]. For many decades, transportation has been an essential link between many facets of life around the globe. In most cases, the state of the natural environment, society's welfare, and the economy's advancement rely upon the state of the world's transportation networks. Therefore, transport systems that are generally safe, clean, sustainable, and egalitarian are beneficial to the economic development of countries, particularly in cities and other metropolitan centers. Nevertheless, a significant body of research demonstrates that the transportation systems used in most cities and urban regions must be more sustainable. Moreover, there are a few of these transportation systems that are seen to pose a risk to the environmental, social, and economic elements of future generations. According to this point of view, it is necessary to involve various regional, national, and worldwide stakeholders to change such transportation patterns [32]. Encouraging environmentally responsible modes of transportation is another essential component of environmentally responsible regional planning. This may involve the development of public transit networks, the creation of bicycle lanes and pedestrian walkways, and using traffic-calming measures to alleviate congestion and enhance air quality. These programs can lessen reliance on fossil fuels, cut down on levels of air pollution, and encourage citizens to lead more physically active and health-conscious lives. Consequently, mobility refers to a person's capacity to move about on their own or with the assistance of vehicles. The mode and means of transportation that someone chooses are determined, in part, by his requirements or preferences and physical and financial capabilities. Mobility as a service is a relatively new revolutionary transportation idea projected to bring about substantial shifts in how transportation is now practiced. However, there is a need for greater clarity surrounding the notion; it is still being defined what the essential qualities of mobility as a service are and how they may be addressed at this time [33]. Unquestionably, having a car opens up new doors of possibility for the individual driving it. Despite this, the continually expanding number of vehicles causes the road network to become overloaded, damages its state, endangers the lives of people and the environment, and leads to a rise in noise and harmful emissions. In addition, the stress it produces in society and the time it wastes make the work of urban emergency response agencies more difficult (ambulance, fire, emergency, and other units). Experts from all over the world agree that the personal automobile era is drawing to a close. Yet, the creation of a new generation of mobility is one of the most challenging issues of our time. The emergence of the new term "sustainable mobility" is inextricably linked with sustainable development, which connotes social progress, economic growth, environmental conservation, and concern for future generations. The term "sustainable development" was coined by the United Nations in 1987. About 30 years ago, urbanists in Western Europe first proposed the concept of environmentally responsible transportation. At this time, it was institutionalized at the legislative level and became a vital component of the state policy of many nations around the globe. The potential for equality of opportunity and the constitutional guarantee that every individual has the freedom to travel freely are at the heart of sustainable mobility. The 2030 Agenda for Sustainable Development (Agenda 2030) was approved in 2015 at the UN World Summit with the involvement of 193 nations, including the Republic of Belarus. This document specified 17 sustainable development goals (SDGs), which became humanity's primary guidelines [34].

3.6 Harmony with the natural environment

Humanity's destruction of the natural world has reached a tipping point, and now every living thing on Earth must begin training its consciousness toward a more sustainable way of existence. To live sustainably means to provide for the needs of the present without jeopardizing the ability of future generations to do the same. Unfortunately, only a few of our actions have been centered on how those objectives may be met, with no thought given to the consequences to the natural world. The government should begin with urban and regional planning. Still, the environmental catastrophe we have produced compels us to reassess our strategy for growth [35].

Forests encompass almost one-third of the Earth's land surface, play an essential part in the global carbon cycle, and are home to much of the world's terrestrial biodiversity. Forests also provide various other ecosystem services, such as nutrient cycling and soil formation, provisioning services such as food, timber, and medicinal plants, and regulating services such as erosion control, flood mitigation, water and air purification, pollination, and pest and disease control [36, 37]. Sustainable regional planning frequently includes policies for protecting natural resources, including wetland and forest ecosystems and the habitats of various animal species. These solutions may involve the establishment of conservation easements, protected areas, and other safeguards to shield these resources from the effects of development and different types of deterioration [38]. To achieve harmony with nature and environmental sustainability, it is necessary to reduce the negative impact of action on the local ecosystem, boost energy efficiency, and reduce the use of finite resources. However, researchers, politicians, residents, and others all find managing materials waste difficult to solve [39, 40]. In addition, it draws attention to the distinction between the usage of resources and the exploitation of such resources.

3.7 Maintain an equilibrium with the past

The foundations of this approach are the protection of historical monuments and careful attention to the distinctive cultural and social iconography of the areas, including its signs and symbols. Before establishing the legal framework for protecting historical monuments, and religious and cultural heritage, it is essential to define the extent of this protection by identifying the same parts of this legacy. The range of possible elements—both tangible and intangible—that the notion of architectural, religious, and cultural heritage might encompass is rather extensive; this may include complexes of buildings, sites of archeological or historical significance, ancient works of art, ethnographic itemizes, and so on [41]. It is necessary to plan and build to preserve the cultural heritage and the historical urban fabric by the established building techniques and style, pursuing goals to link the old and the new cultural values. This can be accomplished by planning and building by the established building techniques and style [42]. As a result of the fact that the architectural environment of a city is a subject information carrier that is created by architects and transmitted to the consumer by a one-of-a-kind system of signs of urban architecture, the preservation of the historic environment is a conscious measure that arises from the principle of cultural and historical continuity.

3.8 Appropriate technology implementation

The most significant challenge to sustainable development is how to sustainably and fairly feed the world's growing population. Projects to encourage the dissemination of technology and information among communities to support sustainable development have been initiated by the United Nations Educational, Scientific, and Cultural Organization (UNESCO), along with other government initiatives and NGOs [43]. A global problem, especially in developing nations lacking recycling availability and waste management infrastructure, is combating the environmental disaster generated by improper waste management. This crisis has been caused by improper disposal of trash. Slow pyrolysis, which converts municipal garbage into fuel used in cooking, is one strategy that communities might implement to address this situation [44]. Appropriate technology uses locally available materials and energy resources whose tools and processes are maintained and operationally controlled by the local population. Additionally, the right technology must be compatible with local, cultural, and economic conditions (i.e., the economy's human, material, and cultural resources). Therefore, technology is considered "appropriate" only if it is consistent with the preexisting norms and institutions of the community in which it is implemented. This is the only criterion that determines whether technology is suitable. The essential concept of the "right technology" is employing building materials, construction techniques, and project management suited to the location's requirements. Establishing a strong connection between the available resources and the many technologies that can solve environmental problems is vital. Every ecological crisis has the potential to be solved by a diverse range of technologies [45].

A modern city's infrastructure includes the fundamental information technologies, organizational structures, and associated services and facilities required for a business or industry to operate in a smart city. Implementing new technological tools within a modern city's infrastructure is known as "smart city infrastructure." A particular urban infrastructure can be connected to the infrastructures of other cities and nations, creating what may be referred to as regional, national, or even global infrastructures in the process [46]. In the business world, one may refer to industrial or corporate digital infrastructures [47, 48].

3.9 Friendliness and space attractiveness

The realm of place creation that encompasses urban planning and design defines the term "attractiveness" by the issues they face [49]. The difficulties faced by society, the possibilities presented by the community, and the experiences gained through daily life all contribute to the conception of beauty held by the local populace. The production of beauty is influenced in part by day-to-day experiences and in part by the assimilation of the knowledge regarding the location and the community through the consumption of media reporting and government papers about urban planning. The cityscape needs to be designed in such a way that it encourages people to communicate with one another and strikes up conversations with their neighbors. This is something that can be accomplished using urban design and public areas. The advocates of intelligent urbanism assert that a dynamic, interactive society that enables its members to connect and meet may be crafted with the assistance of design and can be realized *via* its implementation [50]. The use of the concept necessitates the production of the following categories of space.

4. Socioeconomic and managerial interaction in regional planning

Planning for sustainable regional development typically also involves activities to encourage the growth of sustainable economies [51]. This might entail fostering the expansion of industries reliant on renewable energy sources, advancing environmentally friendly technology, and promoting sustainable tourism. In addition, sustainable regional planning typically involves the promotion of sustainable agriculture. This may include the preservation of farmland as well as the encouragement of organic and sustainable agricultural techniques. The ability of regional and local authorities to implement the goals in their respective contexts is critical to the success of Sustainable Development Goals (SDGs), which must be incorporated into action plans and planning tools [52]. This will require involvement, collaboration, and development work across sectors and authority levels, as well as the development of guidelines for how this can be accomplished. Coordinating efforts between various fields and industries is essential for environmentally responsible regional planning. This indicates that the planning efforts should involve a wide variety of stakeholders, including officials from the government, representatives from the private sector, members of the community, and specialists in transportation, housing, environmental protection, and economic development. Sustainable regional planning can guarantee that all voices are heard and that choices are made with a wide range of issues in mind by bringing together a variety of viewpoints and putting them into dialog with one another [53]. There is a pressing need to solve the challenge of transitioning to sustainable economic growth so that developing nations can catch up to and master the VI technology mode. Developing natural resources, maintaining ecological balance, improving the economy's territorial structure, and guaranteeing population employment are all a key to effective regional economic strategy. In addition, new, effective methods of influencing the economy are essential if regions are to realize their potential for modern socioeconomic development. This is especially true of program-targeted forecasting, which enables expert-level solutions to the strategic problems facing specific regions' development within a unified state policy on the area.

- *Challenges in a rural district*

The ecological, demographic, economic, historical, and cultural potential of the world's rural regions is enormous. Rural areas make up the majority of the world's landmass. Their state is a significant factor in determining the dynamic growth of the country's economy and enhancing the standard of living for people worldwide [54]. There, for many decades, the primary method of developing rural areas, regardless of their agricultural potential, was reduced to a narrow-sectoral agrarian approach. This approach was distinguished by maintaining the "functional purity" of the countryside and preventing the development of other industries other than agriculture.

5. Conclusion

A complete method of developing and managing a region's physical, economic, and social growth is known as sustainable regional planning. Sustainable regional planning may assist in building dynamic, livable, and resilient communities for current and future generations by balancing economic growth and development with the conservation of natural resources, the promotion of social fairness, and mitigating adverse environmental consequences. The world as we know it right now is at a defining moment, one at which we must look ahead to the future and identify those investments that may provide people with a better living and a better lot while also being suited to the new situation caused by the epidemic, in a world where Sustainable Regional Development must meet the difficulties of the twenty-first century, from post-pandemic recovery to the digital and green transition. One thing we can learn from this crisis is the extent to which the international community stands behind and helps to develop certain areas, as well as the businesses and industries located therein. So that the site maintains its vitality in terms of economy, jobs, public transit, health infrastructure, and education, public authorities must carefully plan and prioritize future investments. In the years ahead, novel initiatives that promote a carbon-neutral economy and prioritize the needs of residents will be the primary engines of regional growth. However, future investments cannot be put to good use unless local public bodies strengthen their administrative competence and engage regional and local actors and residents in a genuinely democratic partnership. Encouraging resilient community development is a crucial component of environmentally responsible regional planning. This includes planning and developing communities that can resist and recover from crises, including economic downturns, natural catastrophes, and other disruptions. For example, this might involve supporting economic diversification to decrease the impact of economic downturns, developing buildings and infrastructure to be more robust to harsh weather, and providing green spaces that can function as natural buffers against floods.

It is essential to be aware that sustainable regional planning is an ongoing process that has to be routinely examined and modified to take into account changes in the region, such as increases in population, shifts in land use, and the introduction of new technologies. Setting up monitoring and evaluation mechanisms and including stakeholders in the planning process are viable options for accomplishing this goal. A complete and holistic strategy for planning and managing a region's physical, economic, and social growth is what we mean when discussing sustainable regional planning. It seeks to strike a balance between economic expansion and development on the one hand and the preservation of natural resources, the advancement of social fairness, and the mitigation of adverse effects on the environment on the other hand. It is a continuous process that should be frequently examined and updated. It involves the integration of many sectors and disciplines, the use of data and technology, the promotion of resilient communities, and sustainable economic growth.

Author details

Amjad Almusaed[1*] and Asaad Almssad[2]

1 Department of Construction Engineering and Lighting Science, Jönköping University, Jönköping, Sweden

2 Faculty of Health, Natural Sciences, and Technology, Department of Engineering and Chemical Sciences Construction Technology, Karlstad University, Karlstad, Sweden

*Address all correspondence to: amjad.al-musaed@ju.se

References

[1] Almusaed A, Almssad A. Urban social sustainability - Case study; Gellerupparken–Denmark. In: Almusaed A, Almssad A, Hong LT, editors. Sustainability in Urban Planning and Design. London, UK: IntechOpen; 2020. DOI: 10.5772/intechopen.93124

[2] Kosolapova NA. Modeling resource basis for social and economic development strategies: Water resource case. Journal of Hydrology. 2017;**553**:438. DOI: 10.1016/j.jhydrol.2017.08.007

[3] Paglia E. The Swedish initiative and the 1972 Stockholm Conference: The decisive role of science diplomacy in the emergence of global environmental governance. Humanities and Social Sciences Communications. 2021;**8**:2. DOI: 10.1057/s41599-020-00681-x

[4] Wan M. The United States, Japan, and the European Union: Comparing political economy approaches to China. The Pacific Review. 2007;**20**(3):397-421. DOI: 10.1080/09512740701461520

[5] Kruk ME et al. High-quality health systems in the sustainable development goals era: Time for a revolution. The Lancet Global Health. 2018;**6**(11):e1196-e1252. DOI: 10.1016/S2214-109X(18)30386-3 Epub 2018 Sep 5. Erratum in: Lancet Glob Health. 2018 Sep 18; Erratum in: Lancet Glob Health. 2018 Nov;6(11):e1162. Erratum in: Lancet Glob Health. 2021;9(8):e1067

[6] Almusaed A, Almssad A, Najar K. An innovative school design based on a biophilic approach using the appreciative inquiry model: Case study Scandinavia. Advances in Civil Engineering. 2022;**2022**:18. Article ID 8545787. DOI: 10.1155/2022/8545787

[7] Aall C. Municipal environmental policy in Norway: From 'mainstream' policy to 'real' agenda 21? Local Environment. 2000;**5**(4):451-465. DOI: 10.1080/713684886

[8] Bonnet J, Coll-Martínez E, Renou-Maissant P. Evaluating sustainable development by composite index: Evidence from French Departments. Sustainability. 2021;**13**:761. DOI: 10.3390/su13020761

[9] Bukhari SAA, Hashim F, Amran A. Green banking: A strategy for attainment of UN-sustainable development goals 2030. International Journal of Environment and Sustainable Development. 2022;**22**(1):13-321. DOI: 10.1504/IJESD.2023.127419

[10] Grainger-Brown J, Malekpour S. Implementing the sustainable development goals: A review of strategic tools and frameworks available to organisations. Sustainability. 2019;**11**:1381. DOI: 10.3390/su11051381

[11] Almusaed A, Almssad A. Quality assurance in a sustainable architecture creation process. Technology Reports of Kansai University. 2020;**62**:13. [Tokyo, Japan] Available from: https://www.kansaiuniversityreports.com/volume/TRKU/62/07/quality-assurance-in-a-sustainable-architecture-creation-process-5f18aec69b700.pdf

[12] Mateoc-Sîrb N et al. Sustainable tourism development in the protected areas of Maramureş, Romania: Destinations with high authenticity. Sustainability. 2022;**14**:1763. DOI: 10.3390/su14031763

[13] Brokking P, Mörtberg U, Balfors B. Municipal practices for

integrated planning of nature-based solutions in urban development in the Stockholm region. Sustainability. 2021;**13**:10389. DOI: 10.3390/su131810389

[14] Oliveira E, Tobias S, Hersperger AM. Can strategic spatial planning contribute to land degradation reduction in urban regions? State of the art and future research. Sustainability. 2018;**10**:949. DOI: 10.3390/su10040949

[15] Almusaed A, Alasadi A, Almssad A. A Research on the Biophilic Concept upon School's Design from Hot Climate: A Case Study from Iraq. Advances in Materials Science and Engineering. 2022;**2022**:12. Article ID 7994999. DOI: 10.1155/2022/7994999

[16] Almusaed A. Introductory chapter: A general reading process on landscape architecture. In: Landscape Architecture - The Sense of Places, Models, and Applications. London: IntechOpen; 2018. DOI: 10.5772/intechopen.77971

[17] Torelli R. Sustainability, responsibility, and ethics: Different concepts for a single path. Social Responsibility Journal. 2021;**17**(5):719-739. DOI: 10.1108/SRJ-03-2020-0081

[18] Almusaed A, Almssad A. City phenomenon between urban structure and composition. In: Almusaed A, Almssad A, Hong LT, editors. Sustainability in Urban Planning and Design. London, UK: IntechOpen; 2019. DOI: 10.5772/intechopen.90443

[19] Yan S, Growe A. Regional planning, land-use management, and governance in German metropolitan regions—The case of Rhine–Neckar metropolitan region. Land. 2022;**11**:2088. DOI: 10.3390/land11112088

[20] Almusaed A, Almssad A. Introductory chapter: Overview of sustainable cities, theory and practices. In: Almusaed A, Almssad A, editors. Sustainable Cities - Authenticity, Ambition, and Dream. London, UK: IntechOpen; 2018. DOI: 10.5772/intechopen.82632

[21] Yiftachel O, Hedgcock D. Urban social sustainability: The planning of an Australian city. Cities. 1993;**10**(2):139-157. DOI: 10.1016/0264-2751(93)90045-K

[22] Nijkamp P. Global environmental change: Management under long-range uncertainty. Journal of Contingencies and Crisis Management. 1994;**2**(1):1-9. DOI: 10.1111/j.1468-5973.1994.tb00021.x

[23] Pozoukidou G, Angelidou M. Urban planning in the 15-minute city: Revisited under sustainable and smart city developments until 2030. Smart Cities. 2022;**5**:1356-1375. DOI: 10.3390/smartcities5040069

[24] Chen X, He B-J. IOP Conference Series: Earth and Environmental Science. 2022;**1122**:012005. DOI: 10.1088/1755-1315/1122/1/012005

[25] Almusaed A, Almssad A. Introductory chapter: Sustainable housing – Introduction to the thematic area. In: Almusaed A, Almssad A, editors. Sustainable Housing. London, UK: IntechOpen; 2022. DOI: 10.5772/intechopen.101968

[26] Coscia C, Mukherjee S, Palmieri BL, Quintanal Rivacoba C. Enhancing the sustainability of social housing policies through the social impact approach: Innovative perspectives form a "Paris affordable housing challenge" project in France. Sustainability. 2020;**12**:9903. DOI: 10.3390/su12239903

[27] Almusaed A, Almssad A. Lessons from the World Sustainable Housing (Past Experiences, Current Trends, and Future Strategies). London, UK:

IntechOpen; 2021. DOI: 10.5772/intechopen.100533

[28] Almusaed A, Almssad A. Introductory chapter: Housing policy matters. In: Almusaed A, Almssad A, editors. Housing. London, UK: IntechOpen; 2018. DOI: 10.5772/intechopen.81622

[29] Bibri S, E, Krogstie J, Kärrholm M. Compact city planning, and development: Emerging practices and strategies for achieving the goals of sustainability. Developments in the Built Environment. 2020;**4**:100021. DOI: 10.1016/j.dibe.2020.100021

[30] Hegazy I et al. Assessment of urban growth of Jeddah: Towards liveable urban management. International Journal of Low-Carbon Technologies. 2021;**16**(30):1008-1017. DOI: 10.1093/ijlct/ctab030

[31] Wang L, Xue X, Zhao Z, Wang Z. The impacts of transportation infrastructure on sustainable development: Emerging trends and challenges. International Journal of Environmental Research and Public Health. 2018;**15**(6):1172. DOI: 10.3390/ijerph15061172

[32] Bamwesigye D, Hlavackova P. Analysis of sustainable transport for smart cities. Sustainability. 2019;**11**:2140. DOI: 10.3390/su11072140

[33] Jittrapirom P et al. Mobility as a service: A critical review of definitions, assessments of schemes, and key challenges. Smart Cities – Infrastructure and Information. 2017;**2**(2):13-25. DOI: 10.17645/up.v2i2.931

[34] Vinuesa R, Azizpour H, Leite I, et al. The role of artificial intelligence in achieving sustainable development goals. Nature Communications. 2020;**11**:233. DOI: 10.1038/s41467-019-14108-y

[35] McPhearson T, Raymond CM, Gulsrud N, et al. Radical changes are needed for transformations to a good Anthropocene. NPJ Urban Sustainability. 2021;**1**:5. DOI: 10.1038/s42949-021-00017-x

[36] Nugroho H et al. Mainstreaming ecosystem services from Indonesia's remaining forests. Sustainability. 2022;**14**:12124. DOI: 10.3390/su141912124

[37] Almusaed A. Introductory chapter: Overview on grass topic. In: Almusaed A, Al-Samaraee SMS, editors. Grasses - Benefits, Diversities and Functional Roles. London, UK: IntechOpen; 2017. DOI: 10.5772/intechopen.70407

[38] Almusaed A, Almssad A. Environmental reply to vernacular habitat confirmation from vast areas of Scandinavia. Renewable & Sustainable Energy Reviews Journal. 2015;**48**:ss.825-ss.834. ISSN: 1364-0321. Available from: http://www.sciencedirect.com/science/article/pii/S136403211500283X

[39] Almusaed A, Almssad A. Building materials in eco-energy houses from Iraq and Iran. Case Studies in Construction Materials Journal. 2015;**1**:ss. 42-ss. 54. ISSN: 2214-5095. Available from: https://www.sciencedirect.com/science/article/pii/S2214509515000029

[40] Kumar R et al. Impacts of plastic pollution on ecosystem services, sustainable development goals, and need to focus on circular economy and policy interventions. Sustainability. 2021;**13**:9963. DOI: 10.3390/su13179963

[41] Tsivolas T. The legal foundations of religious cultural heritage protection. Religions. 2019;**10**:283. DOI: 10.3390/rel10040283

[42] Almssad A, Almusaed A, Homod RZ. Masonry in the context of sustainable

buildings: A review of the brick role in architecture. Sustainability. 2022;**14**:14734. DOI: 10.3390/su142214734

[43] Bobek V, Slanovc M, Horvat T. The role of NGOs in protecting and preserving cultural heritage in the EU: The case of Slovenia-Austria cross-border program. Global Perspectives on Non-Governmental Organizations [Internet]. 2023 Jan 4. DOI: 10.5772/intechopen.108138

[44] Foong Y, Browning S, Seay J. An appropriate technology approach for utilizing plastic waste derived cooking fuel to reduce indoor air pollution from solid fuel cooking. Journal of Appropriate Technology. 2022;**8**(3):101-108. DOI: 10.37675/jat.2022.00157

[45] Almusaed A, Almssad A. Introductory chapter: Overview of a competent sustainable building. In: Almusaed A, Almssad A, editors. Sustainable Buildings - Interaction between a Holistic Conceptual Act and Materials Properties. London, UK: IntechOpen; 2018. DOI: 10.5772/intechopen.77176

[46] Almusaed A, Almssad A. Sustainable wooden skyscrapers for the future cities. In: Du G, Zhou X, editors. Wood Industry - Past, Present and Future Outlook. London: IntechOpen; 2022. DOI: 10.5772/intechopen.105809

[47] Almusaed A, Almssad A. Blending human ware with software and hardware in the design of smart cities. In: Sustainable Smart Cities: A Vision for Tomorrow. IntechOpen; 2022. DOI: 10.5772/intechopen.109053

[48] Almusaed A, Almssad A, et al. Coherent investigation on a smart kinetic wooden Façade based on material passport concepts and environmental profile inquiry, published by MDPI energy in construction and building. Materials (Switzerland). 2021;**14**:3771. DOI: 10.3390/ma14143771

[49] Ali MM, Al-Kodmany K. Tall buildings and urban habitat of the 21st century: A global perspective. Buildings. 2012;**2**:384-423. DOI: 10.3390/buildings2040384

[50] Allam Z, Bibri SE, Jones DS, Chabaud D, Moreno C. Unpacking the '15-Minute City' via 6G, IoT, and digital twins: Towards a new narrative for increasing urban efficiency, resilience, and sustainability. Sensors. 2022;**22**:1369. DOI: 10.3390/s22041369

[51] Yigitcanlar T, Kamruzzaman M. Planning, development and management of sustainable cities: A commentary from the guest editors. Sustainability. 2015;**7**:14677-14688. DOI: 10.3390/su71114677

[52] Enayat A et al. Achieving the sustainable development goals requires transdisciplinary innovation at the local scale. One Earth. 2020;**3**(3):300-313. DOI: 10.1016/j.oneear.2020.08.006

[53] Roberts P. Sustainable regional planning. Regional Studies. 1994;**28**(8):781-787. DOI: 10.1080/00343409412331348666

[54] Bongaarts J. Human population growth and the demographic transition. Philosophical Transactions of the Royal Society of London. Series B, Biological Sciences. 2009;**364**(1532):2985-2990. DOI: 10.1098/rstb.2009.0137

Chapter 2

Sustainable Land Development: Biodiversity, Natural Disasters, and Topographic Gradient

Kazutoshi Fujihira

Abstract

Inappropriate land development has been causing various problems, such as environmental destruction, biodiversity loss, climate change, and increased natural disaster risks. Aiming to avoid such issues and achieve sustainable land development, this study shows a method of dividing a region or municipality into development-restrictive areas and allowable areas. First, it presents three significant land attributes related to judging the appropriateness of development: (1) biodiversity, (2) natural disaster risk, and (3) topographic gradient. Then, regarding these land attributes, the following three sections illustrate ways to define problematic areas for development. Section 3 examines how to delineate sites contributing more to biodiversity, considering both significant areas for biodiversity and conservation practices. The fourth section outlines ways to avoid high-risk areas from predicted climatic and tectonic hazards, aiming to reduce natural disaster risks. Section 5 examines topographic gradient standards for determining steep-sloping areas. Finally, this study demonstrates how to integrate these three kinds of spatial data in the Geographic Information System (GIS) and delineate development-restrictive areas.

Keywords: biodiversity, natural disaster risk, topographic gradient, development-restrictive area, GIS

1. Introduction

Inappropriate land development, such as uncontrolled exploitation and urban sprawl, has been causing various problems worldwide. Such issues generally include deforestation, environmental destruction, biodiversity loss, climate change, and increased natural disaster risks.

For example, recently in Japan, a large number of problems occurred due to hasty land development. After 2012, the new policy for expanding renewable energy utilization accelerated the construction of photovoltaic power plants nationwide [1]. Following this rapid expansion, many troubles occurred in most prefectures of the country [1, 2], as exemplified in **Figure 1**. In addition, the number of problematic cases reported by the media reached 159 as of August 2021 [2]. According to Japan's Ministry of the Environment, major issues with such solar power plants

A solar power plant in Hyogo Prefecture. This plant partially collapsed during the Western Japan Heavy Rain Disaster of 2018. ©Nikkei BP.

A large-scale solar power plant in Kagoshima Prefecture. A landslide occurred in this plant site under a concentrated downpour in July 2022. ©Effort Inc.

Figure 1.
Examples of solar power plants constructed in deforested or steep-sloping areas in Japan.

were as follows: (1) erosion and its impact on water quality, (2) landslides, (3) harmful impacts on animals, plants, and ecosystems, and (4) scenic destruction [1]. Meanwhile, the problematic cases mostly occurred in newly deforested areas or sloping sites [1].

As the above example suggests, developing problematic areas for development leads to cause issues. Therefore, in order to avoid such issues and pursue sustainable land development, it is beneficial to define development-restrictive areas in advance.

Focusing on this point, this study illustrates how to divide a region or municipality into development-restrictive areas and allowable areas. First, the next section shows three main land attributes related to judging the appropriateness of development: (1) biodiversity, (2) natural disaster risk, and (3) topographic gradient. After that, concerning these three land attributes, Sections 3, 4, and 5, respectively, illustrate ways to define problematic areas for development. Lastly, Section 6 demonstrates how to integrate these three sorts of cartographic information in the GIS and delineate development-restrictive areas.

2. Strategy for sustainable land development

2.1 Sustainable urban design guidelines

We are producing sustainable urban design guidelines, based on the system control approach. As the United Nations "Sustainable Development Goals" implies, achieving sustainability is a goal-oriented challenge. The science of system control can be applied to all goal-oriented tasks [3, 4]. Moreover, that science has produced remarkable results in many fields, in particular engineering. Therefore, by applying system control, we have been progressing in research on sustainable design [3, 4].

The main process of producing sustainable urban design guidelines is shown in **Figure 2**. The production process starts with identifying environmental, social, and economic problems related to cities. Next, based on the specified problems, requirements for the sustainable urban design are also identified. After that, the requirements are converted into design guidelines, so that users can efficiently conduct spatial planning and urban design on local maps [4]. The sustainable urban design guidelines consist of three steps: (1) development-restrictive/allowable areas, (2)

DOI: http://dx.doi.org/10.5772/intechopen.110235

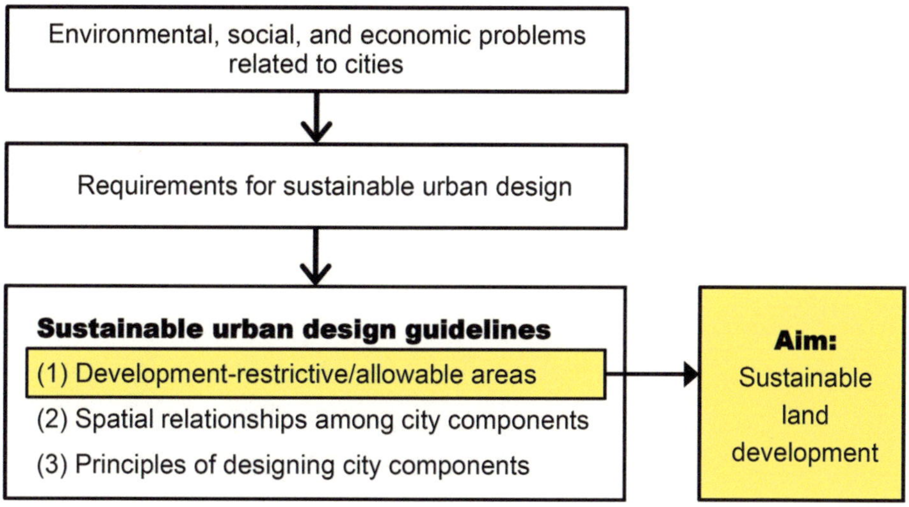

Figure 2.
The main process of producing sustainable urban design guidelines and the aim of the first step.

spatial relationships among city components, and (3) principles of designing city components [4].

This study focuses on the first step of sustainable urban design guidelines. The first part is intended to solve or prevent problems related to land development and meet requirements for sustainable land development. Examples of such issues have been illustrated in the introduction. Meanwhile, the requirements for sustainable land development include nature and biodiversity conservation, environmental protection, avoiding natural disaster danger areas, and consideration for the landscape.

2.2 Main land attributes related to judging development

In order to define development-restrictive and allowable areas, this study examines the relationship between land attributes and development. As demonstrated on the left side of **Figure 3**, there are three significant land attributes related to judging the appropriateness of development: (1) biodiversity, (2) natural disaster risk, and (3) topographic gradient [4].

2.2.1 Biodiversity

"Biodiversity" is a key area when seeking nature conservation and environmental preservation [5]. Efforts to conserve biodiversity also help meet global goals for greenhouse gas emissions to mitigate climate change. Moreover, such biodiversity and nature-based solutions help protect us from climate change impacts, securing ecosystem services [5–7].

2.2.2 Natural disaster risk

"Natural disaster risk" is connected with one of the requirements, "avoiding natural disaster danger areas." Geomorphologically, the Earth's surface is shaped by tectonics and climate [8]. As this simple statement implies, natural disasters can

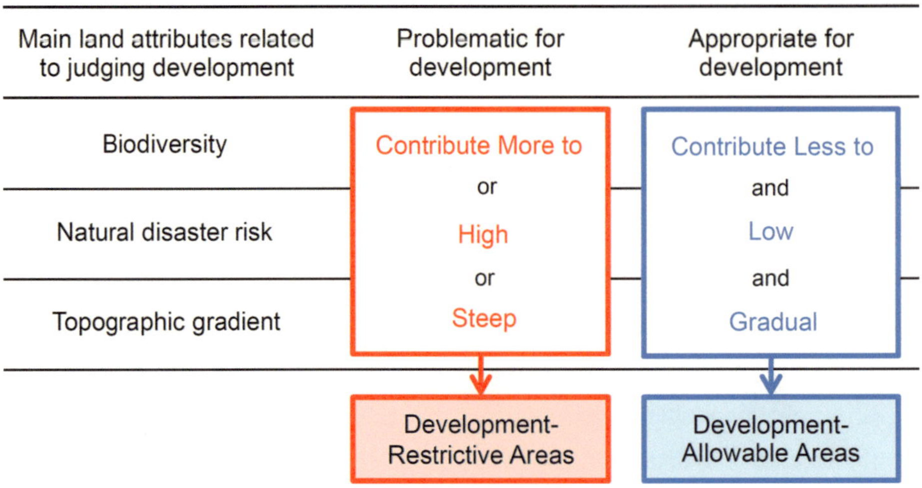

Figure 3.
Strategy for delineating development-restrictive areas and allowable areas.

be divided into two categories: "tectonic" and "climatic" [9, 10]. Tectonic disasters include catastrophes caused by earthquakes and volcanic eruptions. Meanwhile, examples of climatic disasters are serious damage resulting from floods and droughts.

2.2.3 Topographic gradient

"Topographic gradient" is mainly associated with two requirements for sustainable design, "environmental protection" and "accessible and universal design" [4]. Vegetation removal on steep-sloping sites often triggers significant environmental issues, such as soil erosion, landslides, increased downstream runoff, and flooding [11–13]. Meanwhile, as slopes become steeper, the provision of infrastructure and accessible design becomes more difficult and expensive [12]. Furthermore, "topographic gradient" is also associated with another requirement, "consideration for the landscape," because developing steep-sloping areas frequently leads to serious disfiguration of the scenic landscape [11, 12].

2.3 Development-restrictive and allowable areas

This study shows a method of dividing a region or municipality into development-restrictive and allowable areas, based on the strategy demonstrated in **Figure 3**. "Development-restrictive areas" means sites where land development, including vegetation removal and construction, should be prohibited. Development-restrictive areas are the unions, or amalgamations, of the following sites: (1) areas contributing more to biodiversity, (2) high-risk areas from predicted natural disasters, and (3) steep-sloping areas [4].

On the other hand, "development-allowable areas" means sites where land development can be permitted. Development-allowable areas are the intersections, or overlaps, of the following: (1) areas contributing less to biodiversity, (2) low-risk areas from predicted natural disasters, and (3) gradual-sloping or non-gradient areas [4].

3. Biodiversity

Area-based conservation efforts are the cornerstone for halting the global biodiversity crisis [14, 15]. As shown on the left side of **Figure 4**, site-based biodiversity conservation can be divided into two main categories: (1) identification of significant areas for biodiversity and (2) conservation practices [16]. Currently, there has been a steady increase in the coverage of both identified significant areas and conservation areas. However, further and urgent expansion is necessary [14, 15]. The main coverage targets are listed on the right side of **Figure 4**.

3.1 Identification of significant areas for biodiversity

The most authorized approach to designate important areas for biodiversity is the identification of Key Biodiversity Areas (KBAs). KBAs are nationally identified sites that contribute significantly to the global persistence of biodiversity, in terrestrial, freshwater, and marine ecosystems [17]. The KBA approach builds on a methodology originally developed for specifying Important Bird and Biodiversity Areas (IBAs). After that, it was expanded to cover a wider range of taxa and conservation initiatives, such as the Alliance for Zero Extinction (AZE) [18]. In 2016, the International Union for Conservation of Nature (IUCN) published "A Global Standard for the

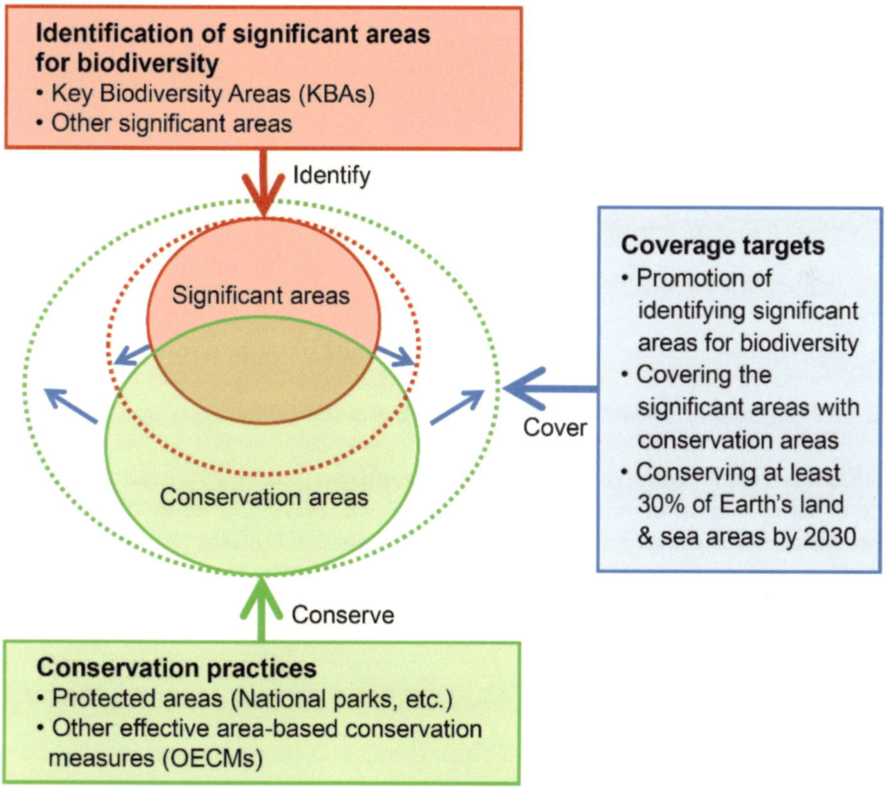

Figure 4.
Current situation of area-based biodiversity conservation and main coverage targets.

Identification of Key Biodiversity Areas," providing a set of globally standardized criteria and quantitative thresholds [17].

In addition to sites of global biodiversity significance, there must be areas of national or regional biodiversity importance. Such sites should also be identified as significant areas for biodiversity [19].

3.2 Conservation practices

Area-based conservation practices consist of "protected areas" and "other effective area-based conservation measures (OECMs)" [20].

3.2.1 Protected areas

Protected areas form the foundation of national biodiversity conservation strategies [21]. IUCN defines a protected area as "A clearly defined geographical space, recognized, dedicated and managed, through legal or other effective means, to achieve the long-term conservation of nature with associated ecosystem services and cultural values" [21]. The definition is expanded by the following management categories [21]:

- Category Ia: Strict nature reserve.
- Category Ib: Wilderness area.
- Category II: National park.
- Category III: Natural monument or feature.
- Category IV: Habitat/species management area.
- Category V: Protected landscape/seascape.
- Category VI: Protected areas with sustainable use of natural resources.

3.2.2 Other effective area-based conservation measures (OECMs)

OECM is a conservation designation for areas that are achieving effective *in situ* biodiversity conservation outside of protected areas [20]. Those concerned with conservation have recently recognized the importance of OECMs [20]. In addition, while protected areas must have a primary conservation objective, this is not necessary for OECMs. OECMs may be managed with conservation as a *primary* or *secondary* objective. Otherwise, long-term conservation may simply be the *ancillary* result of management activities [20].

3.3 Coverage and its targets

Efforts to identify significant areas and conservation practices are in progress. However, the coverages of KBAs and conservation areas are not yet sufficient. The world needs a further and effective increase in the coverage of such areas to halt biodiversity decline [14, 15].

3.3.1 Key biodiversity areas

Global KBA coverage has been increasing. More than 16,000 KBAs have been mapped worldwide [22]. KBAs have been identified in over 200 countries. KBA sites cover approximately 4% of the world's surface [18]. However, defining and delineating KBAs is an ongoing process [18]. Further efforts in KBA identification are necessary for many countries, taxa, and ecosystems [15, 18, 22].

3.3.2 Coverage of KBAs by conservation areas

In January 2018, 21% of KBAs were estimated to be completely covered by protected areas. However, 35% of KBAs had no protection through systems of conservation areas [18]. Currently, the average coverage of KBAs by conservation areas is 43.2% [22]. Therefore, prioritization of conservation efforts is necessary to cover all KBAs with conservation areas, namely protected areas or OECMs [15, 19].

3.3.3 Global coverage of conservation areas

Between 2010 and 2019, the world's protected areas expanded from 14.1% to 15.3% of the terrestrial areas and from 2.9% to 7.5% of the marine areas [14]. According to the official document on the UN biodiversity conference released at the end of 2022, currently, 17% and 10% of the world's terrestrial and marine areas, respectively, are under protection [23].

In December 2022, the UN Convention on Biological Diversity agreed to a new plan to preserve nature with the latest Global Biodiversity Framework. This framework has four overarching global goals and 23 targets for 2030. "Target 3," one of the most prominent targets, calls for a significant increase in the areas covered by conservation status. This target requires conserving at least 30% of Earth's land and sea areas by 2030 through protected areas or OECMs [23].

4. Natural disaster risks

4.1 Natural hazards and disaster risks

Figure 5 shows the relationship between natural hazards and disaster risks. Natural hazards are extreme natural events that can harm humans and their activities. Natural hazards can be divided into two main categories, "climatic" and "tectonic" [8–10]. Climatic hazards occur when a region has certain weather conditions, for example, heavy precipitation can lead to flooding. Tectonic hazards, such as earthquakes and volcanic eruptions, are caused by the movement of tectonic plates.

Disaster risk is widely recognized as the consequences of the interaction between a hazard and the characteristics that make people and places exposed and vulnerable [24]. Accordingly, there are three key elements that determine disaster risks: (1) intensity of natural hazards, (2) exposure to natural hazards, and (3) vulnerability to natural hazards. As demonstrated on the left side of **Figure 5**, "climate change impacts" intensify climatic hazards.

Meanwhile, as shown on the right side of **Figure 5**, there are two disaster risk reduction strategies, "reducing exposure to hazards" and "reducing vulnerability to

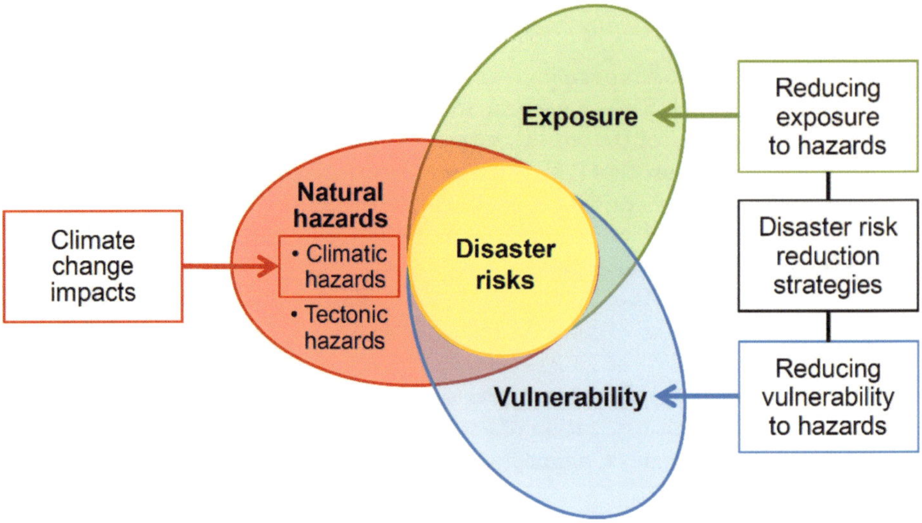

Figure 5.
Factors influencing natural disaster risks.

hazards." Exposure reduction is essential to reduce natural disaster risks, especially when considering land use. Theoretically, if neither humans nor their activities exist within the impact range of natural hazards, disasters will be less likely. Therefore, locating or relocating human activities outside hazardous areas reduces disaster risk. On the other hand, vulnerability reduction strategies include structural and nonstructural disaster prevention measures. Structural measures are physical methods, such as dams, flood levees, ocean wave barriers, earthquake-resistant construction, and evacuation shelters [25]. Nonstructural measures use knowledge, practice, or agreement to reduce disaster risks and impacts, such as public awareness raising, training, and education [25].

4.2 Climatic hazards

Climatic hazards occur when a region is in extreme weather conditions. For instance, as shown in the central part of **Figure 6**, heavy rainfalls and cyclones can cause flooding and landslides [9, 26]. Meanwhile, dryness and heat waves can trigger droughts and wildfires [27].

Recently, the frequency and intensity of climatic hazards have been increasing. According to the Intergovernmental Panel on Climate Change (IPCC), human-induced climate change is already affecting many weather and climate extremes in every region across the globe [27]. Evidence of changes in extremes, such as heat waves, droughts, heavy precipitation, and tropical cyclones, has strengthened since the 1950s [27].

Continued global warming will further intensify the global water cycle, the severity of wet and dry events, and global monsoon precipitation [27]. For example, at 2°C global warming and above compared to 1850–1900, heavy precipitation and associated flooding events are projected to become more intense and frequent in the Pacific Islands, across many regions of North America and Europe, and in some regions in Australasia and Central and South America [27]. Region-specific changes also include the intensification of cyclones and increases in river floods [27]. Meanwhile, several

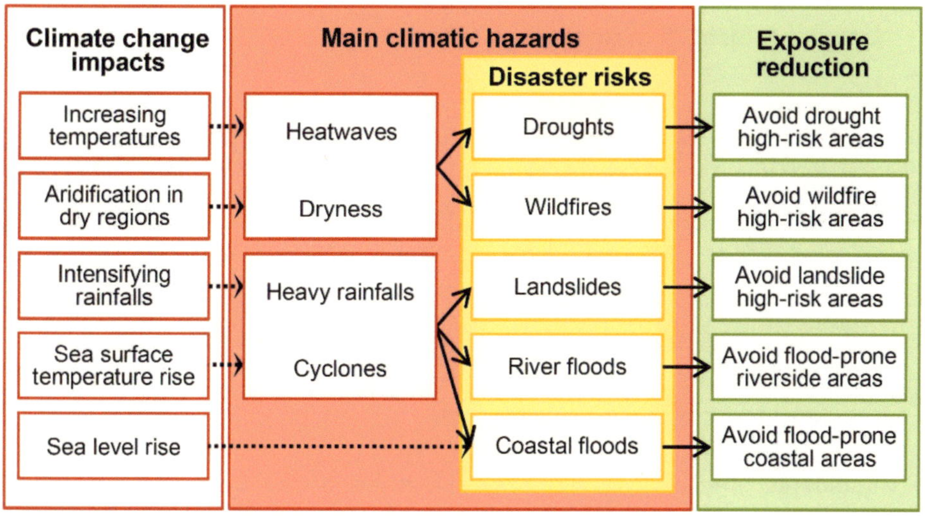

Figure 6.
Current main climatic hazards, disaster risks, and exposure reduction.

regions in Africa, South America, and Europe are projected to experience an increase in the frequency and/or severity of droughts. Such increases are also projected in Australasia, Central and North America, and the Caribbean [27]. Warming and increased dry conditions are also projected to increase wildfires [27].

Sea level rise has accelerated due to increasing rates of ice loss from the Greenland and Antarctic ice sheets, as well as continued glacier mass loss and ocean thermal expansion [28]. Sea level rise will continue throughout the twenty-first century [27]. Increases in cyclone winds and rainfall, and increases in extreme waves, combined with sea level rise, exacerbate extreme sea level events and coastal hazards [28].

As shown in the right side of **Figure 6**, when considering land use, exposure reduction to climatic hazards is crucial to reduce climatic disaster risks. For instance, efforts to avoid flood-prone riverside areas contribute to reducing damage from river flooding.

4.3 Tectonic hazards

As demonstrated in **Figure 7**, two major tectonic hazards are earthquakes and volcanic eruptions. Tsunamis are mainly caused by submarine earthquakes. Tsunamis are also generated by underwater volcanic eruptions. In addition, earthquakes and volcanic activity can initiate landslides [26], although these relations are omitted from the diagram.

An earthquake originates in a sudden release of energy in the Earth's lithosphere that produces seismic waves. When seismic waves reach the Earth's surface, they cause ground shaking. In general, the severity of ground shaking increases as the magnitude increases, and it decreases as the distance from the causative fault increases [29].

Underwater tectonic events generate tsunamis by displacing a substantial volume of water or perturbing a sea [30]. Tsunamis cause destruction both locally and at very distant locations from the area of tsunami generation [29]. Tsunami waves encroach upon the land and destroy things.

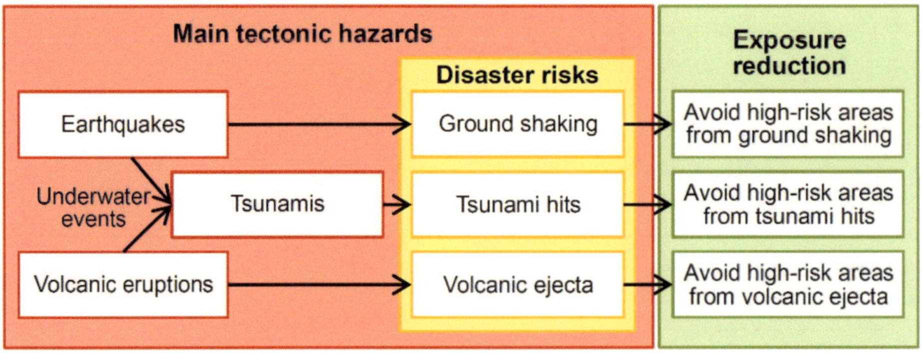

Figure 7.
Main tectonic hazards, disaster risks, and exposure reduction.

Volcanoes are often found where tectonic plates are converging or diverging. Volcanic eruptions occur when hot materials from the Earth's interior are thrown out of a volcano. Such materials, namely "volcanic ejecta," include lava, rocks, and gas compounds [31].

As shown on the right side of **Figure 7**, exposure reduction to tectonic hazards primarily requires people to consider avoiding high-risk areas from ground shaking, tsunami hits, and volcanic ejecta. For example, it is effective for municipalities to anticipate future tsunami risks and locate facilities on higher ground for safety.

5. Topographic gradient

5.1 Categorization of slopes

As outlined in Section 2.2.3, developing steep-sloping areas causes various problems, such as adverse effects of removing vegetation. In order to avoid issues caused by developing steep slopes, it is necessary to consider judging standards for regulation. How do we define a steep slope? There is no common definition of "steep slopes" [13]. However, many municipalities, especially in North America, define steep slopes as areas with 15% or more gradients [11, 12, 32]. Moreover, the Southern Tier Central Regional Planning and Development Board in New York State classifies slopes into three categories by gradients: (1) gradual (10% or less), (2) moderate (10–15%), and (3) steep (15% or more) [11]. Based on these examples, this study has divided slopes into four categories, as shown in **Figure 8(a)**.

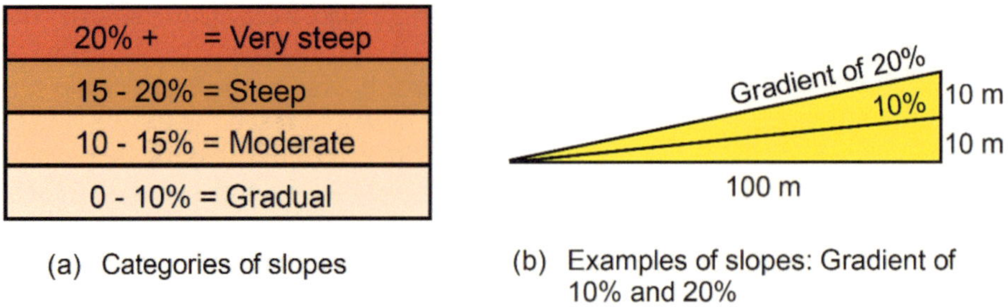

Figure 8.
(a) Categories of slopes, (b) examples of slopes.

In addition, examples of slopes with a 10% and 20% gradient are demonstrated in **Figure 8(b)**. For instance, a 10% slope has a 10 m vertical rise over 100 m of horizontal run.

5.2 Topographic gradients and development restrictions

This section examines the relationship between topographic gradients and development restrictions from four aspects of slope development issues: (1) adverse effects of removing vegetation, (2) grading and construction burdens, (3) accessibility, and (4) landscape.

5.2.1 Adverse effects of removing vegetation

Removing vegetation and disrupting natural drainage patterns on steep slopes can trigger various problems. Potential consequences include soil erosion, landslides, an increase in downstream runoff, flooding, and decreased water purity [11–13, 33]. In order to minimize such adverse effects, reasonable control of development is necessary. Soils with a slope of 15% or more should always involve severe limitations to development [34]. In addition, land development on slopes of 10% or more in environmentally sensitive areas will have similar potential to increase issues, such as surface runoff and soil erosion [34].

5.2.2 Grading and construction burdens

As slopes become steeper, grading and infrastructure provision become more complex and expensive [11, 12]. For example, constructing roads on steep slopes often calls for substantial grading and extra-wide rights of way to accommodate road slopes [12]. As shown in **Figure 9,** retaining walls are often necessary to support soil laterally when slopes are steep. Sewer and water systems are especially challenging and costly to plan and construct on steep slopes [11].

When the gradient exceeds 15%, the cost of building on the site begins to increase significantly because construction becomes more challenging [35]. Typically, the development requires deeper excavation, more concrete, retaining walls, and specialized solutions for drainage and sewer systems [35].

Figure 9.
A residential site developed on a steep-sloping area in Tokyo.

5.2.3 Accessibility

As topographic features become steeper, incorporating accessibility and universal design into the developed areas needs greater planning. As shown in **Figure 9**, the built environment in hilly areas often includes inclined roads and long stairways, which are hard to go up and down, especially for physically challenged people.

In addition, Japan has recently faced mobility issues, resulting from the age of the slope dwellers. In many hilly residential areas developed on the outskirts of megacities, aging has hindered the physical abilities of many residents. As a result, such older dwellers have started suffering from limited physical and social activities due to the steep slopes and long stairways [36].

According to a study on housing land development and accessibility, the maximum slope for general residential land should be 1 in 5, or 20% [37]. Meanwhile, the maximum gradient advisable for pedestrian ramps, which is also suitable for prams and trolleys, is 1 in 10, or 10% [37]. This figure significantly influences the design of pedestrian networks in various housing projects. In addition, ramp gradients are becoming more important with the increasing provision of ramps for the physically challenged in new projects [37].

5.2.4 Landscape

Steeply sloping areas often offer significant views of hills and valleys. Thus, developing such areas frequently leads to serious disfiguration of the scenic natural beauty [11, 12].

Generally, as the topographic gradients of the developed area become steeper, the more easily the landscape changes come into view from many places in the city or town areas. Therefore, in order to maintain scenic hillside landscapes, municipalities should take measures to control development and construction on steep-sloping sites. For example, Cardinia Shire Council in Victoria, Australia, manages development and construction in areas with a slope of greater than 10%, mainly for retaining scenic landscapes and amenity values [38]. Moreover, the Council strictly limits land development in areas with slopes in excess of 20% [38].

The above examinations from the four aspects can be integrated into the following. Generally, steep slopes, namely areas with a slope of 15% or more, should always involve restricting land development. In particular, land development in slope areas in excess of 20% should be strictly limited. Meanwhile, restrictions on developing areas with environmental sensitivity, landscape importance, and walkability-oriented design should start at 10% of a topographic gradient.

6. Integration: defining development-restrictive areas

The previous three sections illustrate how to define areas contributing more to biodiversity, natural disaster-prone sites, and steep-sloping areas. Based on these preparations, this section shows how to integrate the three sorts of cartographic information and delineate development-restrictive areas. In addition, nowadays, the Geographic Information System (GIS) has become increasingly common in handling spatial data; accordingly, the following proceeds with the explanation on the assumption that a GIS is used.

Figure 10 illustrates an example of combining the three sorts of geographical information in a GIS and defining development-restrictive areas. The left side

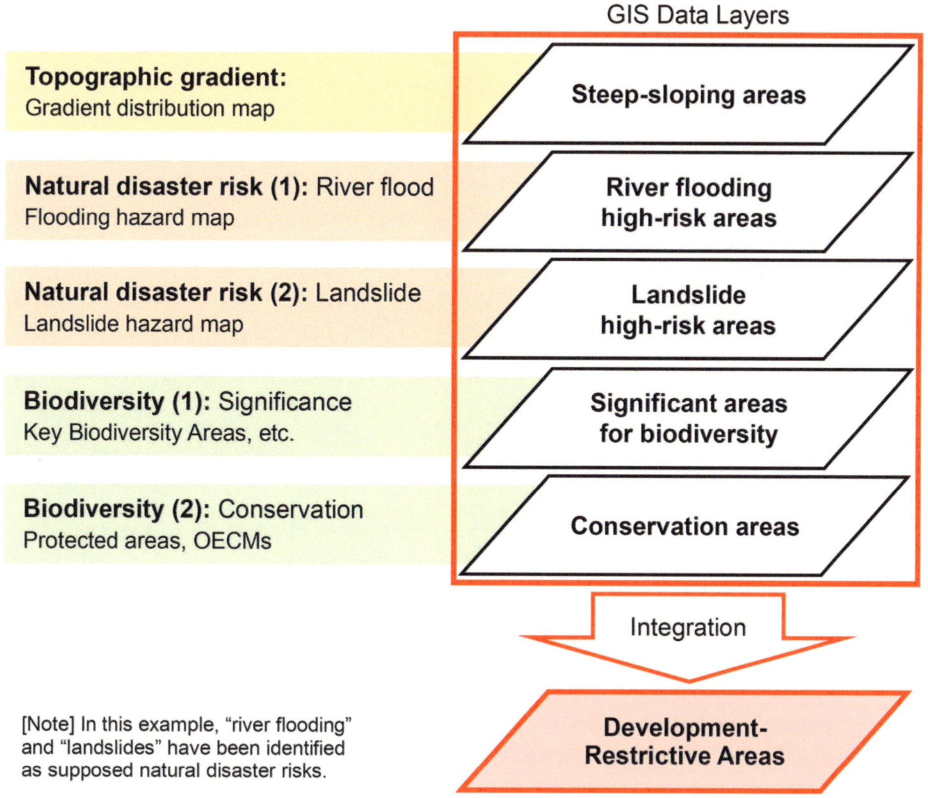

Figure 10.
An example of the process for delineating development-restrictive areas.

includes necessary spatial data, such as hazard maps and relevant specified areas. The right side demonstrates how the prepared GIS data layers are overlaid and integrated.

Concerning "topographic gradient," planners draw up a map layer on which steep-sloping areas are shown. In this case, it is helpful to color the map of the relevant region or municipality according to gradient classifications: 0–10% (gradual), 10–15% (moderate), 15–20% (steep), and 20% + (very steep). An example of such coloring has been demonstrated in **Figure 8 (a)**.

As for "natural disaster risks," the planners create data layers by natural hazard type. In the case of **Figure 10**, two types of hazards, namely river flooding and landslides, are predicted in the region or municipality. Accordingly, a data layer showing river flooding high-risk areas and one displaying landslide high-risk areas are prepared. If reliable hazard maps are open to the public, such maps can be utilized.

When handling data on "areas contributing more to biodiversity," it is helpful to divide them into "significance" and "conservation." This division is based on the categorization illustrated in **Figure 4**. The data layer of "significance" represents significant areas for biodiversity, such as Key Biodiversity Areas (KBAs). Meanwhile, the "conservation" layer displays conservation areas, namely protected areas and OECMs.

After preparing the necessary data layers, the planners place them on top of one another and integrate the spatial data. In this case, the unions, or amalgamations, of the relevant areas shown on the data layers correspond to development-restrictive areas.

The last part of this section compares development-restrictive areas with the coverage targets of conservation: (1) covering KBAs with conservation areas, and (2) conserving 30% of Earth's land and sea areas by 2030. These two coverage targets have been shown in Subsections 3.3.2 and 3.3.3, respectively. Through the integration process, significant areas for biodiversity, such as KBAs, are taken in development-restrictive areas. Therefore, if the development-restrictive areas are given conservation status, all KBAs are naturally covered with this status. Meanwhile, development-restrictive areas contain natural disaster-prone sites and steep-sloping areas, in addition to areas contributing more to biodiversity. Thus, if development-restrictive areas are actually preserved, the coverage ratio of conservation areas to the Earth's surface must expand greatly. In short, defining development-restrictive areas substantially contributes to meeting the two coverage targets of conservation.

7. Conclusion

In order to pursue sustainable land development, this chapter demonstrated a method of dividing a region or municipality into development-restrictive and allowable areas. First, Section 2 showed three land attributes closely related to judging the appropriateness of development, namely "biodiversity, "natural disaster risk," and "topographic gradient." Then, regarding these land attributes, the following three sections illustrated ways to define problematic areas for development, respectively. Finally, Section 6 demonstrated how to integrate these three kinds of spatial information in the GIS and delineate development-restrictive areas.

The second section illustrated how to produce guidelines for sustainable land development, based on the system control approach. In order to deal with various land development-related problems inclusively, this section identified the three significant land attributes mentioned above. After that, it demonstrated a strategy for defining development-restrictive and allowable areas, as shown in **Figure 3**. In addition, development-restrictive areas are the amalgamations of the following: (1) areas contributing more to biodiversity, (2) high-risk areas from predicted natural disasters, and (3) steep-sloping areas.

Section 3 presented the overall picture of current area-based biodiversity conservation efforts. **Figure 4** concisely illustrated how to define sites contributing more to biodiversity, dividing them into two categories: (1) significant areas for biodiversity and (2) conservation areas. The diagram also showed the main coverage targets of these site-based conservation efforts.

Section 4 demonstrated the whole picture of area-based natural disaster risk reduction measures in the three diagrams. **Figure 5** portrayed the relationship between natural hazards and disaster risks, dividing them into "climatic" and "tectonic." Next, **Figure 6** demonstrated how to avoid risky areas from main climatic hazards, also considering climate change impacts. Meanwhile, **Figure 7** showed how to avoid high-risk sites from main tectonic hazards.

Section 5 examined topographic gradient standards for determining problematic areas for development. It first classified slopes by gradient into categories, such as gradual and steep. After that, this section considered the relation between gradients and development restrictions from four aspects: (1) adverse effects of removing vegetation, (2) grading and construction burdens, (3) accessibility, and (4) landscape.

Lastly, **Figure 10** in the sixth section illustrated how to prepare necessary GIS data layers and integrate them. Displayed sites on the data layers are steep-sloping areas, high-risk areas from predicted natural hazards, significant areas for biodiversity, and conservation areas. The unions of these sites are equivalent to development-restrictive areas. In addition, preserving development-restrictive sites helps satisfy the world's conservation coverage targets, as well as avoid various problems concerning land development.

Our next work is case studies based on this method. Through such practical work, we are aiming to refine this method and contribute to sustainable land development.

Author details

Kazutoshi Fujihira
Institute of Environmentology, Inagi, Tokyo, Japan

*Address all correspondence to: fujihira@kankyogaku.com

References

[1] Ministry of the Environment Government of Japan. Report from the Investigated Commission on Fundamental Views of Environmental Assessments on Solar Power Generation Plants and Others (in Japanese). 2019. Available from: https://www.env.go.jp/content/900513059.pdf [Accessed: November 29, 2022]

[2] Yamashita N, Maruyama Y. Troubles with solar power generation and local governments' responses to them. In: Maruyama Y, Nishikido M, editors. How Energy Shifts Can be Achieved Smoothly? (in Japanese). Tokyo: Shinsensha; 2022. pp. 24-47

[3] Fujihira K. Basic schemes: Preparations for applying control science to sustainable design. In: Fujihira K, editor. Sustainable Home Design by Applying Control Science. Rijeka: IntechOpen; 2017. DOI: 10.5772/intechopen.71325. Available from: https://www.intechopen.com/chapters/57401

[4] Fujihira K. How to design sustainable structures. In: Sarvajayakesavalu S, Charoensudjai P, editors. Environmental Issues and Sustainable Development. London: IntechOpen; 2021. pp. 277-297. DOI: 10.5772/intechopen.95012. Available from: https://www.intechopen.com/chapters/74360

[5] European Commission. Biodiversity and Nature-Based Solutions: Analysis of EU-Funded Projects. Luxembourg: Publications Office of the European Union; 2020. p. 48. DOI: 10.2777/183298

[6] The Royal Society. Climate Change and Biodiversity: Interlinkages and Policy Options. 2021. Available from: https://royalsociety.org/topics-policy/projects/biodiversity-climate-change-interlinkages/ [Accessed: December 3, 2022]

[7] Cohen-Shacham E, Walters G, Janzen C, Maginnis S, editors. Nature-Based Solutions to Address Global Societal Challenges. Gland, Switzerland: IUCN; 2016. xiii + 97p. DOI: 10.2305/IUCN.CH.2016.13.en

[8] Whittaker AC. How do landscapes record tectonics and climate? Lithosphere. 2012;**4**(2):160-164. DOI: 10.1130/RF.L003.1

[9] NASA Solid Earth Science Working Group. How do Tectonics and Climate Interact to Shape the Earth's Surface and Create Natural Hazards?. 2021. Available from: https://solidearth.jpl.nasa.gov/PAGES/tect01.html [Accessed: December 4, 2022]

[10] BBC. Natural Hazards. 2022. Available from: https://www.bbc.co.uk/bitesize/guides/zxh4wxs/revision/1 [Accessed: December 3, 2022]

[11] Southern Tier Central Regional Planning and Development Board. Steep Slopes and Land Use Decisions. 2012. Available from: https://www.stcplanning.org/wp-content/uploads/2020/09/SteepSlopes_LandUse.pdf [Accessed: December 26, 2022]

[12] Lehigh Valley Planning Commission. Steep Slopes: Guide, Model Regulations. 2008. Available from: https://www.lvpc.org/pdf/SteepSlopes.pdf [Accessed: December 26, 2022]

[13] Land-of-Sky Regional Council. Mountain Ridge and Steep Slope Protection Strategies. Asheville: Land-of-Sky Regional Council; 2008. p. 76

[14] Maxwell SE et al. Area-based conservation in the twenty-first century.

Nature. 2020;**586**:217-227. DOI: 10.1038/s41586-020-2773-z

[15] Kullberg P, Minin ED, Moilanen A. Using key biodiversity areas to guide effective expansion of the global protected area network. Global Ecology and Conservation. 2019;**20**:e00768. DOI: 10.1016/j.gecco.2019.e00768

[16] UNEP-WCMC. Biodiversity A-Z, Areas of biodiversity importance. 2020. Available from: https://www.biodiversitya-z.org/themes/areas?category_id=23 [Accessed: December 26, 2022]

[17] IUCN. A Global Standard for the Identification of Key Biodiversity Areas, Version 1.0. First ed. Gland, Switzerland: IUCN; 2016. p. 46

[18] UNEP-WCMC. Biodiversity A-Z, Key Biodiversity Areas (KBA). 2020. Available from: https://www.biodiversitya-z.org/content/key-biodiversity-areas-kba [Accessed: December 26, 2022]

[19] Langhammer PF et al. Identification and Gap Analysis of Key Biodiversity Areas: Targets for Comprehensive Protected Area Systems. Gland, Switzerland: IUCN; 2007. p. xiv + 116

[20] IUCN-WCPA Task Force on OECMs. Recognising and Reporting Other Effective Area-Based Conservation Measures. Gland, Switzerland: IUCN; 2019. p. x + 22

[21] Dudley N, editor. Guidelines for Applying Protected Area Management Categories. Gland, Switzerland: IUCN; 2008. p. x + 86

[22] KBA Partnership. KBA Programme Annual Report 2021. 2022. p. 26

[23] UN Convention on Biological Diversity. COP15: Nations adopt four goals, 23 targets for 2030 in landmark UN Biodiversity Agreement. 2022. Available from: https://www.cbd.int/article/cop15-cbd-press-release-final-19dec2022 [Accessed: December 20, 2022]

[24] UNDRR. Understanding Disaster Risk. 2022. Available from: https://www.preventionweb.net/understanding-disaster-risk/component-risk/disaster-risk [Accessed: December 4, 2022]

[25] UNDRR. Structural and non-structural measures. Available from: https://www.undrr.org/terminology/structural-and-non-structural-measures [Accessed: December 23, 2022]

[26] United States Government Geological Survey. What is a Landslide and What Causes One?. 2021. Available from: https://www.usgs.gov/faqs/what-landslide-and-what-causes-one [Accessed: December 23, 2022]

[27] IPCC. Summary for policymakers. In: Masson-Delmotte V et al, editors. Climate Change 2021: The Physical Science Basis. Contribution of Working Group I to the Sixth Assessment Report of the Intergovernmental Panel on Climate Change. Cambridge and New York: Cambridge University Press; 2021. pp. 3-32. DOI: 10.1017/9781009157896.001

[28] IPCC. Summary for policymakers. In: Pörtner HO, et al, editors. IPCC Special Report on the Ocean and Cryosphere in a Changing Climate. Cambridge and New York: Cambridge University Press; 2019. pp. 3-35. DOI: 10.1017/9781009157964.001

[29] U.S. Geological Survey. What are the Effects of Earthquakes?. Available from: https://www.usgs.gov/programs/earthquake-hazards/what-are-effects-earthquakes [Accessed: December 23, 2022]

[30] Haugen K, Lovholt F, Harbitz C. Fundamental mechanisms for tsunami generation by submarine mass flows in idealised geometries. Marine and Petroleum Geology. 2005;**22**(1-2):209-217. DOI: 10.1016/j.marpetgeo.2004.10.016

[31] Dunkley PN, Young SR. Volcanic Hazard Mapping for Development Planning. Nottingham: British Geological Survey; 2000. p. 163

[32] Pennsylvania Land Trust Association (WeConservePA). Steep Slope Ordinance. 2019. Available from: https://conservationtools.org/guides/59 [Accessed: December 28, 2022]

[33] US Department of Housing and Urban Development. HUD Exchange Environmental Assessment Guide: Land Development - Slope. 2022. Available from: https://www.hudexchange.info/programs/environmental-review/environmental-assessment/guide/land-development/slope/ [Accessed: December 28, 2022]

[34] Randolph Township, New Jersey. Land Development Code: 15-44.1. Available from: https://library.municode.com/nj/randolph_township/codes/land_development_code?nodeId=LAND_DEVELOPMENT_CODE_ARTIVSUZORE_15-44REDEWISTSLAR_15-44.1PUFIFA [Accessed: December 29, 2022]

[35] Legal Eagle Contractors. What are the Problems of Building a House on Sloped Land?. Available from: https://legaleaglecontractors.com/building-house-on-slope-challenges/ [Accessed: December 26, 2022]

[36] Jeon S, Ohara K, Lee Y, Fujioka Y. The role of the built environment for active aging in hilly area: Community resource and social activity among older adults in I district (in Japanese). Architectural Institute of Japan's Journal of Architecture and Planning. 2016;**81**(726):1621-1629. DOI: 10.3130/aija.81.1621

[37] Abbott D, Pollit K. Hill Housing: A Comparative Study. London and New York: Granada; 1980. 308 p

[38] Urban Design & Management Pty Ltd. Guidelines for Slope Management in Subdivisions: Cardinia Shire Council – Pakenham East Precinct Structure Plan. 2017. Available from: https://vdocument.in/guidelines-for-slope-management-in-subdivisions-guidelines-for-slope-management.html?page=1 [Accessed: December 31, 2022]

Chapter 3

Conceptual Framework of Adaptive Social Planning for Sustainable Urban Transformation

Barakat Tiamiyu and Nancy Odendaal

Abstract

The notion of social planning has a long history as a method used by social planners to tackle societal problems. Programmes and policies implemented during this act of providing social amenities, developing policies and programmes, and other developmental activities are frequently rendered obsolete with time. Because social requirements and community elements are always changing, thereby creating the need for continual, collaborative, and inclusive efforts to deliver or adapt social development activities to meet people's current needs. This chapter seeks to develop an adaptive social planning framework that will provide a guideline for the adaptation of social needs and ensure the continuous integration of social dynamism into urban transformation. The non-adaptability of social planning and the lack of integrative social planning were extracted from the literature, which forms the basis for the development of the conceptual framework. Findings revealed the social transformation needed to achieve sustainable urban development, which were integrated to establish a holistic view of social needs. With this, a direct connection between the adaptation of social planning activities and sustainable urban transformation was established. This study provides a guideline for social planners and policy-makers to continuously monitor and evaluate the credibility and impact of any program/policy through empirical findings.

Keywords: social planning, adaptive social planning, sustainability development, urban transformation, conceptual framework

1. Introduction

Social planning connotes the activities of social planners towards community development efforts, to combat issues arising from social, economic, and physical complications in the urban environment. The term is conceptualised in diverse fields because of its applicability to a variant of human activity, which reflects the expectations by groups in promoting community well-being. Social planning is the process by which policymakers, legislators, government agencies, planners, and funders try to solve community problems or improve conditions in the community by devising and implementing policies and programmes intended to have certain results [1].

This entails a wide range of spectrum of uncertainty and complexity relating to program or policy design for different community character [2]. It is an important developmental activity for community advancement that aims to address the existing problem or introduce new initiatives to meet the need of the people and the community at large.

However, it is important to note that planning itself is an iterative and adaptive process [3, 4]. Adaption in the context of social planning is the application of fit-for-use project scope for a range of policies and programmes. Fit-for-use is the designing of appropriate programmes or policies that will yield maximum public acceptance and impact. Adaptation is a devised not to be optimal, for the best estimated future, but robust across a range of plausible futures and responds to changes over time, as well as makes explicit provision for learning [5]. This means that adaptive social planning acknowledges the need for knowledge acquisition for risk reduction through data, consultation, and design processes to reduce uncertainty and assumptions by planners or policymakers. There is a high degree of interconnectedness and dynamic consequences in a world where unpredictability and the presence of unknown are the underlying traits [6]. Hence, the plan should be designed to evolve and allow for a continuous flow of information that is generated from the feeds of the conceptualisation, planning, implementation, evaluation, and monitoring process.

There are diverse areas in which development activities by the government, private sector, or individuals are directed such as health, infrastructure, services, programmes, policies, education, and security, among others. The adaptation of these activities to the specific host communities is intricate and has its inherent complexity, which sometimes seems cumbersome with public interests, which often leads to a disjointed project objective. The human-centric nature of social planning programmes creates a dynamic and unsatisfying quest for rapid transformation, from the people and government. However, the adaptive nature of social planning will enable planners to incorporate data and knowledge accumulated within a particular context and to fine-tune spatial management arrangements to fit the dynamic social-ecological systems [7]. The activity of a social planner is an unending process of priority identification, adaptation, and implementation of social planning programmes, by acquiring data, engaging stakeholders in a consultative process, and through the design phases [5]. Stakeholders in this sense, refer to those who the program or policy will influence or whose influence has a direct effect on the program.

The outcome of social planning is a sustainable urban transformation and spatial gratification of the community and beneficiaries [8–10]. As explained by McCormick et al. [8], sustainable urban transformation refers to structural transformation processes—multidimensional and radical change—that can effectively direct urban development towards ambitious sustainability goals. Sustainable urban transformation, according to Ernst et al. [9] encompasses both sustainable urban structures and environments and (radical) economic, social, cultural, organisational, governmental, and physical change processes. Ernst et al. [9] explained that sustainable urban transformation relates to a multitude of urban sustainability issues, ranging from poverty, over-population, unhealthy housing conditions, inadequate infrastructure, hygienic problems, poor water quality, and uncontrolled pollution in developing countries to segregation and growing social tensions, local traffic problems, solid waste generation, and the large consumption of energy and material in developed countries.

This implies that urban transformation is a structural transformation process, multidimensional, and radical change that can effectively direct urban development towards ambitious sustainability goals [6]. This also implies that ensuring a sustainable urban transformation requires a continuous process of ensuring urban developmental projects meet their target impact for a long period. In other words, in the process of ensuring the adaptability of social planning programmes through a continuous approach to ensuring impact attainment, there is a residual effect on community development. The residual effect is the sustainability of urban transformation [11, 12]. As noted by Galbiati et al. [13], a sustainable urban transformation will unlock various economic potentials for the host community. Likewise, Itasca [14] concluded that sustainable urban transformation will impact peoples' living standards, physical development, and other socio-economic pull factors, such as those from investors and investments, economic growth, and population growth.

Social development initiatives often attract a lot of interest and attention at the early stage, but these interests fade even before the projects are completed. The concept and proposed impact create a buzz in the beneficiary community; however, these initiatives are later abandoned due to uninformed social planning procedures in a dynamic context. As a result, non-adaptive social planning undermines the sustainability of urban transformation, especially in developing countries. Also, the non-adaptability of social planning makes the urban transformation to be faced with many challenges that have complicated the social structures and impaired the delivery of social and economic growth. Besides, there is a dearth of adaptive social planning frameworks which would have aided the policymakers and social planners in integrating the flow of data and communicating strategies to stakeholders for better decision-making, outcome, and impact metrics in social development.

As a result of the non-adaptability of social planning and lack of adaptive social planning framework, policymakers and social planners are making assumptions of what people need or want; while the prevailing social planning approach is generic and complicated by disparities in the economic, social, and physical context. These shortcomings have caused many social development projects to fail; either not being completed or abandoned after completion. Hence, there is a need to establish the importance and elements of adaptive social planning towards the development of a participatory approach to the adaptation and development of sustainable urban transformational initiatives. The intended aim of this paper is to develop an adaptive social planning framework that will positively impacts sustainable urban transformation.

2. Literature review

The importance of the city to individual lives cannot be overemphasised because the majority of people live in cities and urban areas [15]. The city, as the metropolitan area plays a dominant role in alleviating poverty through economic opportunities, improving social status, and providing good living standards [14]. John [1] describes a city as a growth centre for existing and emerging global consumption, production, pollution, economic, and social activities. In the opinion of [16], a city is a source of growth for domestic and global economies. Smith [17] observed that city or urban transformation is essential to ensure that urban areas are resilient, safer, healthier, more equitable; and better able to absorb, recover from, ad successfully adapt to future adverse events. However, the growth or transformation of a city is highly dependent on social investment [8].

This indicates that the transformative process of any community is dependent on social investment or development by the municipal government or private bodies.

Social planning is one of the activities involved in social development [18]. It entails effective change management within a dynamic population and expectations. It scans across all social aspect of community needs that is assessed at the grassroots for a strong public social planning process. Community needs as contained in social planning varies from geographic location to another, economic and political atmosphere, and social literacy [18]. These needs may take the form of laws, regulations, incentives, media campaigns, programmes or services, information, and a wide range of possibilities [1]. Svend [19] notes that social planning encompasses a wide range of development activities, which cannot be defined specifically because of the ever-changing social needs such as health service, education, infrastructure, services, programmes, and policies.

There is a long history of social planning and the need for adaptive social planning in many parts of the world. In the United States, Franklin Roosevelt's New Deal contains various public health programmes in the 1940s and 1950s. During this time, social planning was used as a tool for economic and political reasons, which created a biased approach to development initiatives [17]. Consequently, there were many failed programmes (such as the Boston west-end neighbourhood levelling program) as planners assume what was best for the people instead of engaging the people in the selection of appropriate projects [20]. In South Africa, social development was designed to meet specific pressing needs with little or no consideration for the sustainability of the development [21]. The transition of the government of South Africa to democratic rule created an economic system that promotes gross income disparities and a widening gap between the rich and poor [22].

In China, changes in the political economy and deficiency in development activities of the local communities, contradict the ideas of social planning and have led to the regulation of social and economic development [23]. In Nigeria, social welfare emanates during the colonial rule by missionaries; and the post-colonial era by philanthropists and the Social Development Division in the Federal Ministry of Labour and Social Welfare [24]. Several development plans need assessment programmes, political visions and structural programmes were introduced by the different government to meet the need of the public in the area of housing, education, and social alleviation program have since being developed heuristically. Deedam et al. [25] noted the enormous failure of achieving the respective aim and objective of social development programmes, due to inadequacy of planning process and implementation. Other researchers, political analysts and social professionals, and foreign observers have identified sources of social planning failure from corruption, non-implementation of the plans to the fullest, political interference, or coup and counter-coup d'ëtat [25]. This indicates that the public administrators in Nigeria have no defined concept of social planning or the need to adapt social planning for social development.

Non-adaptability of social planning has been linked with a disruption in the social structure of a community [1]. Studies such as [26] have also reported that the non-adaptability of social planning affects the established community's sense of place which was what made the city colourful and full of life. Non-adaptability of social planning can occur when planners assumed that a luxury complex residence profiling for residents is best to improve their living standards when an affordable residence is the best option to improve the living standards of the community [18]. The negative impacts of the non-adaptability of social planning on the sustainable transformation of urban areas have led to the making of community participation a notable requirement in project planning and implementation [18].

This implies that adaptive social planning is necessary to ensure that plans remain relevant and impactful, project objectives are evaluated, and there is feedback on the effectiveness of implemented management actions [27]. Adaptive social planning accounts for communities with peculiar characteristics based on culture, environment, religion, climate, socio-economic and political attributes; by ensuring plans or policies are made with the right people, for the right people, and by the right people to achieve sustainable development [28]. Similarly, Lei and Jianfa [23] pointed out that the idea of ensuring plans stay impactful and relevant is what sustainable development is about [23].

However, implementing an adaptive social planning initiative for sustainable urban transformation is quite challenging, due to the enormous time and resources that will be required [7]. According to Grantham et al. [27], the development of an adaptive plan is a complex process as a result of the workload. Therefore, it becomes imperative to develop a framework that will provide guidelines for the implementation of the adaptive social planning process in the generation of social policies, programmes, projects, and other urban transformation activities in countries like Nigeria that are lacking in social planning and development [29–31].

Thus, the focus of this research is on the development of an adaptive social planning framework that will provide a guideline for the adaptation of social needs and ensure the continuous integration of social dynamism into urban transformation.

3. Research methods

A research methodology framework for this study is shown in **Figure 1**. A literature review on the application of social planning and the historical challenges with its application was first examined. A systematic overview of the concept of social planning as it applies in different countries [17, 23] shows that social planning was used for economic, social, political, and cultural reasons. However, a majority of these social planning initiatives fail because they do not meet the needs they were created for. From the literature review, two distinctive problems were identified: the non-adaptability of social planning and the lack of an integrative adaptive social planning framework for social and sustainable development.

Munasinghe [32] depicts in sustainable urban transformation framework that sustainable development can only be possible if there is interdependency and correlation between the demographic, economic, social and spatial transformation. Ojoko et al. [33] also present Munasinghe's principle of sustainability as an important development tool to drive social, economic, and physical systems. These two

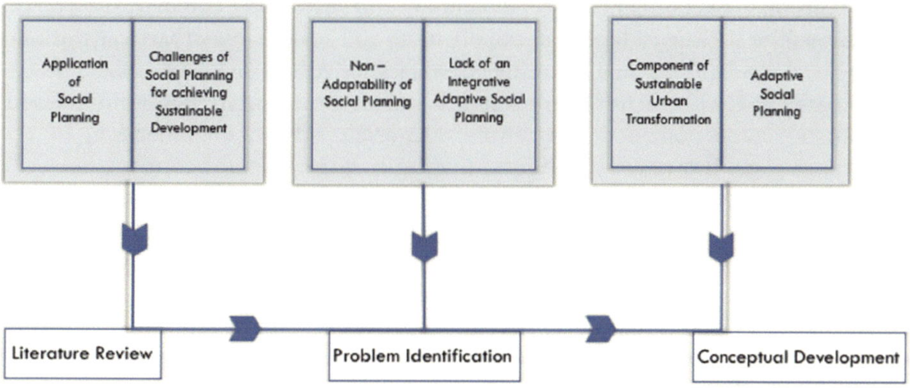

Figure 1.
Research methodology framework.

viewpoints provide crucial elements to achieve urban transformation, which lead to the development of modified sustainable urban transformation components. Social transformation, economic transformation, environmental transformation, demographic transformation, and spatial transformation encompass all the driving forces to address development.

To create an integrative and simplified process for adaptive social planning to achieve sustainable urban transformation, this chapter developed a conceptual framework to assist social planners and policymakers to continuously adapt and rethink social initiatives and developments through an empirical and systematic approach.

4. Development of conceptual framework

Social issues that are associated with urbanisation in developing countries include housing demands, technology advancement, health, employment, and transportation [29]. In addition, social needs are ambiguous while the majority of the developmental initiatives are ineffective and unable to meet social needs [33]. This suggests that adaptive social planning is required to ensure the sustainability of urban transformation by continuously addressing social issues and effectively meeting social needs. However, social planning will only be adaptive and bring sustainable urban transformation when developmental programmes are selected based on social goals, policymaker perspective, and grassroots viewpoint. This research proposes a conceptual framework of adaptive social planning for sustainable urban transformation.

The conceptual framework of this research is shown in **Figure 4**. It depicts a modification of Wei's sustainable urban framework (**Figure 2**) with Munasinghe's principle of Sustainability (**Figure 3**), to investigate the impact of adaptive social planning on sustainable urban transformation. **Figure 2** shows the sustainable urban framework proposed by Wei et al. [34]. The framework explained that demographic transformation, economic transformation, social transformation, and spatial transformation interact closely and interdependently with each other to bring about a sustainable urban framework. The three viewpoints of the Munasinghe sustainability triangle present social, economic, and physical systems as a distinctive driving force with objectives that address development, which has been widely adopted and in operational (see **Figure 3**).

The conceptual framework for this research (as shown in **Figure 4**) proposes a positive relationship between adaptive social planning and sustainable urban transformation. The relationships between adaptive social planning and sustainable urban transformation suggest that adaptive social planning is a collaborative and inclusive social development initiative that will ensure the consideration of environmental, economic, and social characteristics before policies or programmes are developed and implemented.

The conceptual framework in **Figure 4** underlines the need for continuous adaptation and rethinking of social development with new empirical findings to make development more sustainable. As illustrated in the conceptual framework, the elements of adaptive social planning processes for a sustainable urban transformation necessitate the need to create a centralised workplace and integrative process for planners to carry out effective and efficient social development. The adaptive social planning process as shown in **Figure 4** was conceptualised as a procedural-based framework with distinctive and connective elements. This provides an overview and guidelines for planners to engage, develop, plan, evaluate, and implement adaptive programmes and policies that meet social needs.

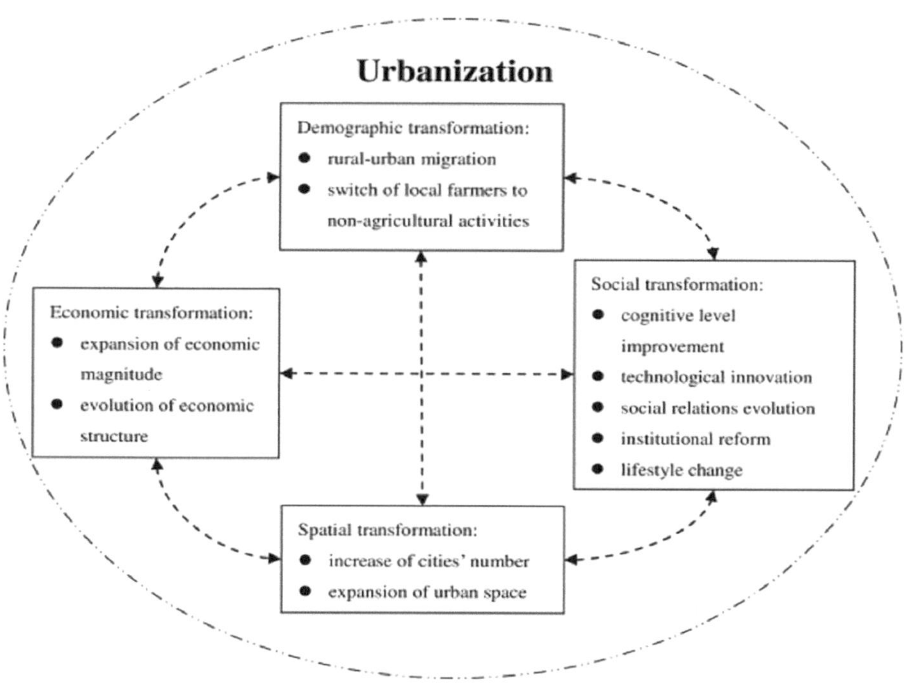

Figure 2.
Sustainable urban transformation framework [34].

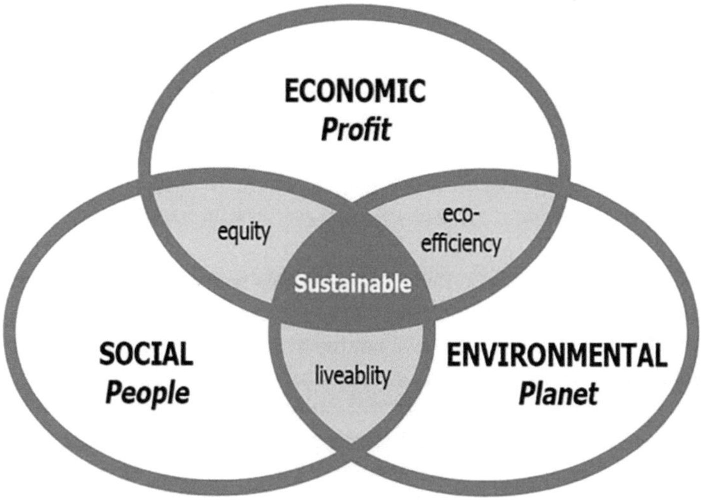

Figure 3.
The dimensions of sustainability by Munasinghe [32].

Also, the conceptual framework in **Figure 4** outlines the proposed criteria and components that a social development initiative must fulfil and capture before it can be said to have adopted adaptive social planning. The criteria suggest that development programmes or policies in the area of health, technology, institutional, relief, and financial will only ensure sustainable urban transformation if they fulfil all of the criteria for adaptive social planning. Also, the criteria suggest that social planners

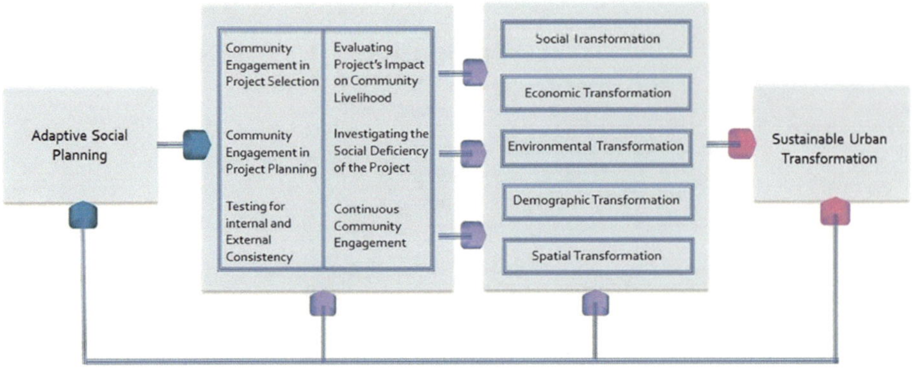

Figure 4.
A conceptual framework of adaptive social planning for sustainable urban transformation.

should first engage in open consultation with stakeholders in the community on possible programmes or projects that are of importance to them. This can be done using a set of toolkits that will allow a wide range of open suggestions and ideas from the selected beneficiaries themselves. Also, the planner must prioritise the selected programmes into a set of three alternatives, having reconciled them with the available resources, regional policy, and allocations to test the feasibility of the project to be implemented.

Lastly, these sets of alternatives must be presented back to the community in a brief and a final project or policy is selected. Community engagement as conceptualised in **Figure 4** will create a sense of ownership, enable the understanding of the social needs from the peoples' perspective, reduce uncertainty and assumptions about urban transformation, and create community responsiveness to urban transformation. The impact of the transformation through various developmental projects must be measured through community assessment of project impact on their personal and community livelihood. Whatever be the impact of the projects, both positive and negative outcomes of projects are very important to policymakers and funders to inform subsequent development activities.

In addition, the conceptual framework proposes that adaptive social planning will lead to social transformation, economic transformation, environmental transformation, demographic transformation, and spatial transformation. Social transformation, as an element of sustainable urban transformation, is geared towards enriching communities, relationships, and individual aspirations. Economic transformation is concerned with an increase in the consumption of goods and services, which in turn improves human welfare. Social transformation occurs in the framework as cognitive level improvement, information exchange, cultural communication, technological innovation, social relations evolution, institutional reform among others, which could further influence people's behaviour and lifestyle. The environmental transformation is concerned with the protection of the integrity and resilience of ecological systems.

5. Conclusions

Prior studies have shown that social planning and social development are intentional community development actions to curb issues arising from social, economic,

DOI: http://dx.doi.org/10.5772/intechopen.109653

and physical complexities. These activities are often carried out by social planners, policymakers, and other organisations, help create a community sense of place, and attract economic development and urban expansion. However, the approach to carrying out these development efforts, also known as social planning, has raised concerns over the years. As cities expand, it creates an increase in demographics, puts a strain on the physical and environmental components, and introduces a complex policy framework for planners and policymakers. Consequently, adaptive social planning is needed for a sustainable urban transformation.

This chapter proposes a paradigm for social planning that is integrative, collaborative, and inclusive in order to achieve sustainable urban transformation. It illustrates the direct correlation between adaptive social planning and sustainable urban transformation. The framework would help guide the implementation of social development, as well as continuously monitor its effectiveness, get public perception, and evaluate its impact. The impact of its application would ensure projects and plans remain relevant through time and social change. Further research is needed on the application and viability of the conceptual framework using a practical social development initiative.

Author details

Barakat Tiamiyu* and Nancy Odendaal
School of Architecture, Planning, and Geomatics, University of Cape Town, South Africa

*Address all correspondence to: barakatbidemi96@gmail.com

References

[1] John F. Towards sustainable neighborhoods: The role of social planning in China-A case study of Ningbo, Zhejiang Province. 2011

[2] Aelenei L, Ferreira A, Monteiro CS, Gomes R, Gonçalves H, Camelo S, et al. Smart city: A systematic approach towards a sustainable urban transformation. Energy Procedia. 2016;**91**:970-979

[3] Collie JS, et al. Marine spatial planning in practice. Estuarine, Coastal and Shelf Science. 2013;**117**:1-11

[4] Ehler C, Douvere F. Visions for a sea change: Report of the first international workshop on marine spatial planning, intergovernmental oceanographic commission and the man and the biosphere programme UNESCO Headquarters. Paris, France. 8-10 November 2006. 2007

[5] Walker W, Rahman S, Cave J. Adaptive policies, policy analysis, and policymaking. European Journal of Operational Research. 2001;**128**:282-289

[6] Swanson D, Barg S, Tyler S, Venema H, Tomar S, Bhadwal S, et al. Aspect of the rural-urban transformation of countries. Journal of Economic Geography. 2005;**5**:23-42

[7] Morena M, Rebecca W, Robert L, Mary G, Rose-Liza E, Amanda T, et al. Real-world progress in overcoming the challenges of adaptive spatial planning in marine protected areas. Biological Conservation. 2015;**181**:54-63

[8] McCormick K, Anderberg S, Coenen L, Neij L. Advancing sustainable urban transformation. Journal of Cleaner Production. 2013;**50**:1-11

[9] Ernst L, de Graaf-Van Dinther RE, Peek GJ, Loorbach DA. Sustainable urban transformation and sustainability transitions; conceptual framework and case study. Journal of Cleaner Production. 2016;**112**:2988-2999

[10] Yang Y. Sustainable urban transformation driving forces, indicators, and processes [Doctoral dissertation]. ETH Zurich. 2010

[11] Hellström R, Maria, et al. 2012. Advancing sustainable urban transformation through living labs: Looking to the Öresund region. International Conference on Sustainability Transitions, Sustainability Transitions (IST). Copenhagen, Denmark: 2012;**2012**

[12] Huang W, Cui S, Yarime M, Hashimoto S, Managi S. Improving urban metabolism study for sustainable urban transformation. Environmental Technology and Innovation. 2015;**4**:62-72

[13] Galbiati ML, Piredda F, Bertolotti E. Envisioning the city: A design-oriented communication process for a sustainable urban transformation. In: ESA research network sociology of culture midterm conference: Culture and the making of worlds. SSRN: Social Science Research Network. 2010. pp.1-12

[14] Itasca SI. Sustainable urban transformation in small cities in Egypt: An UN-habitat perspective. Journal of Cleaner Production. 2013;**50**:200-204

[15] Chima OO. Causes of failure and abandonment of projects and project deliverables in Africa. PM World Journal. 2017;**6**(1):1-16

[16] Sukhdev P. Putting a price on nature: The economics of ecosystems and biodiversity. Solution Journal. 2010;**1**(16):34-43

[17] Smith A. Empowerment series: Foundations of social policy: Social

justice in human perspective. Google Books. 6th ed. Cengage Learning; 2018. pp. 302-05. https://doi.org/ISBN: 978-1-305-94324-7

[18] Dalton H, Maurice J, Wandersman A. Community psychology: Linking individuals and communities. 2nd ed. Belmont, Tennessee, USA: Wadsworth/ Thomson Learning; 2007

[19] Svend R. Social planning and social organization. American Journal of Sociology. 1947;**52**(6):508-516

[20] Meredith C, Dunham M. Real Clout: A How-to Manual for Community-Based Activists Trying to Expand Healthcare Access by Changing Public Policy. Boston: : The Access Project; 1999

[21] Midgley J. The critical perspective in social development. Social Development Issues. 2001;**23**(1):42-50

[22] Mel G. The progress of social development in South Africa. International Journal of Social Welfare. 2006;**15**(Suppl. 1):53-64

[23] Lei W, Jianfa S. Changing spatial planning in China's five-year planning system. Urban Design, and Planning. 2014;**167**(DP5):189-195

[24] Abigail O. The evolution of social welfare and social work in Nigeria. The Journal of Consumer Research. 2011;**8**(3):238-252

[25] Deedam GD, Akpe CI, Juliana OE. Nigeria development plans and its challenges: The way onward. International Journal of Advanced Academic Research/Social and Management Sciences. 2019;**5**(12):1-12

[26] Abubakar A, et al. Smallholder farmer's adaptation strategies to drought in the Sahelian Zone of Jigawa State, Nigeria. International Journal of Advanced Academic Research. 2018;**4**(10)

[27] Grantham HS, Bode M, McDonald-Madden E, Game ET, Knight AT, Possingham HP. Effective conservation planning requires learning and adaptation. Frontiers in Ecology and the Environment. 2009;**8**:431-437

[28] Ken S. Strategic planning and community involvement as contributors to sustainable tourism development. Current Issues in Tourism. 2001;**4**(1):3-41

[29] Okafor EE. Child labour dynamics and implications for sustainable development in Nigeria. Journal of Sustainable Development in Africa. 2010;**12**(5):8-21

[30] Ajomale O. Country report: Ageing in Nigeria–current state, social and economic implications. Summer Newsletter of the Research Committee on the Sociology of Aging of the International Sociological Association. Oxford: Oxford Institute of Ageing; 2007:15-20

[31] Ajala OA, Sanni L, Adeyinka SA. Accessibility to health care facilities: A panacea for sustainable rural development in Osun State Southwestern, Nigeria. Journal of Human Ecology. 2005;**18**(2):121-128

[32] Munasinghe M. Basic concepts and principles of sustainomics. 2007. Available from: http://www.eoearth.org/view/article/51cbed0b7896bb431f68f477

[33] Ojoko EO, Abubakar HO, Ojoko O, Ikpe EO. Sustainable housing development in Nigeria: Prospects and challenges. Journal of Multidisciplinary Engineering Science and Technology. 2016;**3**(5):4851-4860

[34] Wei Y, Huang C, Lam PT, Yuan Z. Sustainable urban development: A review on urban carrying capacity assessment. Habitat International. 2015;**46**:64-71

Chapter 4

Smart and Sustainable Cities: A New Urban Transformation

Jorge Carro-Suárez, Susana Sarmiento-Paredes and Doroteo Nava

Abstract

The "smart city" concept was born with the aim of providing an urban complex that allows society to achieve greater well-being and a better quality of life through greater efficiency of public services and its infrastructure through digital technologies. However, today society not only demands to improve its current well-being, but also cares about that of future generations, in such a way that smart cities must consider sustainability as an integral part of their development and evolution. Therefore, the aim of this chapter is to determine the impact of digital technologies implemented in smart cities on the social, environmental, economic and institutional dimensions of sustainable development and thus promote urban development that is not only smart but also sustainable, taking as reference dimensions evaluated by indices worldwide.

Keywords: smart cities, digital technologies, dimensions of sustainable development, sustainable cities, Industry 4.0

1. Introduction

United Nations predict that 70% of the world population will live mainly in urban areas and cities in 2050, which means that the main social problems that affect society will also increase [1]. Population growth, environmental pollution, traffic congestion, lack of physical infrastructure and new environmental and energy requirements are today the main challenges facing large cities [2]. That is why cities around the world are seeking to achieve progress and improvements through the adoption of smart and innovative technologies that address these problems and that not only improve people's quality of life, but also lead them to be more sustainable spaces, energetically more efficient and environmentally less polluting [1]. This situation gave rise to the concept of "smart cities" given the need to implement modern urban development that takes advantage of the exponential growth of digital technologies [3].

The concept related to the smart city was born in the United States in 1992 in the book "The Technopolis Phenomenon", which was used to represent the growth of information technology and its application in modern infrastructures of the 1990s [3]. Today, it is contextualized more as a relationship between new digital technologies and the physical infrastructure of cities to solve various urban challenges [4], in addition to maintaining a high quality of life, improving economic productivity and generating a safe environment in the event of emergencies or unexpected situations such as the COVID-19 pandemic that devastated the world in 2020 [5].

However, despite the fact that the development of smart cities has achieved acceptance by a large part of society as a means to mitigate various problems derived from growing urban development, there is also concern that with the arrival of new digital technologies cities will have to face other types of problems, such as the leakage of confidential information [6], the social risks that may arise such as social exclusion [7], the permanent rejection of technology by certain sectors of society and discontent in the face of an inequitable social impact [8].

So, the new trend is that smart technology and social innovations should be combined and be the base of any smart city [4]. In such a way that for the development of a smart city, the development of an urban green economy must also be promoted through the reduction of polluting emissions, energy consumption and greater regional development [9], taking from now on the necessary measures to take the next step and achieve a transformation, that is, to go from being smart cities to being smart and sustainable cities [10].

Currently, there are cities in the world that have begun to adopt these measures, such is the case of Singapore with the launch of its Smart Nation initiative in 2014, which contemplates a new eco-intelligent city completely free of vehicles supported by clean public transport and efficient, with more safe areas for pedestrians and cyclists [11]. Helsinky, Finland, has set itself the goal of becoming carbon neutral by 2035 by gradually reducing its emissions from traffic by implementing various measures, such as transitioning its urban service bus fleet from internal combustion systems to electric systems, incorporating and promoting the use of renewable energy for 100% of the city's buildings [12]. Zurich (Switzerland) is innovating in its public lighting system with smart lamps that adapt to traffic levels through sensors, gradually increasing its energy savings, based on efficient management of heating, electricity and cooling in the city through of smart buildings. At a strategic level, initiatives such as “Zurich 2035 Strategies”, “2000 Watt Society”, “Urban Transport 2025” and the “Open Data Strategy” have been developed [13]. Due to the rapid growth of its population, New York (United States) implemented a smart city pilot program in 2020, accumulating data to help manage its waste management and collection services more efficiently, enabling smart centers with contactless technology and services of car sharing to reduce traffic congestion, with the goal of reaching net zero emissions by 2050 [14]. Seoul (South Korea) started these campaigns in 2014 based on data management. Through the analysis of urban patterns such as speed and traffic flow, as well as air quality measured by sensors, it has formed a solid infrastructure with intelligent services. On the other hand, it has focused its technology on the care of its adults who live alone and to detect and prevent criminal patterns. The Pilot National Smart City program was implemented to freely integrate Industry 4.0 related technologies by creating an innovative industrial ecosystem that can drive more creative business models [15]. Finally, Oslo (Norway) has as its main objective to reduce polluting emissions by 95% in 2030. To this end, it is making efforts and creating opportunities in areas that include electric vehicles with their special charging equipment and services, smart grids, renewable fuels, green buildings and circular economy [16].

It is clear that, in the world, these efforts are carried out from different perspectives, which leads to the need to do so in a sustainable and comprehensive way. Therefore, the objective of this chapter is to determine the impact of digital technologies implemented in smart cities on the social, environmental, economic and institutional dimensions of sustainable development, taking as reference specific dimensions evaluated by various indices worldwide on smart cities and promote an urban development that in addition to being intelligent, is also sustainable.

2. Smart city performance indices

Globally, there are various smart city indices that assess different factors. Among the most prominent are the following:

2.1 Cities in motion index

Published by the Center for Globalization and Strategy and the Strategy Department of the IESE Business School. It is a platform that connects a global network of experts in cities and companies whose objective is to promote changes at the local level and develop innovative ideas and tools to make cities more sustainable and intelligent. The last report issued in 2022 evaluated 183 cities through 9 dimensions under the following criteria [17]:

- Human capital

  It believes that the main objective of every city should be to improve its human capital. A city with intelligent governance must have the capacity to attract and retain talent, create plans to improve education and promote both creativity and research. Although human capital includes factors that make it broader than what can be measured with certain indicators, there is an international consensus that educational level and access to culture are irreplaceable components for its measurement. The Human Development Index (HDI) published annually by the United Nations Development Program (UNDP) mentions that education and culture are the determining pillars for the development of human capital. Therefore, Cities in Motion Index (CIMI) 2022 considered the percentage of the population with higher education, the number of public and private schools, the number of internationally recognized universities, annual spending on education, spending on leisure and recreation, the international movement of students and number of museums, galleries and theaters as indicators that explain the differences in the human capital of the cities.

- Social cohesion

  Social cohesion is a sociological dimension of cities that can be defined as "the degree of consensus of the members of a social group" or also as "the perception of belonging to a common project or situation". It is a measure of the intensity of social interaction within the group. In the urban context, social cohesion refers to the level of coexistence between groups of people who live in the same city and have different incomes, cultures, ages or professions. Concern for the social environment of the city requires the analysis of factors such as immigration, community development, care for the elderly, the efficiency of the health system, and citizen security and inclusion. In these times of COVID-19, the capacity and universality of health systems are evaluated and measured in a special way, with the hope that the health crisis has served to understand the importance of these systems and strengthen them.

- Economy

  This dimension includes all those aspects that promote the economic development of a territory: local economic promotion, transition and strategic industrial plans; cluster generation; innovation and entrepreneurial initiatives. If we consider that the dimension intends to measure through multiple indicators the future sustainability of the main cities of the world and the quality of life of its inhabitants, the real GDP becomes a real measure of the economic power of the city and of the income of those who inhabit it. In fact, in numerous studies, GDP is the only measure—or the most important—of the performance of a city or country.

- Governance

  Term commonly used to designate the efficiency, quality and proper orientation of State intervention. Given that the citizen is the meeting point to solve all the challenges that cities face, factors such as the level of citizen participation and the ability of the authorities to involve business leaders and local agents must be taken into account, as well as the application of e-government plans. Likewise, this dimension includes all those actions aimed at improving the efficiency of the administration, which include the design of new organizational and management models. In this section, great opportunities open up for private initiative, which can provide greater efficiency.

- Environment

  The sustainable development of a city can be defined as "development that meets the needs of the present without jeopardizing the ability of future generations to meet their own needs." In this sense, factors such as the improvement of environmental sustainability through anti-pollution plans, the promotion of ecological buildings and alternative energies, efficient water and waste management and the existence of policies that help to counteract the effects of climate change are essential to guarantee the sustainability of cities over time.

- Mobility and transportation

  The cities of the future have to face two great challenges in the field of mobility and transport: facilitating movement (many times, in large territories) and access to public services. Mobility and transport—both in terms of road and route infrastructure, the car park and public transport as well as air transport—affect the quality of life of the inhabitants of a city and can be vital for sustainability of it over time. However, perhaps the most important aspect is the externalities that are generated in the production system, either due to the need to move the labor force or the need to exit production.

- Urban planning

  The urban planning of cities has always been considered a driver of development and poverty reduction. Today, it is also a collective exercise that must involve all stakeholders, such as citizens, civil society organizations, the public and private sectors, multilateral organizations and academics.

- In turn, urban planning is closely related to sustainability. To improve the habitability of any territory, it is necessary to take into account local master plans and the design of green areas and spaces for public use, in addition to committing to intelligent growth. New urban planning methods should focus on creating compact cities, with good connections and accessible public services.

- International projection

  Cities that want to progress must achieve a privileged place in the world. In this sense, maintaining the global projection involves improving its brand and its international recognition through strategic tourism plans, attracting foreign investment and representation abroad. Cities may enjoy a greater or lesser international projection, even if they belong to the same country, but it is not independent of the degree of national openness. This dimension aims to reflect these differences and measure the international projection of cities.

- Technology

  Information and communication technologies (ICTs) are part of the backbone of any society that seeks to achieve the status of "intelligent". Technology is an aspect of society that improves the current quality of life, while the level of development is an indicator of the achieved or potential quality of life. In addition, technological development allows cities to be sustainable over time, as well as maintain or expand the competitive advantages of their production system and the quality of employment. A technologically backward city has comparative disadvantages compared to others, both from the point of view of security, education or health—all of which are essential for the sustainability of society—and from the perspective of the productive system. As a result, production functions have become anachronistic, and competitiveness without protectionism is diminished, which has a negative impact on the city's capacity for consumption and investment, as well as reducing labor productivity.

Based on these dimensions, CIMI presents London (United Kingdom) as the first place, followed by New York (United States), Paris (France), Tokyo (Japan) and Berlin (Germany) as the first five cities in the ranking. 2022.

2.2 E-government development index

The e-government development index (EGDI) presents the state of development of the electronic government of the Member States of the United Nations. Along with an assessment of the factors of Web site development in a country, the EGDI incorporates access characteristics such as infrastructure and educational levels to reflect how a country is using information technologies to promote the access and inclusion of all their society. The index is a composite measure of three important dimensions of e-government [18]:

- Online service (OSI)

  The OSI component is a composite indicator that measures the use of ICTs by governments for the provision of public services at the national level. The OSI values are based on the results of a comprehensive survey covering multiple aspects of the online presence of all 193 Member States. The survey evaluates the technical characteristics of national Web sites, as well as the electronic government policies and strategies applied in general and by special sectors in the provision of services. In the 2022 edition, for the first time, the OSI was calculated based on five weighted sub-indices. Specifically, Member States are assessed for service provision, technology, the institutional framework supporting e-government development, content provision and e-participation.

- Telecommunication infrastructure

  Consider deployment for home Internet access, individual Internet usage and mobile phone penetration.

- Human capital

  It collects information on the educational level of the population, with the literacy rate and average years of schooling.

The 10 countries with the highest evaluation of the 2022 survey corresponding to the EGDI are Denmark, Finland, South Korea, New Zealand, Sweden, Iceland, Australia, Estonia, the Netherlands and the United States.

2.3 Smart city index IMD

The Smart Cities Index prepared by the Smart Cities Observatory of the IMD Global Competitiveness Center in collaboration with the Singapore University of Technology and Design (SUTD) analyzes and rates 102 cities in the world. Evaluates indicators such as health and safety, mobility, activities and opportunities together with the level of infrastructure of cities, benefits and technological services for its inhabitants, these data integrate the two pillars on which the perception of residents is requested, which are focused in measuring how citizens perceive the efforts that their city makes to become a smart city. This takes economic and technological aspects into account, but from a social approach adapted to human dimensions, as explained by the observatory itself [19]:

- Attitudes

  It shows and compares responses to three key aspects of privacy (willingness to give up data, comfort with facial recognition and whether online information has increased trust in authorities).

- Structures and technologies

  It collects key data from the structures and technologies survey in five key areas: health and safety, mobility, activities, opportunities and governance.

According to the index, being a globally recognized smart city is today critical to attracting investment and talent, creating a virtuous circle. The 2021 results show that the top ten smart cities in the ranking are Singapore, Zurich, Oslo, Taipei City, Lausanne, Helsinki, Copenhagen, Geneva, Auckland and Bilbao, highlighting aspects such as public transport, basic sanitation in areas humbler, green spaces, the provision of medical services and recycling services in terms of structures. Also noteworthy is the improvement in the organization of online medical appointments, as well as its use in buying tickets for shows.

2.4 Smart city strategy index

Index issued by the German consultancy Roland Berger measures the efficiency and management of urban centers in light of the key digital applications that identify a smart city, providing a global vision of strategies applied by various cities. In its last edition in 2019, it identified 153 cities with an official smart city strategy, considering Vienna Austria in the lead, followed by London in the United Kingdom and St. Albert Canada. Its evaluation framework is based on 3 smart city dimensions, 12 criteria and 32 sub-criteria [20].

- Planning

  Consider the time it takes to apply your strategic plans, your viable objectives, forms of financing, priority and execution, administrative coordination and the acceptance of all your stakeholders and partners.

- Infrastructure and policy

  Consider the political and legal framework both in its regulation, financial and innovation support and secure data management. In infrastructure, technological connectivity, transparency of information and the speed of their Internet networks are evaluated.

- Action fields

  It evaluates the constructions from their ease of management, applications in the home and construction, and in energy and environment, it considers the management of energy, water and waste. In mobility, it evaluates the management of urban traffic and logistics. In education, it takes into account the capacity of educational platforms, learning models and digital skills. In health, it analyzes information systems, assisted medicine and telemedicine, finishing with the government through its e-services and its digital public administration.

As a complementary part, in this report it considers seven essential points as urban planners for the design of a perfect smart city [20]:

- Take stock

  This exercise requires taking stock of existing plans and projects, being a strategy that presents the perfect opportunity to reassess the role of the city and clearly define its digital services and its portfolio of current and future projects.

- Involve all stakeholders

  To ensure greater efficiency, all stakeholders, including citizens and external providers, must participate in the strategies proposed by the city, as well as in their implementation.

- Think integrated

  A smart city must be based on clearly defined fields of action, supported by solid actions in strategic digitization areas.

- Get private support

  The private sector is the key to providing financing and expertise. Therefore, it is crucial that companies not only have as part of their business model the development of technologies, but also contribute to the financing of new and better solutions to the country's problems.

- Foster innovation

  As such, there is no perfect smart city model, so cities and their society should not limit companies to try new solutions. Innovation laboratories, technical and financial support must be determining points to implement this strategy.

- Establish (open) urban data platforms

  Data flow is the soul of a smart city. Digital solutions work better when they have more information, which means that data platforms are key points, hence the importance of developing solid cybersecurity policies to prevent the leakage information.

- Make concrete action plans

  The action plans defined by each city must be reviewed and updated periodically to keep their results up to date and guarantee proper efficiency with respect to their funding sources.

2.5 Smart cities index report

Issued by Ifm Engage in conjunction with Yonsei University and University of Cambridge, it represents a comprehensive study of 31 cities around the world, which reviews the effectiveness of local innovation initiatives, service development and infrastructure projects. It highlights trends and provides information on the impact of smart city development globally for better future strategic planning. In his 2022 report, it presented a baseline evaluation in eight dimensions [21]:

- Service innovation

  It is essential for the development of a smart city that the urban infrastructure is improved through the application of advanced ICTs. In this way, smart cities drive sustainable innovation as new industrial sectors emerge that offer services that are synergistic with a city's own information provision. Since the health crisis due to the COVID-19 outbreak, ICTs mediated services have become more essential than ever to people's lives. Many leading cities around the world are investing heavily in innovative advanced technologies with the aim of providing citizen-centric smart city services while enhancing the competitiveness of cities by driving various smart city policies. New York, Barcelona, Seoul, London and Amsterdam lead this dimension.

- Urban intelligence

  Cities around the world are actively working on how to solve various urban problems using Industry 4.0 technologies, such as the use of big data and blockchain, through services based on Web applications, infrastructure services and specific projects. On the other hand, the use of artificial intelligence (AI) technology to promote infrastructure services is also recognized, as well as in the fields of health care, social services, crime and disaster prevention, transportation, and energy management. In this dimension, Seoul, Barcelona, Amsterdam, Helsinki and Lisbon stand out.

- Urban sustainability

  At this point, smart city projects are classified by service areas, identifying the type of development (services, infrastructure and projects) that has the greatest impact on urban sustainability, with special attention to the energy and environment sector, as well as to Web application-based services. Amsterdam, Copenhagen, Helsinki, Berlin and London lead this dimension.

- Urban openness

  It represents one of the key factors that determine the competitiveness of a city together with the development of a smart city. An open city is innovative and usually takes the form of an ecosystem, within which city government offices work together with large, medium and small companies, as well as young start-ups. Two parameters are basically considered: (1) the opening of urban data and (2) informed citizen participation under the concept of open innovation. Taipei, Seoul, New York and Singapore lead this dimension.

- Infrastructure integration

  The infrastructure of a smart city is a complement of Industry 4.0 technologies that interconnect the physical urban infrastructure (building and public lighting) with cloud applications and data, fostering innovation and improving services for society. It also represents the concept of a smart city as a way to bridge the digital divide, responding preemptively to the vision of a hyper-connected and hyper-intelligent society by making infrastructure fully accessible within the city through various digital technologies. In this dimension, Seoul, Incheon, Barcelona, Taipei, Busan and Dubai stand out as leaders in infrastructure integration.

- Urban innovativeness

  A smart city is innovative to the extent that it undertakes the task of creating and designing an ecosystem in which Industry 4.0 technologies can be implemented and prosper, including aspects such as construction and zoning, promoting innovative projects with the aim of generating greater socioeconomic value through the development or commercialization of services through technological innovations.

- Collaborative partnership

  It is defined as the mutual and cooperative relationship to create and activate ecosystems for the efficient development of services and infrastructures in smart cities. Its application is useful in overcoming budgetary and resource constraints. Local governments, research institutions, private companies, Small and Medium-sized Enterprises (SMEs), citizens and other neighboring cities are included. It identifies the establishment of policies so that associations can be a driving force that increases synergy in the development of infrastructure and quality services, fostering innovation and agility in the development of resources. In this context, Seoul, Shanghai and Singapore are the leading cities in this dimension.

- Smart city governance

  The term refers to the organizational structures in which a city brings together multiple stakeholders to solve common problems that occur during the delivery of public services. Through governance systems, governments and their collaborating partners share a vision, identify priority problems, propose strategies and methods and determine political decisions to promote projects with social impact, systemically establishing a digital transformation in key sectors, such as transport, energy and environment. New York, Amsterdam, Vienna, Singapore and Dubai are the leading cities in this dimension.

Cities around the world have accepted these challenges and have adopted various factors and strategies as part of their response to the problems that afflict society in general. Likewise, through the development of innovative ideas and approaches, transformations have been shaped that have taken advantage of new digital technologies not only to improve people's quality of life, but have also identified that the environment and the planet require the same attention. The next step is to determine how a smart city can promote more sustainable and sustainable urban habitats from an integral point of view, that is, that its impact is of a social, economic, environmental and institutional for the benefit of all.

3. Smart cities and comprehensive sustainability

3.1 Industry 4.0 digital technologies

Improving processes and optimizing available resources by perfecting and/or replacing human action was one of the main causes for which the concept of Industry 4.0 was born based on disruptive digital technologies that boosted the growth and development of companies and organizations around the world. Nine main pillars are highlighted [22]:

- Big data
- Autonomous robots
- Simulation and augmented/virtual reality
- System integration
- Internet of the things
- Cybersecurity
- The cloud
- Additive manufacturing

- AI.

These technologies allowed the industrial sector to achieve greater operational efficiency, better management of its resources and improvement of its processes, resulting in greater productivity based on the creation of value of its products and services. Therefore, the new trend recognizes that these new technologies are the way for not only the business sector to develop, but also for smart cities to better manage their resources and more appropriately face and solve their main current and future problems.

3.2 Integral sustainability

Implicitly sustainable development plays a determining role for smart city service management, as part of a complete innovation cycle [21]. Despite the fact that digital technologies under the concept of Industry 4.0 were developed due to the need to convert normal machinery into intelligent machinery with learning capacity to improve its performance [23]. The World Economic Forum's (WEF's) Network Readiness Index 2021 described that technology and people need to be integrated into effective governance to have the right impact on the economy, society and the environment. Recognizes that despite the fact that digital technology is changing people's lives in a dizzying way, there are still gaps in access to it in various regions and economies of the world [24].

Likewise, the United Nations 2030 Agenda has objectives that can be achieved more quickly thanks to the development of smart technologies. Ensure healthy lives and promote well-being for all at any age (SDG 3); guarantee inclusive, equitable and quality education (SDG 4); guarantee the availability of water and sanitation (SDG 6); guarantee access to affordable, safe and non-polluting energy (SDG 7); build resilient infrastructure and foster innovation (SDG 9); make cities and human settlements inclusive, safe, resilient and sustainable (SDG 11); taking urgent action to combat climate change (SDG 13) and building effective, accountable and inclusive institutions at all levels (SDG 16) are part of the sustainable development goals that can be directly linked to the development of a smart city [25]. Por lo tanto, un desarrollo sostenible integral requiere una atención específica a cuatro dimensiones [26]:

- Environmental

  In this dimension, it is established that products and processes must be friendly to the environment through the prevention of pollution and good management of natural resources, recognizing the design of green products, from their raw material to the end of production its life cycle.

- Social

  The benefit of people and their environment is sought with an efficient administration of human resources, providing health, safety and economic growth, both to organizations, personnel and to the place where the company has been established.

- Economic

  The proposal of both economic and social profitability is proposed, in the search to generate results and/or benefits based on investment in technology to reduce energy consumption and improve the environmental quality of the processes, also conceiving that its aim is not only to generate economic profits, but also to give something back to society at a social, economic and environmental, level that allows the company to establish permanent roots.

- Institutional

  The creation of a sustainable culture is proposed, with a mission and vision that impacts organizations and society, establishing norms, habits and values so that they are practiced at all times, making them a new form of behavior and education.

Under these principles, it is highlighted that a smart city must not only consider the implementation of technologies to improve the living conditions of its inhabitants, and it must comprehensively cover these four dimensions. The dimensions and strategies presented by the evaluation indices are the best indicator to assess the environmental, social, economic and institutional impact experienced by cities and which, in turn, will serve as a basis to promote them in other cities that are different economic, social and political conditions. Based on the foregoing, **Table 1** shows the relationship of the dimensions proposed by smart city indices analyzed with each of the dimensions of sustainable development.

In the first instance, from the point of view of sustainable development, **Table 1** highlights the following:

In the environmental dimension, mobility and transport problems are determining aspects that every smart city must solve by virtue of reducing pollution rates. The environment and urban sustainability must assess and address the effects on ecosystems of air pollution, water quality, biodiversity, deforestation and climate change. A determining and vitally important point is to implement technology for the efficient management of water resources to really have a sustainable future. Water is a vital

Smart city indices	**Sustainable development dimensions**			
	Environmental	**Social**	**Economic**	**Institutional**
CIMI	Environment Mobility and transportation Technology	Human capital Social cohesion Urban planning	Economy Mobility and transportation Technology	Social cohesion Governance International projection
EGDI		OSIs Human capital	Telecommunication infrastructure	
Smart City Index IMD		Structure and Technologies	Structures and technologies	Attitudes
Smart city strategy index (SCSI)		Action fields	Infrastructure and policy Action fields	Planning
Smart cities index report (SCIR)	Urban sustainability	Service innovation Urban intelligence	Infrastructure integration	Urban openness Urban innovativeness Collaborative partnership Smart city governance

Table 1.
Relationship between dimensions of smart cities with sustainable development.

component for the population, so an intelligent administration that manages access to its supply as a basic element in the development of any city is important. In another context, the solid waste generated by each person and each household in a city represents great potential damage not only to people's health, but also to the environment.

According to **Table 1**, the social dimension represents the main focus of attention for the development of a smart city, emphasizing the importance of improving people's quality of life, from providing safe streets and green spaces to quality hospital care and efficient OSIs. Another determining aspect is the valuation of human capital, so it is necessary to strengthen them from various perspectives, one being education, promoting its evolution through the incorporation of digital technologies, in such a way that in a smart city, its educational institutions. Regardless of the academic level, they incorporate the use of cutting-edge digital elements into their development and performance. From another perspective, efficient and high-quality health services must be provided for the entire population, incorporating cutting-edge technology for better care in the service and medical treatments.

In the economic dimension, the fact that a smart city must present both local and international competitiveness through a new type of economy according to our era, the digital economy is highlighted. In fact, the integration of Industry 4.0 digital technologies can significantly contribute to achieving the sustainable development goals of the 2030 Agenda. However, despite the fact that its ability to change the labor market will imply the inevitable disappearance of certain jobs in the future, it will also promote the creation of other, better-paid jobs, but requiring a new series of knowledge and skills. All this will require greater investment and new learning models in the education sector. Improve urban planning through green designs of public buildings, green cities, green industrial corridors, adoption of Internet of things (IoT) systems, clean and renewable energy sources, application of emerging technologies for urban agriculture, central markets smart technologies, the technological efficiency of public services (such as transport and mobility) and the modeling of digital environments today represent economic alternatives for the search for not only a digital economy, but also a green economy.

In the institutional dimension, smart cities must be part of a shared vision between the government and citizens based on social justice, solidarity and democratic values. These principles are decisive for the image that each smart city presents to the world. Being recognized internationally will lead to a greater attraction of foreign investment, a greater number of tourist visits and a continuous international projection. This requires good strategic planning, in which the government, civil society and public and private companies must align themselves with the same mission and vision according to their own institutional limitations (political and regulatory).

Secondly, according to the WEF, the main challenges facing the cities of the new century are [27]:

- Environmental threats

  Includes affectations caused by meteorological phenomena related to climate change.

- Resources management

  Water, food and energy will be the most determining resources for each city due to constant urban growth.

- Inequality

  The gap between the haves and have-nots will widen and become more pronounced in the megacities of the future. It will be necessary to formulate policies that guarantee that the fruits of progress are distributed equitably in society.

- Technology

  Technology will be used more and more, especially that related to energy, mobility, transportation, health, safety, education and the environment. However, as cities become digital, access must be guaranteed to all of society without discriminating against the economically disadvantaged groups.

- Governance

  For a better future, good governance practices and good urban planning are required. Therefore, its main objectives should be to guarantee equity, habitability and sustainability for all of society in the face of the inevitable increase and diversity of the population.

Therefore, the smart cities of the future must identify these problems, analyze them together with those they are already dealing with and propose solutions to overcome them, relying on the impact of digital technologies as follows:

- Big data

  Given the exorbitant amount of data and information that it is possible to obtain from the dynamics of large cities, the collection and evaluation of data in real time will be decisive for decision-making. Medical, social, transportation, weather, academic histories, etc., will be valuable information to better understand each problem and thus provide increasingly better results based on continuous learning.

- Autonomous robots

  Robotic systems will no longer only be used in industrial systems. In the new smart cities, they will have various applications for the benefit of society. Companions, protectors and both domestic and public workers are some of the new tasks that they will carry out in the future for the greater well-being of society and the community.

- Simulation and augmented/virtual reality

  The analysis of urban problems through computerized models and virtual environments will allow large cities to optimize their resources and provide solutions more quickly and efficiently. Transportation, pollution and public safety are some of the issues that have already begun to be addressed with these technologies.

- System integration

  The exchange of information through networks and applications will be the best way to integrate various technologies in real time for the safety and well-being of the population regardless of their social or economic condition.

- Internet of the things

  Communication by means of protocols and devices through a digital network will allow the interconnection of countless reading, measurement and control devices located in strategic positions for the management and control of buildings, transport, security and public services in a more efficient and safe.

- Cybersecurity

  Given the need for data management and management, the protection of systems, platforms and servers becomes a critical point that must be addressed with a high priority. In particular, governments and public and private organizations must guarantee the protection of all the information they have under protection, relying on high-security digital technologies.

- The cloud

  The constant growth of the communications infrastructure requires a greater availability of computers and servers. Given this, it is essential that a smart city has as part of its services digital repositories for the storage, access and use of digital and computer resources that are available without the need for physical equipment, with sufficient capacity to share information from any point in the city.

- Additive manufacturing

  Despite the fact that this technology is more focused on industrial aspects, within smart cities it has found a space in the development and construction of houses made with environmentally friendly materials, being a new alternative to correct habitat problems in big cities.

- AI

  Efficient access to public services, crime prevention, intelligent transport networks, development of communication networks, protection of the environment and care and health of the population are some of the issues in which AI is being developed to be implemented in the big cities. The design of AI algorithms will create expert systems that will make predictions based on data analysis and continuously learn to make and improve their decisions in solving urban problems, just as a human would, but with greater precision and speed.

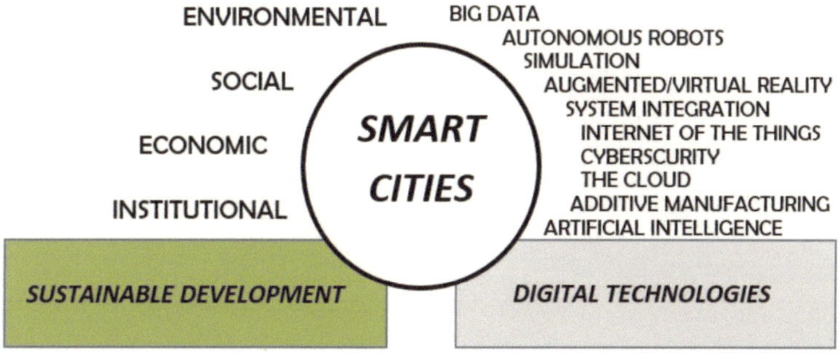

Figure 1.
Pillars of the smart cities of the future.

Figure 1 shows this relationship, in which the smart cities of the future will be supported by two determining pillars: sustainable development and digital technologies.

Both pillars represent the new bases and trends that today's society requires and demands to really obtain a better quality of life, greater economic growth and the guarantee of prosperity for future generations.

4. Conclusions

Digital technologies have evolved, and their media impact on the economic development of companies, society, countries and cities has become evident, boosting their competitiveness and innovation. They have also implicitly generated an impact on the social and environmental dimensions. But, to achieve integral sustainability, it is crucial that cities point towards the institutional dimension through new policies and regulations that regulate their actions more efficiently, thus implementing a new sustainable culture throughout their society.

It is clear that technological development grows along with the needs of society. Therefore, urban problems such as environmental pollution, housing, unemployment, public services, sanitation, health, transportation, energy, and public safety, to name a few, require greater attention from government agencies, and digital technologies are the best tool for your solution. Smart cities should be an example of a life system for those who aspire to be, even for those cities that are developing. The next step is to take advantage of digital technologies to go beyond just addressing these urban issues.

Today's society not only demands to improve their quality of life, now it also demands care for the planet and its natural resources. The Brundtland Report in 1987 mentioned the importance of efficiently and rationally managing our resources so that it is possible to improve the current well-being of our society without compromising the quality of life of future generations [28]. Since then, society has slowly been moving towards that goal. Today, smart cities represent one more step in this development. Therefore, the next step is to transcend time and take another evolutionary leap, to move into the era of smart and sustainable cities.

Conflict of interest

The authors declare no conflict of interest.

DOI: http://dx.doi.org/10.5772/intechopen.110234

Author details

Jorge Carro-Suárez[1*], Susana Sarmiento-Paredes[2] and Doroteo Nava[2]

1 Polytechnic University of Tlaxcala, Tlaxcala, Mexico

2 Autonomous University of Tlaxcala, Tlaxcala, Mexico

*Address all correspondence to: jorge.carro@uptlax.edu.mx

References

[1] Lai O. Too 7 Smart Cities in the World in 2022 [Internet]. 2022. Available from: https://earth.org/top-7-smart-cities-in-the-world/ [Accessed: 05 December 2022]

[2] Patrao C, Moura P, de Almeida AT. Review of Smart City Assessment Tools [Internet]. 2020. Available from: https://www.mdpi.com/2624-6511/3/4/55 [Accessed: 08 December 2022]

[3] Lai CMT, Cole A. Measuring progress of smart cities: indexing the smart cities indices. Urban Governance. 2022 [In Press]. DOI: 10.1016/j.ugj.2022.11.004

[4] Quan X, Solheim MCW. Public-private partnerships in smart cities: a critical survey and research agenda. City, Culture and Society. 2022 [In Press]. DOI: 10.1016/j.ccs.2022.100491

[5] Alshamaila Y, Papagiannidis S, Alsawalqah H, Aljarah I. Effective use of smart cities in crisis cases: a systematic review of the literature. International Journal of Disaster Risk Reduction. 2022 [In Press]. DOI: 10.1016/j.ijdrr.2023.103521

[6] Pandya S, Srivastava G, Jhaveri R, Babu MR, Bhattacharya S, Maddikunta P, et al. Federrated learning for smart cities: a comprehensive survey. Sustainable Energy Technologies and Assessments. 2022 [In Press]. DOI: 10.1016/j.seta.2022.102987

[7] Shayan S, Kim KP. Understanding correlations between social risk and sociodemographic factors in smart city development. Sustainable Cities and Society. 2022 [In Press]. DOI: 10.1016/j.scs.2022.104320

[8] Twist A, Ruijer E, Meijer A. Smart cities & citizen discontent: a systematic review of the literature. Government Information Quarterly. 2022 [In Press]. DOI: 10.1016/j.giq.2022.101799

[9] Liu K, Meng C, Tan J, Zhang G. Do smart cities promote a green economy? Evidence from a quasi-experiment of 253 cities in China. Environmental Impact Assessment Review. 2022 [In Press]. DOI: 10.1016/j.eiar.2022.107009

[10] Alshuwaikhat H, Aina Y.A, Binsaedan L. Analysis of the implementation of urban computing in smart cities: a framework for the transformation of Saudi cities. Heliyon. 2022:**8**(10):1-10. DOI: 10.1016/j.heliyon.2022.e11138

[11] Smart Nation Singapore. Transforming Singapore Through Technology [Internet]. 2021. Available from: https://www.smartnation.gov.sg/ [Accessed: 12 December 2022]

[12] Helsinki Smart Region. Smart City + Smart Countryside = Smart Region [Internet]. 2022. Available from: https://helsinkismart.fi/ [Accessed: 12 December 2022]

[13] Shuang Z. Zurich Smart City Strategy [Internet]. 2018. Available from: http://icity.ikcest.org/city/project/398 [Accessed: 13 December 2022]

[14] Lai O. How New York Smart City Projects are Leading the Way [Internet]. 2022. Available from: https://earth.org/new-york-smart-city/ [Accessed: 13 December 2022]

[15] Smart City Korea. National pilot smart city [Internet]. 2021. Available from: https://smartcity.go.kr/en/ [Accessed: 14 December 2022]

[16] Smartcity Press. Oslo – A Smart City with a pioneering thinking against climate change [Internet]. 2019. Available from: https://smartcity.press/

climate-change-in-oslo/ [Accessed: 15 December 2022]

[17] IESE. Índice IESE Cities in Motion [Internet]. 2022. Available from: https://media.iese.edu/research/pdfs/ST-0633.pdf [Accessed: 18 December 2022]

[18] United Nations. E-Government Survey 2022 The future of Digital Government [Internet]. 2022. Available from: https://desapublications.un.org/sites/default/files/publications/2022-09/Report%20without%20annexes.pdf [Accessed: 20 December 2022]

[19] IMD. Smart City Index 2021 [Internet]. 2021. Available from: https://imd.cld.bz/Smart-City-Index-2021 [Accessed: 21 December 2022]

[20] Roland Berger Consulting. Smart City Strategy Index 2019 [Internet]. 2019. Available from: https://www.rolandberger.com/en/Insights/Publications/Smart-City-Strategy-Index-Vienna-and-London-leading-in-worldwide-ranking.html [Accessed: 22 December 2022]

[21] Ifm Engage. Smart Cities Index Report 2022 [Internet]. 2022. Available from: https://smartcitiesindex.org/smartcitiesindexreport2022 [Accessed: 24 December 2022]

[22] Carro J, Sarmiento S. The human factor and its role in the transition to Industry 5.0: a systematic review and future perspectives. Entreciencias: Diálogos en la Sociedad del Conocimiento. 2022;**10**(24):1-18. DOI: 10.22201/enesl.20078064e.2022.24.81727e24.81727

[23] Lee J, Kao HA, Yang S. Service innovation and smart analytics for Industry 4.0 and big data environment. In: Product Services Systems and Value Creation. In: Proceedings of the 6th CIRP Conference on Industrial Product-Service Systems; 2014; pp. 3-8

[24] Portulans Institute. Introducing the Network Readiness Index 2021 [Internet]. 2021. Available from: https://networkreadinessindex.org/introducing-the-network-readiness-index-2021/ [Accessed: 28 December 2022]

[25] United Nations. The 17 Goals. [Internet]. 2021. Available from: https://sdgs.un.org/goals [Accessed: 29 December 2022]

[26] Carro J, Sarmiento S, Rosano G, Garnica J, Vega-Lebrún CA. The process of innovation as a determinant factor of sustainable development in companies. Intrenational Journal of Innovation and Sustainable Development. 2021;**15**(1):100-125. DOI: 10.1504/IJISD.2021.111552

[27] World Economic Forum. 5 big challenges facing big cities of the future [Internet]. 2018. Available from: https://www.weforum.org/agenda/2018/10/the-5-biggest-challenges-cities-will-face-in-the-future/ [Accessed: 5 January 2023]

[28] United Nations. Report of the World Commission on Environment and Development: Our Common Future [Internet]. 1987. Available from: https://sustainabledevelopment.un.org/content/documents/5987our-common-future.pdf [Accessed: 8 January 2023]

Chapter 5

Local Development Model as an Element of Regional Sustainable Strategy

Mieczysław Adamowicz

Abstract

Regions constitute key elements of the territorial structure of any country. Internal differences in regions permit to distinguish subregional and local units in them, where economic and social activities are undertaken. Despite the internationalisation and globalisation processes, the importance of local development concept is growing. There are several factors influencing the growing role of local units in socio-economic development of regions. In many countries, local communities have a great autonomy in creating local development strategies and policies. Local endogenic strategies and policies create an important complementary activity to sustainable regional planning. The aim of the work is to present a theoretical background of the local economy and local development. Using the subject literature, the theoretical concepts, definitions and the results of local development were presented. Selected models and paradigms of local development as well as interrelations between dimensions of local sustainability and links with regional planning were presented and discussed as well.

Keywords: local economy, local development, regional policy, sustainable development models, regional sustainable strategy

1. Introduction

A factor of key importance for the efficient functioning of a country is the division of the country into regional systems, which are usually established on a statutory basis. Regions separated in consideration of many historical, geographical, natural, social, and economic criteria, forming the basic spatial and functional structure of the country, have a relatively permanent character and play a big role in the development plan and program formation procedure, in the efficient functioning of economy and society and the general management of the state. Regional territorial systems as a concrete way of expressing the role of the geographical space in the socio-economic development of the country is a concept that becomes particularly significant nowadays when we can observe the development of international economic integration processes and the intensification of globalisation. However, economic processes occur and manifest themselves in local systems, which are a part of higher-level territorial systems – sub-regions and regions. Paying attention to the role of local systems and local development involves the need to increase the role of citizens in the management of countries and regions. Tendencies to include local communities in strategy formation and development planning processes were taken into account in concepts of grassroots development, localism, and local development [1, 2]. The interest in local development issues made it possible to show and become more familiar with the size of and reasons for the differentiation of territorial development and the role of modern development factors and to elaborate new concepts, models, policies, strategies and plans of development. The subject matter of this paper is local economy and local development, with particular regard to its essence, concepts, models, and dimensions.

2. Local economy: concept and elements

Local economy encompasses the entirety of economic resources, managing entities, phenomena and forms of the emergence of business activity that occur in the given local system. Local economy can be defined as socio-economic activities aimed at supporting the development of the local unit, i.e., the local territorial system in which local factors are used and local development barriers are taken into consideration [3, 4]. In addition, local economy can be regarded as a complex process in which local authorities stimulate the economic development of the given territorial unit with the use of their own resources and the commitment of external partners [5, 6]. Local economy is the sum of business activity and economic phenomena in the city, village or commune such as production, service, trading, investment, housing, transport, innovation and any other activity. Economic phenomena manifest themselves in the local market of goods and services, the labour market, and markets of other resources and production factors. However, local economy produces goods not only for the local market but also for external markets, even with a global reach; apart from that, it also receives resources, products and services, information, and innovations from the outside. This connection with external markets determines its scale of closure and opening. Although possibilities of development and the method of functioning of local economy depend mainly on the endogenous sphere, this connection with the external environment constitutes a very significant development factor for it. Various elements of local economy are interrelated and form a sort of coupled system serving the local community; external connections help to make better use of own resources in actions for the environment, and external resources create opportunities for the development of an internal system.

From the perspective of value creation in a given local or regional system, economy can be divided into two groups of branches, one of which encompasses basic branches focusing on production for markets outside the border of the given system, whereas the other encompasses supplementary service branches engaged in production mainly for local markets inside the system, i.e., for the local or regional market. Basic production and service activity in the form of export from the system also maintains supplementary activity consumed inside the system. The division into basic (export) activity and locally consumed service activity becomes useful for the determination of plans and strategies for socio-economic development of the given system. The ability to generate supra-local added value is a sign of the development potential existing in the given system.

Local economy can also be defined as a group of entities managing and participating in management systems in a city or commune, which are interrelated and interdependent in various ways [1]. By making use of available resources, each managing entity and each organisation fulfils certain functions and strives to accomplish its own goals, at the same time accomplishing social goals.

In local systems, a local environment comes into being, in which local economy and the local community function. The environment as a general concept is a relatively permanent system and character of surroundings that encompasses people and things important for human life and behaviour on a limited spatial scale. The local environment consists of a set of natural, economic, social and cultural features determining the distinctness and character of the given place or limited territory. An important element of the local environment that affects local economy is the social environment. The social environment can be defined as a relatively permanent system of individuals, groups and other human collectives producing human activity and having an impact on human behaviour. The local system is characterised by the existence of the local community. The local community is understood as a group of people living in a limited and relatively isolated territory that possesses and appreciates common traditions, values and symbols and service and cultural institutions, aware of their unity and distinctness, is ready to act together and lives in a sense of belonging and internal safety [7]. In the context of the concept of local communities, the idea of a civil society characterised by a pro-active attitude, the self-regulation capacity and the ability to accomplish goals without impulses from the state and its institutions emerge nowadays.

3. Growth and development in local systems

Aiming to satisfy their constantly growing needs, people and communities undertake business activity that gives rise to new economic values embodied in the concept of economic growth. Economic growth is considered mainly in terms of quantity and measured by the total value of the newly created value in conversion into a unit of outlays used for its production. A synthetic measure of all effects of management is the concept of economic growth and development, which is measured most often by the amount of the gross domestic product per inhabitant. The economic development process is a combination of economic growth and qualitative changes occurring in the field of production factors and production process results, and also in the institutional and social sphere of management [8]. Economic growth is not a goal in itself but a method of achieving economic, social and environmental goals in various territorial systems: the national, regional and local system. Integrated growth and development processes are shown in the most versatile and mature way in theories recognising the role of the state as the subject of development policy. The deepening of economic polarisation and the development of international integration and globalisation led to an increased interest in regional development (the 1960s and 1970s) and local development (the 1980s).

Many authors do not introduce a distinction between development in various territorial systems and treat local development as regional development on a smaller scale [9, 10]. In fact, there are very strong dependencies between local development and regional development, as well as between regional development and development on a national scale. These connections are reflected by the actions of territorial governments and state public administration. In practice, it is assumed in Poland that regional development refers to the entire province, whereas development on the scale of a county, city and commune should be regarded as local development [11].

Local development is a specific and very complex category of socio-economic development [12]. Its progress is influenced by many entities and factors operating and existing both within and close to the given territory. The complexity of the phenomenon of local development results both from the internal systemic character of local economy and the location of the given local system, particularly in the system of regional and national economy. The progress of basic economic processes on a local scale is shaped by the general economic and social system, the socio-economic policy of the country, regional policy and local development policy pursued by local authorities [13]. The category of local development has a dual character, so it should be considered in two ways: using the local and supra-local methods [2]. The local method is based on understanding local development as a process of transformation of local system structures and mutual relationships between elements of these structures and their connections with the environment through the mobilisation of mainly endogenous factors. In the supra-local perspective, local development is perceived as a transformation of structures of the aforementioned supra-local systems and relations between these structures and their environment.

The increased interest in local development also results from the manifestation of ethnic and cultural regionalisms existing in countries of integrating Europe. The emergence of paradoxes of globalisation showed that the spread of socio-economic phenomena across state borders required the intensification of actions on local scale and making use of grassroots initiatives. To some extent, it is a reaction to failures of the idea of top-down development or development based on external factors. Local development is associated with grassroots (bottom-up) development based on the endogenous potential supported by available exogenous factors. Local development is a special form of regional development in which endogenous factors used by local and supra-local entities for evoking changes strengthening the local development potential and the effectiveness of its use play a crucial role. Drawing attention to local development issues in the first half of the 1980s had economic and social as well as environmental and political grounds. These grounds existed in the then-current crisis of the welfare state concept and were related to the seeking of development opportunities in the centralisation of state management [14]. In Poland, the interest in local development issues also resulted from the systemic transformation process, which also encompassed a change in the territorial division of the country and the restructuring of territorial government administration.

The interest in local development in the European Union in 1980–2000 manifested itself in the introduction of various innovative actions, pilot programs and community initiatives aimed at the utilisation of spontaneous grassroots activity as the driving force of general economic development. After 2000, the local development policy assumed a slightly different, more dispersed form in instrumental and program terms [14].

4. Essence and criteria of the definition of local development

The conception of local development was analysed by many Polish and foreign authors, who formulated their own definitions and specified its essence, conditions and factors. This conception can be divided into two trends. The first trend, represented by a large group of authors, highlights in its definitions a long-term process of directional changes where objects or items evolve from simpler and less perfect forms and states into forms and states that are more complex and perfect, desirable and favourable from the perspective of specific goals.

Another group of authors believe that development is a process of changes based on progressive or regressive differentiation. Definitions of the first trend, which treat development as a process leading to the quantitative and qualitative improvement of a state of a specific territorial system, are used more frequently. Referring development issues to a local system, we can distinguish three main groups of definitions of local development: aspectual, general and systemic definitions [10].

Aspectual definitions consider local development separately within the scope of separate levels, including particularly the economic level. General definitions determine local development in a general way without referring to individual spheres or levels. Systemic definitions refer to three main development levels: the economic, social and environmental levels. Thus, they are similar to the concept of sustainable and permanent development.

The development process is always connected with a specific territory and a concrete spatial system—a local, regional, national or even broader system. Local development is related to a specific place, a given local system constituting the unity of natural and anthropogenic resources and of the local community and local economy. Local development can be considered in various aspects; social and economic aspects are used more frequently. Socio-economic development in a specific territory of a local system is a complex phenomenon that can be defined in various ways, depending on the aim of the analysis and criteria of assessment. Bagdziński [15] understands local development as 'jointly considered favourable changes in the territory of a local system that originate particularly from local natural and material resources and the characteristics of local communities that foster development, and the results of these changes contribute to the fuller satisfaction of inhabitants' needs and improvement of their prosperity'.

A definition similar to the above one specifies local development as a group of positive socio-economic changes in a qualitative and quantitative sense that occur in territories of cities, communes and counties [16] and make it possible to create new functional advantages and improve the existing ones in a given territorial unit and to create specific conditions for the development of economy [17, 18].

Some authors treat 'local development as a sign of local solidarity creating new social relations, which means expressing the will of inhabitants of the microregion regarding resources and factors of economic development'. Others perceive local development as a process through which the population gains broader autonomy in defining and satisfying their own needs, solving their problems and determining their own destiny. It is an outcome of the formation of collective dynamics in which economic, social, cultural and territorial aspects overlap and where local initiatives are connected with supra-local activities and resources [19].

Local development can also be defined as a process of economic, social, cultural and political changes leading to the improvement of the general level of inhabitants' prosperity. In other words, development is a process of transition in evolving into states that are more advanced and complex or more perfect in a certain respect. Local development is based on the rational use of internal and external factors for development of a given territory. Local development is a multi-aspect process that can be expressed in economic growth indicators or measured by social criteria or qualitative parameters of inhabitants' living standards [13, 20].

Local development is a social process based on stimulating the activity of local communities, highlighting local pro-development attitudes and including activities of social institutions in this development. This process must be preceded by the frequently long preparation of both resources for their economic use and people for joint participation in development [21].

We can distinguish a number of levels – e.g., the economic level, the social level and the ecological level – in local development. The economic level encompasses local economy: private and state-owned production and service enterprises, economic institutions and non-government organisations supporting entrepreneurship. The social level encompasses mainly households and the housing sector, as well as educational, health care, cultural, social care and public safety and order institutions providing services to people. The ecological level includes environmental degradation issues and problems, as well as issues concerning the rehabilitation of the natural environment. Local development occurs when programmed activities of local entities lead to the creation of new functional advantages and the improvement of the existing ones of a local territorial unit, the creation of optimal conditions for the functioning of economy and the local community along with the maintenance of the spatial order and ecological advantages.

Local economic development can also be defined as the quantitative and qualitative development of entities running business activity in a given local environment (commune, county) that can be associated with the improvement and introduction of new products and services, the expansion of market outlets, the modernisation of technology and investment, an increase of employment and the improvement of production effectiveness along with auxiliary financial, capital and innovative activities [13, 20].

Local development encompasses the social environment and is carried out from the perspective of local communities and local economic structures, institutions and bodies of territorial self-government. Local social development is based on the increasingly better fulfilment of the social needs of the population and business entities in a given territory. These needs include mainly the sphere of education and upbringing, health care, administrative support and the satisfaction of needs regarding culture and personal development, social care and public safety.

R. Brol [17] relates local development to changes occurring in a city, rural commune and urban–rural commune or a differently delimited sub-region, that is, a local territorial system characterised by specific spatial, economic and cultural qualities and the local preference of needs and value hierarchy, and defines it as the harmonised and systematic activity of the community, public authority and other entities functioning in a given territorial unit that is aimed at creating new functional advantages and improving the existing ones in a given territorial unit, the creation of favourable conditions for economy and the ensuring of the spatial and ecological order. L. Wojtasiewicz [3, 4] specifies local development as a complex of qualitative transformations concerning a given area with regard to the standard of living of its residents and the conditions of functioning of business entities located here. A. Myna [22] defines local development as processes initiated and created consciously by local authorities, entrepreneurs, ecological lobbyists, social and cultural associations and inhabitants that are aimed at the creative, effective and rational use of non-material and material resources.

Drawing upon definitions made by other authors, M. Ziółkowski and M. Goleń [23] describe local development as a process of quantitative and qualitative positive changes in the standard of living of a local community and in the conditions of functioning of business entities that takes into account the needs, priorities and preferences of inhabitants and entrepreneurs as well as the recognition of their systems of values. This definition highlights the importance of needs, preferences and the system of values as well as economic functions in local development processes in a given territory.

The crucial aim of economic development is to provide the community with high-quality jobs, to achieve local stability by responding to the needs of the local economy sector, to build a diversified economic base and employment base and to promote sustainability [24]. It is not possible to perceive the development of local units and regions in a homogeneous manner, because the nature and meaning of the concept of development differ and change in time and space, and specific concepts of development are socially determined by concrete interests and social groups in a specific geographical context and at a given time [25, 26].

Kuźnik [27] states that local development has an integral and multidimensional character. In his view, its basic dimensions include the socio-cultural dimension concerning the process of demographic, social and cultural changes, the environmental dimension (changes in the natural environment and in the environmental infrastructure), the infrastructural dimension (transport, communication, energy, water management, sewage disposal and heating sectors), the economic dimension (development of the economic base of cities and housing estates) and the spatial dimension (forms of spatial development, public order).

According to Klasik [28], the basic components of local development understood as the creation of new values and treated in an integral manner include economic growth and employment, an increase in prosperity, an improvement of the quality of life and investment attractiveness, technological development and innovations, the restructuring and diversification of business activity, the development of institutional infrastructure, an improvement of environmental quality, the enrichment of identity and integration processes (**Figure 1**).

Parysek [29] defines local development as the conducting of activities aimed at the economic and social development of a given territorial entity with the use of its resources, in consideration of inhabitants' needs and their participation in actions being taken. Local development is connected with the creation of new jobs on the one hand and with the comprehensive creation of the most favourable living conditions

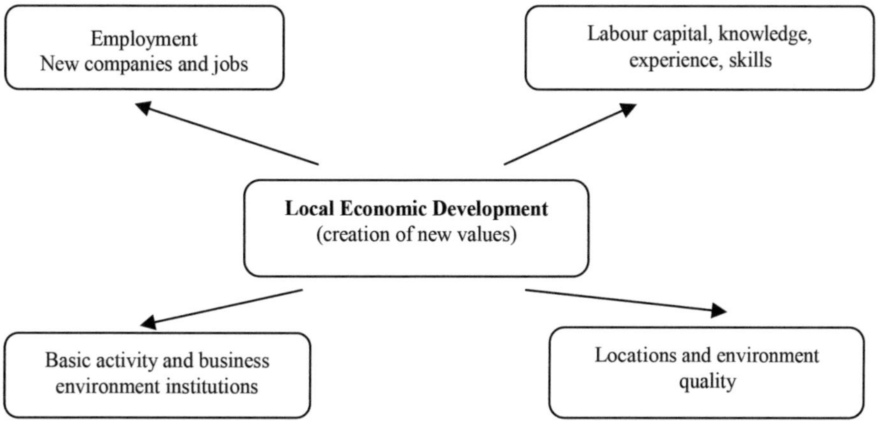

Figure 1.
Basic components of local economic development [28].

for inhabitants on the other hand, which is also expressed by the improvement of the organisation and structure and the functioning of local resources.

The phenomenon of development is also perceived as a process of creating institutions, developing alternative branches of industry and other economic functions, exerting influence on entrepreneurs for the purpose of making better products, helping in the identification of new markets, the transformation of knowledge, the diffusion of innovations and the support of entrepreneurship and others [24].

T. Markowski [30] defines local development as a process of directional changes in the course of which a quantitative increase and qualitative progress occur in the commune, county or region concerned. Local authorities can stimulate this development and make use of relevant development management instruments. According to I. Pietrzyk [31], local development as a process of diversification and enrichment of the business and social activity of a given territory, which has its source in the mobilisation and co-ordination of its resources and energy, is an activity undertaken by the will of local actors (local government officials, business entities, associations, etc.) on the basis of reflections concerning the valorisation of local resources and taking the specificity of the territory into account.

An important feature of local development is that it starts from the grassroots, which does not exclude the need to attract external investments. However, external elements should be included in the network of relations and integrated with the local environment. E. Wojciechowski [32] highlights the fact that local development essentially consists of the creation of new values, and its essence is economic development that encompasses favourable quantitative changes and qualitative and structural transformations, as well as an improvement of living conditions of the population. Referring to various definitions, A. Łuczyszyn [33] notes that the essence of local development is the process of economic and social transformation of the local community, which must be considered on three levels: the economic level, the social level and the ecological and spatial level. By synthesising approaches of various authors, A. Łuczyszyn [33] states: 'to put it simply, local development is a harmonised and systematic activity conducted in the local community with the participation of stakeholders, the results of which serve to satisfy the needs of the local population and contribute to overall progress'.

An attempt to systematise local development issues was undertaken by J. Kot [34], who formulated the following six theories:

- Local development is an autonomous process because it occurs according to principles and rules concerning market mechanisms; moreover, it is also subject to autonomous decision-making by local government authorities being the subject of local policy and development processes;
- Local development is entity-oriented, which means it is a concrete process bringing benefits and advantages to entities participating in it;
- Local development is a result of a compromise of actors and stakeholders operating in the local territorial unit;
- Local development takes place in conditions determined by phenomena and processes occurring both in its internal structures and in its environment; the success and stability of development processes depend on the flexibility of actions of business entities and local authorities;

- Local development is a long-term process conditioned historically by production, political and cultural traditions of the population;

- Local development is ruled by its own logic that results from individual attitudes of the population, organisational behaviours and market mechanisms.

Local development is characterised by autonomy, empowerment, openness, extension in time, flexibility, interactions and subordination to social evaluation and social impact on the change process [6, 17, 35–38].

According to M. Ziółkowski [39], local development should take into account both the needs, priorities, preferences and recognised systems of values of inhabitants and entrepreneurs and the requirements of the natural environment, cultural heritage and rational spatial planning.

Thus, local development is a multi-aspect, multi-thematic and multi-level concept in which values are expressed and which can be shaped through the determination of goals and selection of instruments and evaluated in terms of its outcomes and results. Local development can be perceived from the perspective of an individual local system and from the perspective of the place and role of a given local system in a broader environment—the supra-local environment [2].

As we can see, multi-aspect local development has a number of common elements, including dynamics as a process, the functioning of various entities, among which local government plays the leading role, endogenous local development factors, a broad range of impact (economy, society, environment, culture), the satisfaction of needs and aspirations of inhabitants and local entities, goals, factors and instruments of development that change over time, reducing the significance of 'hard factors' and strengthening of the role of 'soft factors' such as knowledge, intellectual capital and innovations.

Taking into account the analysis of model features of local development perceived from a local perspective, A. Sztando [2] formulates an elaborate definition of this development as a 'long-term, ununified, multidimensional and self-sustaining process of transformations of local system structures and interrelations between them and their connections with the environment that co-forms desirable supra-local processes and is created in a planned, participatory and coordinated manner and in accordance with the idea of integrated order, mainly by the local community and in its interest, through the consensus-based mobilisation of mainly endogenous factors'.

In modern concepts of local development, we can see a distinction between concepts of 'local development' and 'development in a local system'. The first concept stresses the engagement of local entities and development factors, that is, the engagement of the local community, and the stability and self-sustainability of development processes in time (endogenous development, integrated development and sustainable development). Development in local systems can also be a consequence of the impact of external and exogenous factors, considered from the theoretical side as a model for a process specifying the essence, content, requirements, characteristics, connections, internal factors and barriers and connections with the environment, as well as results and methods of their determination, and as a model for a change process [1].

5. The aims, components and outcomes (results) of local development

The object of local development is locally separated social-territorial units possessing a set of specific economic, social, spatial and cultural characteristics and expressing their own needs and hierarchies of values. In the Polish territorial development structure, these are mainly municipal and rural communes and counties established in the administrative division of the country. The commune is a territorial system relatively close to all of its inhabitants, both individuals and local communities and institutions in villages, housing estates and small towns comprising the territorial system of the commune. Therefore, local development concerns also internal (auxiliary) units such as villages, housing estates and other settlement units. This applies both to rural communes, municipal-rural communes and municipal communes. Local systems also include counties constituting a group of neighbouring communes that are relatively homogeneous, at least with regard to certain characteristics. Units separated according to non-administrative thematic criteria, such as a group of nearby communes located in the valley of one river or in a separate geographical area, can also be a form of local system.

Thus, we can distinguish the following elements in a territorial local system:

- sub-local systems (villages, towns/cities and districts),
- primary systems (communes),
- administratively (counties) and thematically integrated systems.

For such systems, development programmes or plans can be prepared. Communes are a primary system of administrative division, stocktaking of resources and the performance of tasks concerning the functioning of the state. The commune carries out many economic, social and political tasks. It is the unit in which basic social needs are fulfilled. The size of a commune is adequately small to enable the integration of inhabitants and local communities ensuring a sense of connection and enabling social activity for the pursuit of common goals and adequately big to achieve a minimum economically justified scale in the provision of primary welfare and social services and the cost-effectiveness of functioning of local markets. The optimum size of a commune in Poland is conditioned regionally and varies from a few to several thousand inhabitants in rural conditions and a few dozen thousand in cities [40–42].

The very complexity of the concept of local development implies the complexity of goals set in development strategies, programmes and policies of this territorial system. The formulation of goals should be perceived through the prism of entities that function in a given system and conduct activity. Each of the business entities (enterprises and households) and institutions functioning in a commune has its own goals specified to a certain extent, which it tries or is obliged to achieve. Their sum ordered according to accepted rules leads to many partial goals common for particular groups of entities. Inhabitants of a commune, as persons forming households and as consumers of goods and services, are interested mainly in the sphere shaping the possibilities and conditions of employment, conditions of material existence, conditions of education and personal, spiritual and cultural development, physical and economic safety, a sense of stabilisation, etc. As can be seen, this group of entities is interested mainly in the efficient labour market and the availability of goods and services, particularly public goods and services, as well as a healthy and friendly environment.

Industrial and commercial enterprises and farms strive to achieve economic goals ensuring their survival and development through the use of available factors in conditions of the local and regional market and the close and distant environment. Social service institutions and units and budgetary establishments seek to maximise their budget and its effective use for the implementation of economic and social tasks. All goals pursued by groups of entities functioning in a given system are the basis of functioning of general development goals of the given local system by the managing authority that exercises territorial government in local systems. In the territorial government of the commune, there is a symbiosis of state authority and inhabitants' self-government.

Local government is responsible for the achievement of general development goals of a given unit and the elaboration of the strategy and policy of their achievement. The primary goal of local development is usually the better fulfilment of needs of local communities and the improvement of the quality of life of inhabitants. The determination of collective needs for various entities in a given unit is not easy and requires the acquisition of knowledge about needs on the one hand and the rational choice of the method of their implementation on the other hand. The fulfilment of the overall goal requires the use of knowledge about resources and their availability and about the possibilities and conditions of their acquisition and mobilisation for socio-economic development. Local development must be based on a solid and diversified economic base used for the fulfilment of social needs without harm to the natural environment.

The economic sphere should ensure the relevant infrastructure and resources for the generation of revenues for entrepreneurs and farmers and income for households and should be characterised by activeness and the entrepreneurial spirit. In the social sphere, inhabitants should have main access to social, educational, health, cultural, sports and other services. What matters, is the social climate, a sense of responsibility and the strengthening of bonds, participation and local identity. However, the difficulty lies in the fact that not all needs can be fulfilled at the same time, which makes it necessary to choose priorities and make compromises. Therefore, the practical determination of the overall goal requires an analysis of goals important for various groups of entities and fields of activity and prioritising within particular spheres and in a general manner in a given system. In the ecological sphere, the fulfilment of the overall goal should ensure the creation and maintenance of spatial order, respect for the values of the natural environment and the protection of the natural and cultural landscape. Consequently, it is necessary to reduce the emission of pollutants by creating relevant technologies and eliminating the sources of environmental degradation and by promoting pro-ecological behaviours of inhabitants and business entities.

The analysis of goals from the perspective of business entities and spheres of activity can be used for the specification and cataloguing of goals of local development. Such a catalogue was prepared, for example, by P. Starosta [43], who distinguished the following goals: 1) economic goals, including economic growth through self-organisation and external aid, 2) social goals, including the engagement and participation of inhabitants, 3) psychosocial goals, reached through the identification of inhabitants with the community and with action programmes, 4) cultural goals, based on the maintenance of norms and values, 5) technological goals, fulfilled through the provision of necessary forms of technical infrastructure, 6) political goals, expressed by the participation of inhabitants in the decision-making process and the creation of democratic power structures and 7) welfare goals facilitating the functioning of the local community.

We must realise that the choice and prioritisation of goals of local development must be adapted to specific local systems. Each commune has its own group and prioritisation of goals corresponding to existing possibilities and expectations. It will be presumably different in rural communes than in municipal-rural communes or municipal communes.

An empirical description of local development processes indicates their huge differentiation and characteristics that can be subject to the evaluation, classification or modelling of these processes. Distinguished forms or models of local development with the use of various criteria always have a very significant element referring to the role of individual entities and the entire local community in the programming, planning and implementation of adopted development programmes, plans and strategies.

The combination and harmonisation of economic, social and ecological development assume the form of sustainable local development. It takes into proper consideration economic, social, technical and environmental aspects of development in a given spatial system. Sustainable development is harmoniously connected with the general condition of economy and society and with regions and macroeconomic development strategies and policies [13, 20].

Local development can involve the following components:

- economic growth – an increase in employment and decrease in the level of unemployment;
- increase in the population's welfare and quality of life – an improvement in the furnishing of households, the state of housing substances, the availability of technical equipment, etc.;
- increase in investment attractiveness, for example, through the provision of access to investment areas, the facilitation of real property trade and investment expansion and the support of local investors and entrepreneurs;
- technological development and implementation of innovations – through the provision of access to information, supporting the development of various innovative companies, consultancy and the creation of local business and enterprise support institutions;
- restructuring and diversification of business activity – through initiation and facilitation of the development of new fields of activity and new organisational and technological solutions in enterprises facing structural difficulties;
- development of public benefit services and resources facilitating the functioning of households and enterprises;
- increase in the professional and social mobility of the population – through training and retraining of employees, the development of the double work system among farmers, the creation of educational and advisory institutions, etc.;
- development of institutional infrastructure – through extension of the scope of services, bringing services closer to places of residence and work, establishment of institutions for the solution of difficult problems and social pathologies and supporting the development of non-governmental institutions;
- rehabilitation of the natural environment, including the conservation of the landscape, the preservation and enrichment of biodiversity, etc.;
- strengthening of local, regional or environmental identity through the support of phenomena integrating the local environment.

All actors of the local community, entities and the entire local community are engaged in the local development process. An important role in this development is played by the territorial government, employees of municipal enterprises and public institutions, activists of non-governmental organisations (NGOs), the professional self-government, chambers of commerce, foundations, general development associations, as well as representatives of individual business entities, social and political organisations and individuals – leaders and persons of authority. Each of these groups or persons can play an important role in the stimulation and implementation of local development.

The local development process requires adequate planning and coordination, and the territorial self-government authority being a subject of local development plays a special part in this development. The main task of the local territorial authority is to determine the direction and strategy of local development in accordance with internal and external conditions and development strategies of higher-level territorial units. The second primary task of the territorial authority is to pursue a local development policy, and the third task is to administer and manage the development and functioning of the local unit on a running basis.

Taking into account views of many authors dealing with local development issues, we can mention a number of characteristics of local development. The most important of them include [10]:

1. Continuity resulting from the performance of successive and interrelated activities aiming at desirable changes,

2. Comprehensiveness and simultaneousness resulting from multiple levels of local development and dependencies between these levels,

3. Complexity resulting from the large number and diversity of needs, goals and conditions of their implementation,

4. Territoriality, which means the connection of various individual activities with a specific location and a specific territory constituting a development base,

5. Commonality and/or co-operation, meaning that actions undertaken for the benefit of local development are approved by the local community and its entities.

6. Endogeneity, meaning the prioritisation of internal factors and forces represented by local related entities in local networks of developing local communities,

7. Feedbacks where individual actions at least partly correspond to the development goal of the entire community, and the implementation of the common goal leads also to the achievement of individual goals.

8. Connection with the environment – local development is an element of development of a broader territory, and conditions and factors of the environment affect the course of the development process in a given local territorial system,

9. Long duration – current development processes are related to the historical course of economic processes concerning a given territorial unit,

10. Autonomy – it is the subject matter of sovereign decisions made not only by individual entities but also by local authorities having specific competencies and funds for development goals.

However, local development has external conditions and limitations resulting from the pursuit of the interregional policy of the state and the intraregional policy of local authorities of the region within which a given local system functions.

According to research conducted by the International Labour Organisation and other agencies of the United Nations, local development [10, 44]:

1. is strongly dependent on the competencies, will and capacities of local actors,

2. concentrates on the strengthening of the local potential,

3. is largely based on the effective functioning of small and medium-sized enterprises everywhere,

4. depends on the ability of business initiatives to cooperate and integrate,

5. requires that the local unit be 'equipped' with relevant instruments,

6. depends on the possibility of active cooperation between systems at the local, regional, national and international levels.

Particularly important is the cooperation with administration and regional government. The region serves as a direct environment for the local system, and the local system is an integral part of the region. Local systems and the regional system are directly integrated and interdependent. There is a belief that local development is not a form of separate development on a small spatial scale, but a sign of grassroots (bottom-up) development in the region that is based on the use of endogenous resources with the support of external stimuli and the skilful control and guiding of production and accumulation processes and on the ability to initiate and generate innovations [45, 46].

The basic issues being separate spheres of local development include:

- Economic development – diagnosis of the level and factors of economic development and entrepreneurship, numbers of employees and employment structures, the unemployment level, sizes of production of goods and services, local and supra-local markets and the flow of capital and investments.

- Social development – the state and growth potential of social welfare and the quality of life of local communities. Income level, working conditions, housing conditions, social services, public safety and functioning of the local government are the parts of this development.

- Spatial-environmental development – features of the local space and the natural environment. Spatial order, spatial and ecological conflicts, space planning, environmental pollution, ecological threats and the aesthetics of the natural and anthropogenic environment.

- Technological and innovative development – investment, housing and tourist attractiveness. Availability of areas, material and human resources, character of local policy and expansion of infrastructure. Institutions supporting the innovative environment.

- Cultural development – the enhancement of image and identity, the strengthening of integrative processes of local communities and integration with the environment.

The outcomes of development processes are subject to assessment and evaluation. A number of measures and indicators can be used for the assessment of each of the aforementioned kinds of development. Some of them can be obtained from the Local Data Bank, whereas others must be sought in local institutions and organisations or by gathering opinions among inhabitants and through other research.

The assessment of the course and consequences of local development must include a number of aspects considered on economic, social, environmental, cultural, political and other levels. The most important determinants of the success of local development include [31]:

- The existence of efficient and effective leadership in the local system that inspires decision-makers to mobilise members of local communities;

- Engagement and active participation of all local social and professional groups in the decision-making process;

- Determination of clear guiding rules and precise assessment of goals before their final acceptance;

- Trust, compromises, cooperation and public-private partnership;

- Inclusion of cultural identity and the social-political structure of the local and regional system;

successive assessment of the course of development and the current updating and adaptation of activities to structural changes occurring in the environment of the given system. Special attention should be paid to innovative phenomena and processes.

6. Models and paradigms of local development

Local development is created with the use of achievements of many sciences: economy, sociology, geography, spatial planning, organisation and management, strategic planning, public finance, regional policy, integration and other sciences providing a theoretical background and organisational, technical and functional aspects.

Research on local development draws upon paradigms elaborated mainly in economic sciences for regional development and spatial planning. Assuming the general way of perceiving reality within a certain scope as a paradigm, we must say that the perception of reality that we understand as local development has changed. Research within this scope led to the formation of many standards and models of this development, and various general concepts and theories generating various problems, concepts, rules of

Conceptions and approaches of development	Specific features and selected advantages of development
Exogenous development and economic base	• Implant development based on external factors and impulses • Development depending on the position occupied within a larger system • Favourable transfer of outside capital for a quick pace of development
Polarised development	• Development occurring in selected privileged places (regions) – created growth poles • Uneven development polarising space at all levels of its organisation • Development accepted and supported financially by decision-making centres
Integrated development	• Multi-aspect and multidimensional development • Development of interdependent fields, requiring integration and coordination • Strategic thinking fosters all development projects
Self-organisation	• Global interdependence of changes in various systems – evolutionary paradigm • Natural autonomy of every system; the skill of using resources of the environment • Integrated local community for the implementation of the eco-development strategy
Endogenous development	• Development based on internal development potentials of cities and regions • Development rooted in the community; self-governance and grassroots initiatives • Lack of motivation and conviction about the relevance of looking for external partners for the financing of development
Self-sustaining development	• Sustainable development that automatically restores the balance on a global and local scale • Clean development eliminating negative side-effects in successive phases of development • Development based on a stable system of values
Sustainable development	• Equal treatment of economic, social and environmental aspects • Management without harm and limitations for future generations • Possibility of including other aspects and stressing various forms of activity • Concept of smart cities • Concept of smart villages • Form of sustainable development • Development of smart specialisations • Creation and implementation of innovations • Building the innovation environment • Development of the cooperation network • Coordination of actions through local governments • Implementation of a circular production system • Recycling and value restoration • Management without waste • Cascading of manufacturing processes

Conceptions and approaches of development	Specific features and selected advantages of development
Smart development	• Concept of smart cities • Concept of smart villages • Form of sustainable development • Development of smart specialisations
Innovative development	• Creation and implementation of innovations • Building the innovation environment • Development of the cooperation network • Coordination of actions through local governments
Circular economy	• Implementation of a circular production system • Recycling and value restoration • Management without waste • Cascading of manufacturing processes
Participating economy	• New approach to the ownership category • Use of factors and products by way of lending or share • Joint consumption
Sustainable consumption	• Minimisation of the use of resources and energy • Waste reduction and recycling • Adaptation of lifestyle to the natural environment
Green economy	• Risk reduction in the natural environment • Prevention of pollution and waste • Maintenance and protection of biodiversity • Reduction of greenhouse gases, climate protection and improvement of social health

Table 1.
Old and new concepts of local and regional development [47–49].

procedure and methods of solving problems have been formulated within the scope of individual models. **Table 1** contains various concepts of local and regional development.

In literature, we can distinguish a number of concepts and typologies of local development models, the most important of which are: exogenous development, endogenous development, grassroots development, top-down (induced) development, sustainable development, self-sustaining development, integrated development, evolutionary development, innovative development, extensive development and intensive development. The size of this chapter is too small to discuss the aforementioned concept, so we will focus only on sustainable development models.

7. Sustainable development

Sustainable development has become a currently leading development model on a global scale in individual countries and in territorial systems within them. The concept that emerged in France and Germany with regard to forestry in the 18th century [50] was popularised and became a basic paradigm of development in the second half of the 20th century. Sustainable development also became the goal of developing strategies and policies that encompassed increasingly broader areas of functioning of the modern society. In the concept restored in the 1970s and 1980s, three dimensions were distinguished: the economic one, the social one and the environmental one. The number of these dimensions is often increased. The primary methods of treating individual dimensions and relationships between them are shown in **Figure 2**.

The first model presents the separate consideration of individual levels, the second model shows jointly connected and interrelated dimensions and the third model is a layout where individual levels are hierarchically dependent and subordinated.

The concept of sustainable development is based on two basic principles. The first principle, resulting from concern about the natural environment and the need for its protection, refers to the equal inclusion of the economic, social and ecological aspects. The second principle includes the aspect of continuity and sustainability responding to the needs of both current and future generations and the retention of the possibility of managing without harm and limitations to the activity of future generations. In aspects of sustainability, the economic, social and environmental levels are sometimes accompanied by other equivalent aspects, such as legal, political, technological, moral or business aspects. The highlighting of the ecological level is connected with the concept of eco-development. The highlighting of the aspect of intergenerational justice as a goal and condition of development is manifested by adopting the name of sustainable development or self-sustaining development. The concepts of sustainability and self-sustaining refer to the preservation of natural and material resources, physical, human and social capital resources and general possibilities of efficient and economic activity for future generations in a given system.

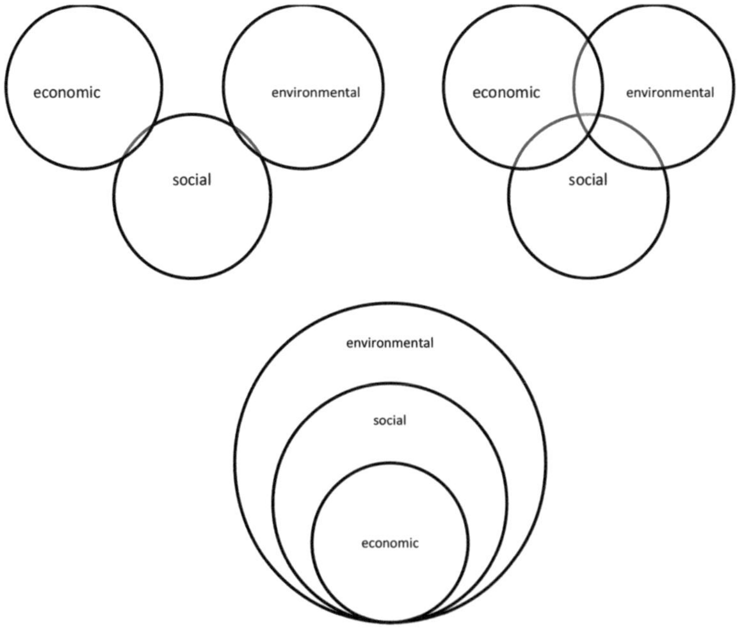

Figure 2.
Models of relationships of sustainable development dimensions [51, 52].

The concept of sustainable development in Poland, adopted as a basis for general development of the country, means an attempt to include both the criteria of effectiveness and social aspirations in laying emphasis on justice and equality in the sphere of spatial and territorial planning. The main core of this paradigm of sustaining general development and balancing the spatial planning of the country assumes the need for constant improvement of the quality of life of local communities through its support with sustainable factors, mainly of economic, but also social and ecological nature, without disturbing the developmental basis of any sphere of this development.

Today, attempts are made to seek various forms of its implementation in compliance with basic principles of sustainable development. This is especially true for specific selected aspects of development, for example, urban development, rural development or development of agriculture or consumption [49].

Sustainable development is sometimes associated with eco-development [53, 54]. The concept of eco-development, the essence of which is expressed by the third model shown in **Figure 2**, lays emphasis on the respect for the advantages of the natural environment and the shaping of spatial order. In this paradigm, attempts are made to find optimum relations between the human being and nature and to improve the quality of the natural environment. As can be seen, this concept highlights the aspect of guaranteeing the survival and development of current and future generations through the prevention, limitation and complete elimination of changes and damages resulting from human activity. Technical progress and innovations are not only used for generating the economic growth of a given territorial system but also for improving the condition of the environment. The authors who separately consider the paradigm of sustainable and ecological development stress that the first one has the character of balanced development on three distinguished levels (the economic, social and environmental levels), and the second one is a type of development that sustains itself in time.

The concept of sustainable development is regularly developed and supplemented. These primary levels are supplemented with new levels and specific aspects forming multidimensional layouts with variously arranged relationships are brought out. The hierarchical layouts of levels under construction are supplemented with human needs and rights, institutional solutions, etc. An example of expanded hierarchical layouts of sustainable development dimensions is shown in **Figure 3**.

The economic dimension encompasses the processes of production and exchange of goods and services that are the basis of the population's incomes. Both quantitative factors of production and qualitative factors, particularly those related to labour factors, as well as conditions are significant for production processes. The measurement of the effects of the economic level currently goes beyond the level of GDP per capita. The UN's Human Development Index is an indicator with a broader range.

The social dimension encompasses a wide range of parameters showing social conditions and the standard of living. The assessment of this level is strongly influenced by connections and relations with the economic and environmental level and the ensuring of fundamental rights and the satisfaction of the population's needs.

The environmental dimension concerns both the environmental impact of human activity and the use of natural resources, as well as conditions created by the environment for the achievement of goals of economic activity and social welfare.

The dimension concerning the possibility of satisfying the needs and guaranteeing legal order with regard to respect for property and individual and group safety is connected with the institutional level that **concerns ensuring** the adequate order of functioning of market economy and democratic society.

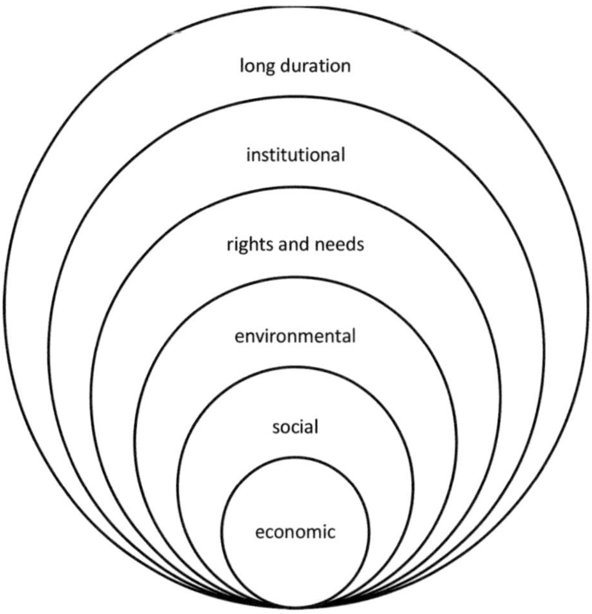

Figure 3.
Six hierarchically subordinated dimensions of sustainable development [1, 55].

Since the emergence of the concept of sustainable development, the dimension of long duration has been an important criterion for avoiding the conflict of interests of the current generation and future generations in development policy and currently serves the need for rational planning in order to maintain the possibility of management by future generations. Thus, successive dimensions are increasingly general and difficult to quantify, but not less important for development processes. These dimensions are relatively concrete and easily measurable in local systems. However, the possibilities of adapting or correcting individual levels are very limited here. These possibilities, if they exist, are larger on the national and regional level, where development policy is shaped and where sources of potential funds for the support of development are located.

8. Conclusions

Over the last three decades, new concepts of development based mainly on the ground of sustainable development emerged. The concept of sustainable development, arising out of the need to prevent the degradation of the natural environment, covers not only social and economic aspects – enriched by contributions from various fields of science but it has also become the basic paradigm of all development programmes, policies and strategies formulated by international organisations, national governments and local and regional authorities [49]. Its essence still lies in ensuring the constant improvement of the life of modern and future generations by shaping rational proportions between various kinds of capital: economic, human, social and natural capital. By adopting the Agenda 2030 in 2015, member states of the UN undertook to pursue 17 Sustainable Development Goals divided into 169 detailed tasks in their domestic policy [56]. The adopted goals introduce new forms of sustainable management, such as closed circuit economy, participating economy, sustainable consumption, the green deal, etc.

By presenting selected old and new concepts of development that can be related to the local scale, we should note that these concepts and models are not separable and alternative towards one another and that the emergence of new concepts and models does not invalidate old ones. The emergence of new concepts, particularly the latest ones, is often an expansion or supplement to the existing ones. This is especially valid with regard to the paradigm of sustainable development, which often assumes a new form depending on the situation or context. This does not mean that a completely new model emerges – it is often a new variety of the basic concept. New names or definitions of development are often introduced, particularly by international global policy-making circles, and new names or slogans are used primarily for the promotion of a certain concept, dimension or factor. Many of these concepts are still too vague and limited to the promotion of behaviours or reactions regarded as desirable. Many of them do not refer only to the local dimension – they are of a general nature.

Conflict of interest

The author declares no conflict of interest.

Author details

Mieczysław Adamowicz
John Paul II University of Applied Sciences in Biala Podlaska, Biala Podlaska, Poland

*Address all correspondence to: adamowicz.mieczyslaw@gmail.com

References

[1] Adamowicz M, Zwolińska-Ligaj M. Lokalne strategie innowacji jako instrument kreowania innowacyjności i wspierania rozwoju. Biała Podlaska: Wydawnictwo PSW; 2021. (in Polish)

[2] Sztando A. Ponadlokalna perspektywa zarządzania strategicznego rozwojem lokalnym na przykładzie małych miast. Wrocław: Wydawnictwo Uniwersytetu Ekonomicznego we Wrocławiu; 2017. (in Polish)

[3] Wojtasiewicz L. Ekonomiczne uwarunkowania rozwoju lokalnego. In: Parysek J, editor. Rozwój lokalny i lokalna gospodarka przestrzenna. Poznań: Wydawnictwo Naukowe; 1996. (in Polish)

[4] Wojtasiewicz L. Czynniki i bariery rozwoju lokalnego w aktualnej polityce gospodarczej Polski. Gospodarka Lokalna w Teorii i Praktyce. 1996;**1**:734. (in Polish)

[5] Blakely EJ. Planning Local Economic Development. Theory and Practice. London: SAGE Library of Social Research; 1989

[6] Parysek JJ. Podstawy gospodarki lokalnej. Poznań: Wydawnictwo Naukowe UAM; 1997. (in Polish)

[7] Pilch T. Zasady badań pedagogicznych. Warszawa: Wydawnictwo Żak; 1995. (in Polish)

[8] Korenik S. Zarządzanie rozwojem lokalnym (gminy). In: Dolata S, editor. Funkcjonowanie samorządu terytorialnego. Doświadczenie i perspektywy. Volume 1. Opole: Wydawnictwo Uniwersytetu Opolskiego; 1998, (in Polish)

[9] Noworól A. Pojęcie rozwoju terytorialnego w kontekście nauk o zarządzaniu, Współczesne Zarządzanie nr 2. Białystok: Fundacja Współczesnego Zarządzania; 2007. (in Polish)

[10] Rakowska J. Fundusze unijne jako czynnik rozwoju obszarów wiejskich w Polsce w świetle teorii rozwoju lokalnego. Warszawa: Wydawnictwo SGGW; 2019. (in Polish)

[11] Szewczuk A. Rozwój lokalny i regionalny – główne determinanty. W: A. Szewczuk, M. Legut-Jaworska, M. Zioło (red.), Rozwój lokalny i regionalny. Teoria i praktyka. Wydawnictwo C.H. Beck: Warszawa; 2011. (in Polish)

[12] Guzal-Dec D. Samorząd gminny w kreowaniu zrównoważonego rozwoju obszarów przyrodniczo cennych województwa lubelskiego, Monografie i Rozprawy Nr 4. Biała Podlaska: Wydawnictwo PSW w Białej Podlaskiej; 2015. (in Polish)

[13] Adamowicz M. Skala lokalna w terytorialnym podziale kraju. In: Adamowicz M, editor. Strategie rozwoju lokalnego tom I, Aspekty instytucjonalne. Warszawa: Wydawnictwo SGGW; 2003. (in Polish)

[14] Jouen M. Local development in Europe. Assessments and prospects after the economic crisis. Policy Brief Notre Europe. 2011:21

[15] Bagdziński SL. Lokalna polityka gospodarcza w okresie transformacji systemowej. Toruń: Wydawnictwo Uniwersytetu Mikołaja Kopernika; 1994. (in Polish)

[16] Szewczuk A, Kogut-Jaworska M, Zioło M, editors. Rozwój lokalny

i regionalny – teoria i praktyka. Wydawnictwo C.H. Beck: Warszawa; 2011. (in Polish)

[17] Brol R. Rozwój lokalny – nowa logika rozwoju gospodarczego. Gospodarka lokalna w teorii i praktyce. Prace Naukowe Akademii Ekonomicznej we Wrocławiu. 1998;**785**:11-15. (in Polish)

[18] Poniatowicz M, Wyszkowska D. Stymulowanie rozwoju lokalnej przedsiębiorczości a konkurencja podatkowa gmin. Zeszyty Naukowe Wyższej Szkoły Bankowej w Poznaniu. 2014;**52**(1):73-93. (in Polish)

[19] Gałązka A. Teoretyczne podstawy rozwoju regionalnego teorie, czynniki i bariery rozwoju regionalnego. Studia BAS. 2017;**49**(1):9-61. (in Polish)

[20] Adamowicz M. Kształtowanie rozwoju lokalnego. In: Adamowicz M, editor. Strategie rozwoju lokalnego tom II, Aspekty instytucjonalne. Warszawa: Wydawnictwo SGGW; 2003. (in Polish)

[21] Parysek JJ. Rola samorządu terytorialnego w rozwoju lokalnym. In: Parysek JJ., editor. Rozwój lokalny: zagospodarowanie przestrzenne i nisze atrakcyjności gospodarczej. Warszawa: PWN; 1995. (in Polish)

[22] Myna A. Rozwój lokalny, regionalne strategie rozwoju, regionalizacja. Samorząd Terytorialny. 1998;**11**:30-48. (in Polish)

[23] Ziółkowski M, Goleń M. Zarządzanie strategiczne rozwojem lokalnym. In: Sochacka-Krysiak H, editor. Zarządzanie gospodarką i finansami gminy. Warszawa: Oficyna Wydawnicza SGH; 2003. (in Polish)

[24] Blakely EJ, Leigh NG. Planning Local Economic Development. Theory and Practice. London: SAGE; 2010

[25] Pike A, Rodriguez-Pose A, Tomaney J. Local and Regional Development. London; New York: Routledge; 2006

[26] Pike A, Rodrigues JA, Tommaney J. What kind of local and regional development and for whom? Regional Studies. 2007;**41**(9):1253-1269. DOI: 10.1080/00343400701543355

[27] Kuźnik F. Regionalne i lokalne aspekty rozwoju społecznego. In: Frączkiewicz-Wronka A, editor. Zarządzanie publiczne w lokalnej polityce rozwoju. Warszawa: WSPTWP; 2007. (in Polish)

[28] Klasik A. Zarządzanie rozwojem lokalnym. In: Kuźnik F, editor. Zarządzanie rozwojem gmin w zespołach miejsko-przemysłowych. Katowice: Wydawnictwo Akademii Ekonomicznej w Katowicach; 1996. (in Polish)

[29] Parysek JJ. Rozwój społeczno-gospodarczy oraz czynniki i uwarunkowania rozwoju. Studia Komitetu Przestrzennego Zagospodarowania Kraju PAN. 1999;**183**:37-56. (in Polish)

[30] Markowski T. Zarządzanie rozwojem miast. Warszawa: Wydawnictwo Naukowe PWN; 1999. (in Polish)

[31] Pietrzyk I. Teoretyczne podstawy rozwoju lokalnego. In: Broszkiewicz R, editor. Związki polityki gospodarczej z polityką regionalną. Wrocław: Akademia Ekonomiczna; 1997. (in Polish)

[32] Wojciechowski E. Zarządzanie w samorządzie terytorialnym. Warszawa: Wydawnictwo Difin; 2003. (in Polish)

[33] Łuczyszyn K. Nowe kierunki rozwoju lokalnego ze szczególnym uwzględnieniem peryferyjnych

ośrodków w metropoliach. Warszawa: Wydawnictwo CeDeWu; 2013. (in Polish)

[34] Kot J. Zarządzanie rozwojem gminy a praktyka planowania strategicznego. Łódź: Wydawnictwo Uniwersytetu Łódzkiego; 2003. (in Polish)

[35] Kot J. Rozwój lokalny – jego istota, cele i czynniki. In: Markowski T, Stawosz D, editors. Ekonomiczne i środowiskowe aspekty zarządzania rozwojem miast i regionów. Łódź: Wydawnictwo Uniwersytetu Łódzkiego; 2001. (in Polish)

[36] Filipiak S, Kogut M, Szewczuk A, Zioło M. Rozwój lokalny i regionalny. Uwarunkowania, finanse, procedury. Szczecin: Fundacja na Rzecz Uniwersytetu Szczecińskiego; 2005. (in Polish)

[37] Kogut-Jaworska M. Instrumenty interwencjonizmu lokalnego w stymulowaniu rozwoju gospodarczego. Warszawa: CeDeWu; 2008. (in Polish)

[38] Romanowski R. Wpływ wsparcia systemów innowacji na rozwój lokalny. Poznań: Wydawnictwo Uniwersytetu Ekonomicznego w Poznaniu; 2015. (in Polish)

[39] Ziółkowski M. Zarządzanie strategiczne w polskim samorządzie terytorialnym. In: Zalewski A, editor. Nowe zarządzanie publiczne w polskim samorządzie terytorialnym. Warszawa: Oficyna Wydawnicza SGH; 2005. (in Polish)

[40] Moseley MJ. Rural Development: Principles and Practice. London, Thousand Oaks, New Delhi: SAGE Publications; 2003

[41] Swianiewicz P. Czy małe jest piękne? Czy duże jest bardziej efektywne? Wielkość gmin w teorii i praktyce. Samorząd Terytorialny. 1995;**7-8**:23-35. (in Polish)

[42] Herbst M, Swianiewicz P. Korzyści i niekorzyści skali w działaniach samorządu terytorialnego. Samorząd Terytorialny. 2002;**6**:14-24. (in Polish)

[43] Starosta P. Poza metropolią. Wiejskie i małomiasteczkowe zbiorowości lokalne a wzory porządku makrospołecznego, Łódź: Wydawnictwo Uniwersytetu Łódzkiego; 1995. (in Polish)

[44] UNDP, ILO, UNOPS, EUR. Las agencias de desarrollo local. New York: UNDP, ILO, UNOPS, EUR; 2022

[45] Siemiński JL. Główne koncepcje rozwoju lokalnego. Człowiek i środowisko. 1994;**18**(3):453-454. (in Polish)

[46] Pietrzyk I. Polityka regionalna Unii Europejskiej i regiony w państwach członkowskich. Warszawa: Wydawnictwo PWN; 2004. (in Polish)

[47] Klasik A, Kuźnik F, editors. Zarządzanie strategiczne rozwojem lokalnym i regionalnym. Wydawnictwo Akademii Ekonomicznej w Katowicach: Katowice; 1998. (in Polish)

[48] Klasik A. Strategie regionalne. In: Formułowanie i wprowadzanie w życie. Katowice: Wydawnictwo AE; 2001. (in Polish)

[49] Adamowicz M. Agricultural development processes in the context of globalization challenges and new approaches to the concept of sustainable development. Zagadnienia Ekonomiki Rolnej/problems of. Agricultural Economics. 2021;**366**(1):24-45. (in Polish). DOI: 10.30858/zer/132395

[50] Baum R. Sustainable development – A modern understanding of the concept.

Annals of the Polish Association of Agricultural and Agribusiness Economists. 2021;**23**(2):9-28. DOI: 10.5604/01.3001.0015.0026

[51] Jourmand R. Durable? Sustainable? Bron: IFSTTAR Researchers; 2011

[52] Pawłowski A. Wielowymiarowość rozwoju zrównoważonego. Problemy Ekorozwoju. 2006;**1**(1):23-32. (in Polish)

[53] Żylicz T. Ekonomia wobec problemów środowiska przyrodniczego. Warszawa: PWN; 1990. (in Polish)

[54] Kozłowski S. Ekorozwój – wyzwanie XXI wieku. Warszawa: Wydawnictwo Naukowe PWN; 2000. (in Polish)

[55] Jourmand R. Environmental sustainability assessment: Towards a new framework. International Journal of Sustainable Society. 2011;**3**(2):133-150. DOI: 10.1504/IJSSOC.2011.039918

[56] Garcia-Feijoo M, Eizaguirre A, Rica-Aspiunza A. Systematic review of sustainable development goal. Development in Business Schools. 2020;**12**:440. DOI: 10.3390/su12010440

Chapter 6

The Impacts of Unsustainable Urbanization on the Environment

Abdulkarim Hasan Rashed

Abstract

Urban areas—cities—are not simply geographic areas for human gathering but are a locus of economic production, cultural and social interactions, and ecological development. Therefore, cities create positive development values when planned and managed on a sustainable footing by considering institutional, governance, environmental, political, economic, coherent policies, cultural, and social conditions and requirements. Sustainable urbanization has multiple benefits including creating more employment opportunities and better incomes, hubs for innovative solutions by attracting competencies, enhancing land utilization efficiency, improving infrastructural performance, providing better services (e.g., education, health, water supply, and electricity), economic growth hub, acting as knowledge centers, better social and cultural life, and providing better living standards. While the impacts of rapid unsustainable urbanization are water stress, scarcity, and high consumption, sanitation wastewater, water pollution, air pollution, climate change, noise pollution, cultivated land depleted, urban sprawl, dust, solid and hazardous wastes, destruction of biodiversity, high energy consumption, traffic congestion, soil pollution, and deforestation. Thus, the 2030 Agenda[1] for Sustainable Development—and its sustainable development goals (SDGs)—and New Urban Agenda are key transformative power toward sustainable urbanization development; this development is not at the expense of the environment while leading to prosperity and improving quality of life.

Keywords: 2030 agenda, SDGs, environmental impacts, sustainable urbanization, sustainable planning, coherent policies

1. Introduction

Urbanization is one of the largest social transformations of the modern era, accompanied by economic, social, physical, and environmental processes [1]; therefore, it has become an important and top priority issue at the global agenda. According to a United Nations report (World Urbanization Prospects: A 2018 Review), the proportion of the world population living in urban areas was 30% in 1950, increased to 55% in 2018; there is a projected increase up to 68% in 2050—i.e., an increased that is more than 126% in 100 years. The period from 1990 to 2018 witnessed an average annual

[1] 2030 Agenda is a global plan of action for the people, planet, and prosperity, and comprises 17 Sustainable Development Goals (SDGs), 169 targets, and 232 indicators.

growth rate reaching 1.8% for world cities that populated more than 300,000 [2]. Moreover, the cities' number and size will continue to increase due to the rate of births being more than the death rate in urban areas, besides that the continuous migration from rural areas to urban areas and from abroad [3].

The transformation process to urbanization should be built on a sound and solid institutional basis—e.g., policies, laws, and regulations—to set up sustainable and resilient urban cities. In this context, the World Cities Report 2020 mentioned several benefits of sound institutions that include: leading to inclusive prosperity, providing a better quality of life, leading urban development, giving the highest benefits to most of the inhabitants, achieving more equitable cities, and motivating inclusive and sustain economy growth [4]. There is a remarkable policy transformation toward adopting the institutional frameworks that entitle coordination and cooperation among urban actors. Thus, there is a need to put the policy purpose into practice through investment in infrastructure development, institutional capacity building, and employment of suitable policy tools in the national context [5].

The impacts of urbanization on the integration between the economy and the environment are non-linear; thus, it does not assist in minimizing environmental pressures [6], and this creates several tangible challenges, such as escalating infrastructure development, housing, economic instabilities, education, health, pollution, transportation, and water projects [7, 8] due to unsustainable urban planning. Uttara et al. reasoned that the core causes of declining environmental quality are planning and management issues [9]. In addition, there is a causal relationship between industrialization and urbanization, where urbanization paved the way for industrialization; the latter accelerated paces of urbanization, and both resulted in challenges that cannot be easily beaten [10].

The next parts of this chapter are structured as follows: Section 2 is an overview of the New Urban Agenda and 2030 Agenda; Section 3 discusses the impacts of urbanization on the environment; Section 4 presents urbanization in the context of Sustainable Development Goals. Section 5 discusses the process toward sustainable urbanization. The last section is the conclusion.

2. New urban agenda and 2030 agenda

Two significant agreements milestones unify the globe; the first in 2015, which was the adoption of the 2030 Agenda for Sustainable Development, and the second ensued in 2016, with the adoption of the New Urban Agenda (NUA) and both agendas need to be implemented in closeness and integration to attain a better future for the people. In this context, the 2030 Agenda provides an operative framework for addressing the issue of urbanization at the global level [11].

The World Cities Report for 2016 clearly revealed that "the current urbanization model is unsustainable in many respects, puts many people at risk, creates unnecessary costs, negatively affects the environment, and is intrinsically unfair" [12]. Thus, NUA considered urbanization as the essence of sustainable development, and this agenda is characterized by a "shared vision of a better and more sustainable future," "a paradigm shift based on urban science," and "new recognition of the relationship between good urbanization and development" [13]. The NUA consists of 175 items but lacks metrics—indicators—for monitoring and measuring their progress [14]. In this context, Sietchiping et al. stated that, there is a need for a substantial transformation to ensure efficient urban planning and integrated policies toward sustainable cities' planning, development, and management at all development levels [15].

Item 9 of NUA states that, "the implementation of the New Urban Agenda contributes to the implementation and localization of the 2030 Agenda for Sustainable Development in an integrated manner, and to the achievement of the Sustainable Development Goals (SDGs) and targets, including Goal 11 of making cities and human settlements inclusive, safe, resilient, and sustainable" [13]. Significantly, urbanization has been acknowledged growingly as a means of achieving the main aspects of the 2030 Agenda, particularly given the efficiency of urban resources and services and economic growth [16].

Notably, SDGs 9 and 11 have primarily covered the infrastructure and cities [14]. SDG11 is a so-called urban sustainable development goal and is explicit about the possibility of cities' transformation and their capacity to realize other SDGs, whereas monitoring the progress of urban SDGs is not a simple task [17, 18], and this goal is also considered a novel tool that enables to reach the resources that have assisted urban improvement [19]. Furthermore, the United Nations initiative on the United for Smart Sustainable Cities (U4SSC) has introduced a more authentic tool intended to assess cities toward achieving smartness, sustainability, and the SDGs [20].

Figure 1 indicates that the synergy between NUA and the 2030 Agenda can generate a successful transformative pathway toward achieving sustainable cities with increased prosperity. Thus, both Agendas—NUA and the 2030 Agenda—must work

Figure 1.
NUA and the 2030 agenda: pathway toward sustainable cities.

Key issues of SDG 11 targets	Linkages to other SDGs	Linkages to NUA items
11.1 By 2030, safe, & affordable housing & upgrade slums	1, 3, 4, 5, 6, 7, 10	31–34, 46, 61, 70, 99, 107, 108, 110, 112
11.2 By 2030, improving road safety, notably by expanding public transport	1, 2, 8, 9, 13	48, 50, 54, 113–115
11.3 By 2030, sustainable human settlement planning & management in all countries	16	29, 39–42, 92, 149, 155–157, 160
11.4 Protect & safeguard the world's cultural & natural heritage	4, 5, 10, 12, 13	38, 63, 66, 121, 122
11.5 By 2030, reduce the number of people deaths & affected; decrease disasters, including water-related disasters	6, 7, 13, 14, 15	65, 68–69, 71, 73–78, 119, 123

Key issues of SDG 11 targets	Linkages to other SDGs	Linkages to NUA items
11.6 By 2030, attention to air quality, municipal & other waste management	6, 7, 13, 14, 15	65, 68–69, 71, 73–78, 119,123
11.7 By 2030, provide universal access to safe, inclusive; accessible, green, and public spaces	1, 5, 8, 9	37, 53, 55–56, 67, 100, 109
11.a Strengthening national and regional development planning	10, 16	87–91,158–159
11.b By 2020, inclusion, resource efficiency, mitigation and adaptation to climate change, resilience to disasters & disaster risk management	12, 13, 16, 17	77–78, 86, 94, 95–98
11.c Support least developed countries financially and technically	10, 12, 13, 16, 17	131–133, 139, 140, 144, 149, 151

Table 1.
Relationship between the SDG11 and the New Urban Agenda.

jointly and concurrently to overcome the exacerbated challenges (e.g., environmental degradation) that face sustainable development toward attaining extensive sustainable urbanization and realizing the essential role of cities in tackling climate change and other issues. Therefore, NUA is a crucial driver for accelerating SDGs' accomplishment and localization. **Table 1** summarizes the relationship between the 2030 Agenda and the New Urban Agenda.

3. The impacts of urbanization on the environment urbanization and environmental

Cities are facing significant challenges due to current unsustainable urbanization patterns that are rapid, unplanned, sprawling, resource management way, and the pressure on the infrastructure to enhance resilience to any slow-onset crises and changes [21]. Adding to that, the global urbanization process is irregular and not uniform [22]. Consequently, urbanization aggravates environmental degradation [23]. The environmental issues drive the recognition of cities on the global agenda [24]; therefore, urbanization and environmental sustainability issues became a top priority global concern [25]. **Figure 2** illustrates some challenges and solutions of urbanization in the context of SDG 11.

The impacts of rapid unsustainable urbanization are water stress, scarcity, and consumption [27, 28], sanitation, wastewater, water pollution [29, 30], greenhouse gases (GHGs) emissions (climate change) [31], air pollution [29], noise pollution [10], cultivated land depleted [32], urban sprawl [33], dust, solid, and hazardous wastes, and destruction of habitats (biodiversity) [9], high energy consumption [34], traffic congestion, soil pollution [35], and deforestation [36]. For instance, urban areas are generated around 70% of GHGs emissions and consumed more than 60% of global energy [37], and the continuously growing urbanization has an impact globally on waste generation and the volume of GHGs emissions [38] because the cities are major contributors to GHGs emissions [5]; in this context, the Ellen Macarthur Foundation predicts that cities will be a source of 60% of both global waste and GHGs emissions by 2050 [39]. Furthermore, the World Health Organization (WHO) has reported that during 2008–2013 global urban air pollution levels increased by 8% [40].

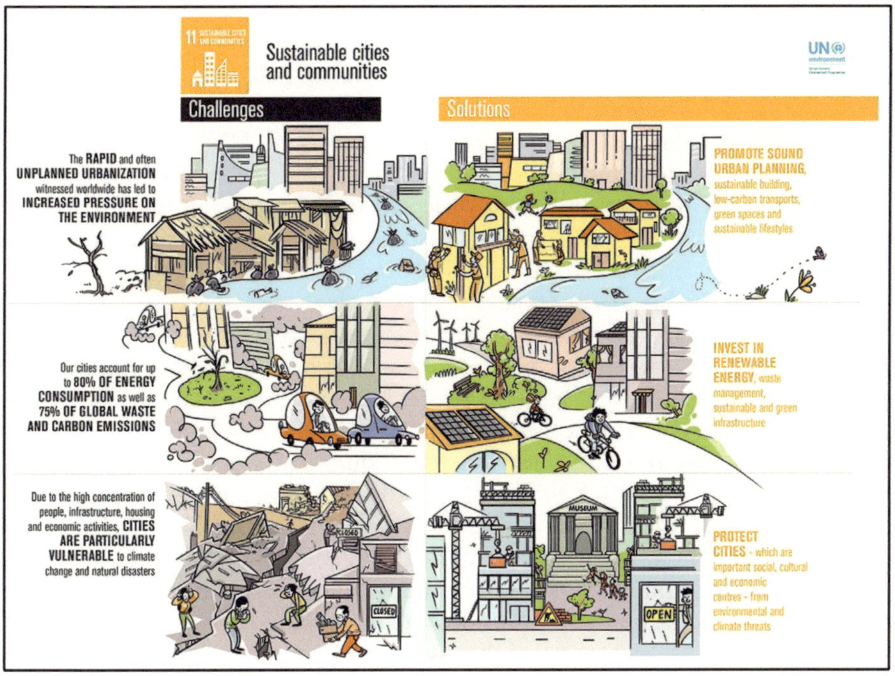

Figure 2.
SDG 11: Some challenges and solutions [26].

In addition, the UN report "The Weight of Cities: resources requirements of future urbanization," [41] reported that in 2010 the material consumption of global cities was 40 billion tonnes, and by 2050 it will increase to nearly 90 billion tonnes where the resource consumption per capita is estimated 8 to 17 tonnes in 2050. Sustainable urbanization is one of the efficient solutions to minimize the environmental footprints of cities. Therefore, the environmental dimension should consider it during planning, designing, and developing cities to reduce their impacts without additional cost to the environment, become resilient to climate change impacts, and contribute to economic growth [21]. Furthermore, the policies should tackle urban environmental issues, such as planning urban spaces and transportation to reduce air pollution [42].

4. Urbanization in the context of sustainable development goals

The 2030 Agenda and its core 17 SDGs is comprehensive of broad multiple significant thematic issues. The core principles of the 2030 Agenda are intimately related to the process and form of urbanization [16]. One of the 2030 Agenda aims is to transform the world by ensuring human well-being [43]. Among the SDGs, SDG 11 explicitly addresses the urbanization issue, aims to provide sustainable living conditions in the city community, and tackles the linkage between urbanization and sustainable development [38, 44]. Therefore, achieving SDG11 will lead to sustainable cities and societies and contribute to accomplishing other goals [44].

There is a strong interlinkage between urbanization and the SDGs. Urbanization has a positive impact on majority of the SDGs [45], thus, urbanization cities should play essential roles in the implementation of all SDGs. The SDGs are harmoniously set and cannot be achieved separately. The large cities have positive and negative impacts on the SDGs, for example, transformation to sustainable cities via SDG11 "sustainable cities and communities," contributing to more profitable economic productivity through SDG8 "Decent work and economic growth," while causing environmental challenges represented in SDG12 "sustainable consumption," SDG13 "Climate Action," SDG14 "Life below Water," and SDG15 "Life on Land" [46]. Therefore, cities should enhance sustainable planning and resource efficiency management through the implementation of the SDGs for achieving sustainable urbanization solutions. **Table 2** summarizes the relationship between the SDGs and the New Urban Agenda items.

To achieve sustainable urban development; sustainable development dimensions should be included in the planning standards of cities, which will lead to gaining several benefits in local and global communities; for instance, it will create more employment opportunities and better revenues, hubs for innovative solutions by attracting competencies, enhance land utilization efficiency and improve infrastructural performance, provide better services (e.g., education, health, water supply, and electricity), enhance the hub of economic growth, act as knowledge centers, better social and cultural life, and providing better living standards [33, 35, 47]. Therefore, integrating the paramount goals of both the 2030 Agenda and the Urban Agenda will create sustainable and innovative solutions and maximize the benefits of urbanization; as a result, the adverse environmental impacts associated with the urban development process on human health and the ecosystem will be avoided or reduced.

SDGs	New Urban Agenda items
6 CLEAN WATER AND SANITATION	73, 120
9 INDUSTRY, INNOVATION AND INFRASTRUCTURE	43, 44, 56–60
11 SUSTAINABLE CITIES AND COMMUNITIES	30, 34, 37, 46, 49–53, 55, 63, 66, 67, 72, 77–78, 80, 87–88, 90–92, 94, 96–103, 107–119, 124–125
12 RESPONSIBLE CONSUMPTION AND PRODUCTION	65, 70–71, 74, 76, 122
17 PARTNERSHIPS FOR THE GOALS	16–19, 21–24, 26, 29, 31–32, 35, 38, 42, 47–48, 81–86, 93, 104–06

Table 2.
The relationship between the SDGs and the New Urban Agenda items.

5. Toward sustainable urbanization

The population growth led to high pressure on the absorptive capacity of cities, which increased the demands on natural resources due to the increase in urban services, which has led to an urgent need to build resilient and sustainable cities. Therefore, to explore the urbanization issue based on sustainable dimensions, different concepts have been manifested in the literature and United Nations reports, such as green cities, eco-cities, eco-friendly cities, smart cities, sustainable cities, and smart sustainable cities. The most supported models of sustainable urbanism are compacted cities and eco-cities [48], while most literature overseen smart cities concepts [35, 38, 49].

The urban context should be emphasized on the dimensions of sustainable development at the early stage of urban planning. According to Bibri, some key challenges facing sustainable cities are the planning and designing the human settlement forms to enhance and improve sustainability [48]. Globally, the problems yielded by human mistakes in the urbanization development process are recurring and similar. Thus, the decision-making mechanism should consider those problems as knowledge lessons learnt to avoid them by acting appropriately to ensure sustainable urbanization results [50]. The contemporary lesson of the COVID-19 pandemic crisis is that there is an urgent need to compile efforts to find integrated solutions to make future development more sustainable and resilient; thus, the role of both policymakers and decision-makers decisions concerning urbanization development and land-use management is critical for better prosperity and quality life [51, 52]. Therefore, cities' competitiveness and sustainability can be achieved when the policies are harmonious and aligned [53].

Considering the 2030 Agenda and the New Urban Agenda, the top requirement is an obligation to set appropriate, effective, and coherent policies and measures into practice to strengthen the sustainable planning and management of urbanization. Further, to strengthen urban governance, evolve infrastructures, enhance institutional collaboration and capacity building based on the urban priority needs, and empower the participation of concerned stakeholders.

6. Conclusion

The 2030 Agenda and the New Urban Agenda are top and foundation stones to adopt practical measures, coherent policies, and effective legal frameworks to improve resilient cities, better human settlements, aspirations life, and resilient infrastructure through sustainable planning, management, and governance to mitigate or prevent the devastating impacts of disasters and pandemics. Further, both Agendas are the transformative powers of sustainable urbanization development, and that development is not at the expense of the environment.

The New Urban Agenda creates a value of sustainable urbanization by encouraging citizens' contribution to economic prosperity, improving environmental quality, stabilizing social justice, and enhancement of social and cultural institutions toward better lives in sustainable cities. All SDGs approvingly support the value of the transformative approach of urbanization toward sustainability, especially Goal 11. Therefore, integrating the paramount goals of both the 2030 Agenda and the Urban Agenda will create sustainable and innovative solutions and maximize the benefits of urbanization; as a result, the adverse environmental impacts associated with the urban development process on human health and the ecosystem will be avoided or reduced.

Cities can respond and contribute effectively to tackle the threats to the environment; such as by employing sustainable water consumption means (SDG 6, e.g., collection and purification of rainwater), including the social value in economic growth (SDG 8, e.g., investing more in the circular economy), adoption of sustainable waste management approach (SDG 12, e.g., increase waste recycling), minimizing climate change footprints, through the adoption of local mitigation and adaptation policies, regulations, and strategies (SDG 13, e.g., circular carbon and renewable energy), and protection of the natural and agricultural areas (SDG 15, e.g., avoid consumption sprawling land). Therefore, cities should enhance sustainable planning and resource efficiency management through the implementation of the SDGs for achieving sustainable urbanization solutions.

Author details

Abdulkarim Hasan Rashed
Environment and Sustainable Development, Bahrain

*Address all correspondence to: kme2001@hotmail.com

References

[1] Bai X, McPhearson T, Cleugh H, Nagendra H, Tong X, Zhu T, et al. Linking urbanization and the environment: Conceptual and empirical advances. Annual Review of Environment and Resources. 2017;**42**(1):215-240

[2] United Nations (UN). World Urbanization Prospects: The 2018 Revision. New York: UNDESA/PD; 2019

[3] Lerch M. International Migration and City Growth. New York: United Nations; 2017

[4] UN-Habitat. World Cities Report 2020: The Value of Sustainable Urbanization. Nairobi: UN; 2020

[5] Bandauko E, Annan-Aggrey E, Arku G. Planning and managing urbanization in the twenty-first century: Content analysis of selected African countries' national urban policies. Urban Research & Practice. 2021;**14**(1):94-104

[6] Wang Q, Wang X, Li R. Does urbanization redefine the environmental Kuznets curve? An empirical analysis of 134 Countries. Sustainable Cities and Society. 2022;**76**:103-382

[7] McPhearson T, Pickett ST, Grimm NB, Niemelä J, Alberti M, Elmqvist T, et al. Advancing urban ecology toward a science of cities. Bioscience. 2016;**66**(3):198-212

[8] Anwar B, Xiao Z, Akter S, Rehman RU. Sustainable urbanization and development goals strategy through public–private partnerships in a South-Asian metropolis. Sustainability. 2017;**9**(11):1940

[9] Uttara S, Bhuvandas N, Aggarwal V. Impacts of urbanization on environment. International Journal of Research in Engineering and Applied Sciences. 2012;**2**(2):1637-1645

[10] Hoi HT. Impacts of Urbanization on the Environment of Ho Chi Minh City. In: IOP Conference Series: Earth and Environmental Science. IOP Publishing; Jul 2020

[11] Ramirez-Rubio O, Daher C, Fanjul G, Gascon M, Mueller N, Pajín L, et al. Urban health: An example of a "health in all policies" approach in the context of SDGs implementation. Globalization and Health. 2019;**15**(1):1-21

[12] UN-Habitat. World Cities Report 2016: Urbanization and Development: Emerging Future. Nairobi, Kenya: UN; 2016

[13] UN-Habitat. New Urban Agenda. Nairobi, Kenya: United Nations Human Settlements Programme; 2017

[14] Diaz-Sarachaga JM, Jato-Espino D, Castro-Fresno D. Evaluation of LEED for neighborhood development and envision rating frameworks for their implementation in poorer countries. Sustainability. 2018;**10**(2):492

[15] Sietchiping R, Reid J, Omwamba J. Implementing the SDGs and the New Urban Agenda. Environment and Urbanization ASIA. 2016;**7**(2):x-xii

[16] McGranahan G, Schensul D, Singh G. Inclusive urbanization: Can the 2030 Agenda be delivered without it? Environment and Urbanization. 2016;**28**(1):13-34

[17] Klopp JM, Petretta DL. The urban sustainable development goal: Indicators, complexity and the politics of measuring cities. Cities. 2017;**63**:92-97

[18] Moreno EL. Concepts, definitions and data sources for the study of urbanization: The 2030 Agenda for Sustainable Development. United Nations Expert Group Meeting on Sustainable Cities, Human Mobility and International Migration. 2017. Available from: https://www.un.org/en/development/desa/population/events/pdf/expert/27/papers/II/paper-Moreno-final

[19] Grossi G, Trunova O. Are UN SDGs useful for capturing multiple values of smart city? Cities. 2021;**114**:103-193

[20] Smiciklas J, Prokop G, Stano P, Sang Z. Collection Methodology for Key Performance Indicators for Smart Sustainable Cities. Geneva, Switzerland: U4SSC; 2017

[21] Kacyira AK. Addressing the sustainable urbanization challenge. UN Chronicle. 2012;**49**(2):58-60

[22] Sun L, Chen J, Li Q, Huang D. Dramatic uneven urbanization of large cities throughout the world in recent decades. Nature Communications. 2020;**11**(1):1-9

[23] Nathaniel S, Anyanwu O, Shah M. Renewable energy, urbanization, and ecological footprint in the Middle East and North Africa region. Environmental Science and Pollution Research. 2020;**27**(13):14601-14613

[24] Kosovac A, Acuto M, Jones TL. Acknowledging urbanization: A survey of the role of cities in UN frameworks. Global Policy. 2020;**11**(3):293-304

[25] Wang Q, Wang X, Li R. Does population aging reduce environmental pressures from urbanization in 156 countries? Science of the Total Environment. 2022;**848**:157-330

[26] United Nations Environment Programme. (n.d.). GOAL 11: Sustainable cities and communities. Available from: https://www.Unep.org/ [Accessed: December 14, 2022]; https://www.unep.org/explore-topics/sustainable-development-goals/why-do-sustainable-development-goals-matter/goal-11

[27] Flörke M, Schneider C, McDonald RI. Water competition between cities and agriculture driven by climate change and urban growth. Nature Sustainability. 2018;**1**(1):51-58

[28] Wang Q, Wang X, Liu Y, Li R. Urbanization and water consumption at national-and subnational-scale: The roles of structural changes in economy, population, and resources. Sustainable Cities and Society. 2021;**75**:103-272

[29] Satterthwaite D. Successful, safe and sustainable cities: Toward a New Urban Agenda. Commonwealth Journal of Local Governance. 2016;**19**:3-18

[30] Strokal M, Bai Z, Franssen W, Hofstra N, Koelmans AA, Ludwig F, et al. Urbanization: An increasing source of multiple pollutants to rivers in the 21st century. Urban Sustainability. 2021;**1**(1):1-13

[31] Zhang Q, Liu S, Wang T, Dai X, Baninla Y, Nakatani J, et al. Urbanization impacts on greenhouse gas (GHG) emissions of the water infrastructure in China: Trade-offs among sustainable development goals (SDGs). Journal of Cleaner Production. 2019;**232**:474-486

[32] Marselis SM, Feng K, Liu Y, Teodoro JD, Hubacek K. Agricultural land displacement and undernourishment. Journal of Cleaner Production. 2017;**161**:619-628

[33] Sitharam TG, Dhindaw J. Benefits and challenges of urbanization and low carbon energy needs in India. Journal of

Sustainable Urbanization, Planning and Progress (Transferred). 2016;**1**(1)

[34] Franco S, Mandla VR, Rao KRM. Urbanization, energy consumption and emissions in the Indian context: A review. Renewable and Sustainable Energy Reviews. 2017;**71**:898-907

[35] Shahidehpour M, Li Z, Ganji M. Smart cities for a sustainable urbanization: Illuminating the need for establishing smart urban infrastructures. IEEE Electrification Magazine. 2018;**6**(2):16-33

[36] Elmqvist T, Fragkias M, Goodness J, Güneralp B, Marcotullio PJ, McDonald RI, et al. Urbanization, Biodiversity and Ecosystem Services: Challenges and Opportunities: A Gobal Assessment. 2013

[37] Watts M. Cities spearhead climate action. Nature Climate Change. 2017;**7**(8):537-538

[38] UN-Habitat. SDG 11 Synthesis Report 2018: The Value of Sustainable Urbanization. Nairobi, Kenya: UN; 2018

[39] Blasi S, Ganzaroli A, De Noni I. Smartening sustainable development in cities: Strengthening the theoretical linkage between smart cities and SDGs. Sustainable Cities and Society. 2022;**80**:103-793

[40] World Health Organization. Air Pollution Levels Rising in Many of the World's Poorest Cities. n.d. Available from: https://www.who.int/ [Accessed: December 14, 2022]; https://www.who.int/news/item/12-05-2016-air-pollution-levels-rising-in-many-of-the-world-s-poorest-cities

[41] UN Environment. The Weight of Cities: Resources Requirements of Future Urbanization. Paris, France: UN; 2018

[42] Kuddus MA, Tynan E, McBryde E. Urbanization: A problem for the rich and the poor? Public Health Reviews. 2020;**41**(1):1-4

[43] UN General Assembly (UNGA). Transforming Our World: The 2030 Agenda for Sustainable Development. 2015. Available from: http://www.un.org/ga/search/view_doc.asp?symbol=A/RES/70/1&Lang=E

[44] Akuraju V, Pradhan P, Haase D, Kropp JP, Rybski D. Relating SDG11 indicators and urban scaling–An exploratory study. Sustainable Cities and Society. 2020;**52**:101-853

[45] Chen M, Chen L, Cheng J, Yu J. Identifying interlinkages between urbanization and sustainable development goals. Geography and Sustainability. 2022;**3**(4):339-346

[46] Brelsford C, Lobo J, Hand J, Bettencourt L. Heterogeneity and scale of sustainable development in cities. Proceedings of the National Academy of Sciences. 2017;**114**(34):8963-8968

[47] Kaika M. 'Do not call me resilient again!': The New Urban Agenda as immunology... or... what happens when communities refuse to be vaccinated with 'smart cities' and indicators. Environment and Urbanization. 2017;**29**(1):89-102

[48] Bibri SE. The underlying components of data-driven smart sustainable cities of the future: A case study approach to an applied theoretical framework. European Journal of Futures Research. 2021;**9**(1):1-14

[49] Pagani RN, Soares AM, da Luz AA, Zammar G, Kovaleski JL. On smart cities and sustainable development goals. Debates sobre innovación. 2019;**3**(1)

[50] Wang J, Shen L, Ren Y, Ochoa JJ, Guo Z, Yan H, et al. A lessons mining system for searching references to support decision making toward sustainable urbanization. Journal of Cleaner Production. 2019;**209**:451-460

[51] Cotella G, Brovarone EV. Questioning urbanization models in the face of Covid-19. TeMA-Journal of Land Use, Mobility and Environment. 2020;**2020**:105-118

[52] Solly A, Berisha E, Cotella G. Toward sustainable urbanization. Learning from what's out there. Land. 2021;**10**:356

[53] UNDP. Leveraging Urbanization and Governance for Growth in Africa: A Framework for Action. Addis Ababa: UNDP; 2017

Section 2

Case Studies

Chapter 7

Nocturnal Developments in the Jos Metropolis, Nigeria

Samuel Danjuma Wapwera

Abstract

This paper seeks to examine the series of development that takes within the Jos metropolis, Nigeria without proper consultation and with permission following the acceptable planning standard. By examining the time of development and examining the process followed to put the development in the location either uphill or low land. The instruments employed were face-to-face interviews and reviews of archived documents on the implementation of master plans and other options to control haphazard development over the past 20 years. The data obtained through the face-to-face interviews and the archive documents were analysed using the thematic and content analyses. The results obtained revealed that the developments have been carried out haphazardly, which was affected by unproved legislation, the plethora of planning authorities covering the planning region. Hence the consideration of the strategies to employ for the implementation of plans concluded with the recommendations to help practitioners, policy makers and academia.

Keywords: nocturnal, developments, Jos metropolis, Nigeria, development control

1. Introduction

Rapid urbanisation in recent times indicates accelerated urban growth and significantly influenced urban form in very profound ways, especially in most exploding uncontrolled cities of less developing countries. Hence, such urbanisation has been characterised by poor housing, poor transportation, inadequate social services, poor environmental quality, extreme poverty and physical dereliction of basic infrastructure amongst others leading to high levels of urban poverty as argued by Wapwera and Dung-Gwom and Bashir [1, 2]. The urban poors are usually unable to obtain land and housing through the formal channels as such have been forced to seek shelter in marginal lands and sometimes even built on lands without following the regular protocol of obtaining permission making them obtained shelter with high environmental and ecological risks [2, 3].

This action has culminated into unprecedented share scale and pace of contemporary urban growth, the most rapid changes in urban form, pattern and structure are taking place where historical roots are weakest and are observed in some settlements with specific peculiarities. Most buildings are constructed in a quake way, casting doubts on their foundation and lacking capacity to withstand harsh weather and environmental conditions any time it comes. This situation shows that the buildings are usually built with substandard materials, boards and zinc sheeting, as well as recycled and mixed materials, which make the houses very vulnerable as argued in Ref. [4].

These buildings are manifesting varying defaults that are very dangerous to life and properties. Recent researchers have shown that these developments have taken place usually at night and at less active hours and times of the day where visibility is not readily available. Based on this observation, it means that some buildings are done at night rather than day time. They could be termed nocturnal buildings or developments.

Nocturnality could be considered as a condition of being active or taking place during the night or the active behaviour of an animal or plant during the night (and then inactive during the day), as observed in this case, having developments of building and structures within some specific areas of the urban metropolis.

Hence, nocturnality refers to the condition of being active or happening during the night and not during the daytime. Worthy of note at this juncture is the fact that nocturnality could be more attributed to behaviour and it could be influenced by environmental factors, which in this instance have forced the people to build at odd hours thereby getting wrong and arbitrary results.

Developments are activities that modified the physical environment; these include mining, minerals or waste disposal matters, excavation, and erection of both temporary and permanent structures as shelter where man works, plays and lives. These activities are expected to be controlled and dealt with by government of county councils in cities and metropolitan areas as argued by Ratcliffe et al. [5]. And if these developments take place without control, it becomes haphazard and hence needs to be checked. Nocturnal development has attracted increased attention in recent years as a result of the default observed from the houses, structures, nature and patterns that have emerged. Hence, a careful assessment of the nocturnal development on household livelihood and income is very low compared to those within planned and organised spaces. Development control (DC), otherwise known as planning control, is a management component of the **United Kingdom**'s planning system for the **town and country planning**, through which local government regulates **land use** and new building control. It relies on a 'plan-led system' whereby urban and regional development plans are formed and the public is consulted.

Ratcliffe et al. [5] Observed that any development requires **planning permission**, which is granted or refused by reference to the development plans as a material consideration. The term 'development control' is often abbreviated to DC. Previous studies have shown that there are 421 **local planning authorities** (LPAs) in the **United Kingdom** while in Nigeria there are 774 local government areas (PAs) and yet they do not have planning departments or sections. Hence, the results obtained are manifested in a chaotic, uncoordinated physical environment lacking in every basic facility.

The efforts put in place to meet one of the goals of the sustainable development Goals [6] especially those that which concerns the Goal 11. Make cities and human settlements inclusive, safe, resilient and sustainable. It is expected that by 2030, people will have access all too adequate, safe and affordable housing and basic services and upgraded slums. This would be achieved by reducing the proportion of urban population living in slums, informal settlements or inadequate housing. Recent studies have shown that there is dearth of need as it relates to physical planning to address the issues relating to slums, informal settlements or inadequate housing. Studies by Dung-

Gwom and Bashir [2] on planning, housing and environmental challenges of Hilltop Settlements in Zinariya and Azurfa Areas of Jos, Nigeria, [7], characteristics and Planning Challenges of Hilltops Settlements in Jos Metropolis, Nigeria and [8], Nature of peri-urban development in Jos, Nigeria show that developments emerge at locations where the provision of basic infrastructure are not readily available and hence the need for the provision by government through their planning authorities.

The physical planning strategies in Jos metropolis, Nigeria are geared towards identifying and targeting such developments, by conserving the available priorities, risks, opportunities and gaps in the urban and regional areas. It is not a static blueprint plan, but an attempt to understand the social, economic and environmental resources within each jurisdiction, linkages within and between urban and regional areas and upgrade of infrastructure. Based on this system of physical planning these developments were observed. It's incumbent upon the government, planning authorities and the communities affected to take proactive measures towards addressing the growing menace.

A strong planning system relies on up-to-date plans, enhanced with current information to address the wicked and dangerous problems that are dynamic in nature. Most African countries lack such systems as these physical planning problems are evident, hence affecting cities and most urban areas as managing them is an uphill task needing the attention of planners and allied experts as well as technocrats in the built environment. Hence, sprawling conditions characterised by fast slums occurring make the problem grow rather than diminish [9]. A plan that identifies and brings forth adequate quantities of land for development into safe and accessible locations is most appreciated as argued by many in recent times.

The era and legacy of failed master plans that are rigid and fixed have left cities, urban areas, residents and the surrounding environment poorly equipped to cope with the exploding urbanisation levels [10]. These cities are important to Africa and Nigeria in particular but are characterized by poor growth and haphazard developments. But due to poor provision of the needed infrastructure such as sophisticated management and effective leadership, it has been robbed of standard and effective plan-led modernisation and growth [11–13].

The choice of the study problem nocturnal developments is based on the fact that areas originally earmarked as buffer zones within a planned environment suddenly are occupied by buildings and settlements emerging without control. Secondly, attempts were made to plan the Zinariya and other areas into a low-density area, after the de-gazetting of the Naraguta Forest Reserve in 1985. The green areas were encroached upon and also sold and 'illegally' occupied by low-income earners, who were unable to go through the cumbersome and unending process of land processing and documentation in the numerous planning authorities as argued by Dung-Gwom and Bashir [2].

Furthermore, there has been no review of the master plan of Jos to incorporate these changes and to ensure proper integration of these areas into the city system. Rough terrain and steep slopes pose great challenges to housing development with inadequate provision of infrastructures, no road designed and its consequent construction amongst others. Due to the high level of poverty and the fact that these areas are mostly inhabited by low-income earners, the residents are unable to provide these facilities, utilities and services on their own. Hence, the quality of housing development is generally poor, lacking in design, inadequate spacing, poor orientation and ventilation exhibiting the weak or non-existence of the presence of development control in the settlements observed to have emerged.

The underlying theories in this chapter are the theory of institutional deficiency and the theory of planning practice which are confirmed by the elements of the non-implementation of the Greater Jos Urban Master Plan due to the following; inadequate man-power, high level of corruption, Bureaucratic Bottleneck and Red Tape, Delay and negligence of responsibility amongst others as argued by Mahalingam and Levitt [14]. The theory of planning practice is also made manifest in all the planning authorities, a good organisation for industrial experience for planners and other professionals within the built environment at various levels, having gone a long way to enriching the town planning courses in tertiary institutions where training takes place and where the planning authorities send employees for industrial experience for different durations.

Also, constraints have been identified with the absence of an appropriate framework to analyse the current institutional framework within which the planning authorities exist in Jos Metropolis, Nigeria and how the constraints have affected the functionality of the framework towards the implementation of the Ref. [15] and other related plans within the Jos metropolis, Nigeria.

Furthermore, the government needs to ensure participation by all and sundry, by working with the less privileged and marginalised groups. A conscious involvement of the public sector by placing much emphasis on traditions, legislation and methods of urban and regional planning, and bringing about the reduction in the dichotomy of the distribution of resources across regions. Good governance brings about good planning because it is vital to the design and implementation of any plan because the market forces usually are too overwhelming. Hence, the solution is always ineffective public sector action, as the government plays a vibrant role in stabilising the economy by providing the needed infrastructure. A good planning approach has a great tendency to contribute to Nigeria's development because it possesses the technical capacity to deliver planning at both the local, state and national levels to bring about sustainable development in Jos Metropolis, Nigeria.

it is against this background that this chapter seeks to examine the series of nocturnal development that takes within the Jos metropolis, Nigeria without proper consultation and permission as well as following the acceptable planning standard. This would be achieved by examining the time of development and the process followed to put the developments in the settlements that have emerged either on uphill or low land with a view to making recommendations to address the physical planning problem.

2. Literature review

The review of literature will be considered under two parts vis; the time of development and the process of development followed to put the development in the locations.

2.1 Time of development

The consideration of the concepts of development and nocturnality has become very pertinent at this juncture to provide vital information as it relates to the development of the urban environment, which has become the major issue at stake as it relates to physical development of the said areas [16].

Development is not purely an economic phenomenon but rather a multi-dimensional process involving reorganisation and reorientation of an entire economic and social system. In related studies such as those by RTPI and Rydin [17, 18] observed that development is a process of improving the quality of all human lives with three equally important aspects vis;

1. Raising peoples' living levels, that is, incomes and consumption, levels of food, medical services and education through relevant growth processes,

2. Creating conditions conducive to the growth of people's self-esteem through the establishment of social, political and economic systems and institutions, which promote human dignity and respect.

3. Increasing people's freedom to choose by enlarging the range of their choice variables, for example, varieties of goods and services.

Development is defined to include mining, minerals or waste disposal matters, excavation, and erection of both temporary and permanent which is dealt with by county councils in non-metropolitan areas [5]. It was further considered under Section 55(1) of the 1990 TCPA defined development as: '*The carrying out of building, engineering, mining or other operations in, on, over or under land or making of any material change in the use of any buildings or other land*'.

From the above definition, two major elements are evident (operation) and material change of use (activity).

2.2 Nocturnal developments

Nocturnality is the act or condition of being active or taking place during the night or the active behaviour of an animal or plant during the night (and then inactive during the day). Hence, nocturnal developments could rightly refer to the condition of being active or happening during the night and not during the day time and for the purpose of this study for building of structures for human habitation. These are developments that occur without obtaining any permission from any regulatory agency making it possible for developments to take place without control showing haphazard developments (building and structures), which needs to be checked. The implication is that these developments occur negating the following development control regulations such as: permitted land uses, set-backs, building heights, plot coverage for different density zones (low, high and medium), permitted fence heights, number of entrances permitted to a plot and available parking spaces amongst others [19].

The aforementioned Nocturnal developments are organised by governments and institutions with intension of improving night-time safety on public transport (Infrastructure) in urban areas, to encourage women and children to travel at night. Even though two-thirds (2/3) of women are afraid to travel on public transport at night and have safety concerns over using car parks after dark [20].

Most planning policies consider bringing activities into city centres at night for safety reasons, hence introducing safety measures, such as better street lighting, improvements to car park design and many other things. The provision of open parking areas is seen as safer for women than multi-storey car parks. Other things considered are improved public transport including evening hail and ride bus services;

shorter train carriages to increase occupancy and a well-lit station entrance. Furthermore, another hopeful sign for the future is the development of light rail systems, which may bring safer travel into inner urban areas [19].

Nocturnal developments are developments usually characterised by lack of basic infrastructural facilities such as water, drainage system, sanitation facilities, inaccessibility and lack of sufficient space amongst others as argued by Wapwera et al. [21]. They are also commonly seen as breeding grounds for social vices and problems such as crime and high rates of psychological maladjustments. The developments generally lack infrastructural facilities, such as good water, electricity, drainages and roads usually caused by the low-income levels of the inhabitants. Governments are not aware of their conditions because they do not contact government for adequate planning regulations and advice. Hence, these nocturnal developments are often faced with the brunt of natural and man-made disasters such as fire outbreaks, flooding, disease outbreaks and the likes [19].

2.3 The process of development

The section seeks to consider the concept of development control, evolution of development control, legacies of development control in Nigeria, philosophy of development control and development control mechanism.

2.4 The concept of development control

Urban areas accommodate more than half of the global population [13]. They are thus centres of high economic and commercial activities. In Africa, urbanisation is surging so fast that the continent is envisaged to account for 21% of the global urban population by the year 2050 [22].

The Nigeria Urban and Regional Planning Degree number 88 of 1992 describes development as "the carrying out of any building, engineering, mining or other operations in, on: over or under any land, or the making of any environmentally significant change in the use of any land or demolition of building including the felling of frees and placing of frees-standing erection used for display or advertisements on the land of the expression "develop" with its grammatical variations shall be constructed accordingly. Meanwhile, control means the process of exercising power over one's area of jurisdiction to check development as argued by Ogundele et al. [23].

Hence, development control is the regulation power exercised by planning agencies to either approve or reject a development application [23]. It is a system by which the use of land and buildings on the land are regulated so that misuses or above of use and nonconforming uses are prevented or checked.

Development control is the cornerstone, the biting tooth of physical planning. Without it, physical planning becomes inactive and impractical. It is the medium through which formulated policy proposals are implemented. Development control in all ramifications is simply put as the way and manner in which land use or spatial development of land is regulated [24]. Dung-Gwom [25] sees it as a cardinal means through which government implement their land policies. The definition of what constitutes development control and its objectives differ from person to person over the years.

Dung-Gwom [25] outline four (4) broad strands to the meaning given to the term 'development control' as means of implementing development plan, facilitating development, ensuring good quality environment and protecting a property owner's immediate environment. In developed countries, development control has been used as a means of involving the people in the development of their environment (public participation) since development control is a tool for plan implementation as used by town planners. Dung-Gwom [25] sees it as a means of facilitating development through the mutual contributions of the private and public sectors. But most developers see development control as an aberration or impediment to achieving their development goals and objectives as it is too rigid or negative to their interest [17, 18].

Hence, Aluko [26] defined Development Control as the power to say "Yes" or "No" to a very wide range of private and (certainly public) proposals".

Development control could be divided into the following:

i. Macro and micro

ii. Normative and monitoring dimension [26]

iii. Negative and Positive instruments of control.

Development control stands at the heart of urban development. It is a technical tool for ensuring that all users of urban land exercise some discipline in land development in conformity with planning rules and regulations. For the purpose of this chapter, development is the 'procedure and substantive rules through which physical modification of the environment is decided upon by planning authorities' hence, whatever the planning authorities decide as the procedures and substantive rules through which development is decided should be done in consultation with developers so they are carried along to improve on the level of awareness and compliances. Consequently, development plan set out the land use policy for an area on which development control is based. It guides the granting and refusal of permission to carry out development. Development plans are considered "as the fulcrum upon which development control oscillates as argued in [13].

2.5 Evolution of development control

The concept of development control started in the early **nineteenth** century, with its protagonist as Todor, Start and Georgian. It evolved as a result of the massive industrialisation of the 19th and 20th centuries, which led to large-scale production and population growth, congestion and overcrowding.

This increase led to alteration in the physical appearance of industrial towns and brought with it its own set of problems. Further studies by several scholars noted that the form and location of development have become haphazard, creating a bewildering mixture of functions and qualities of the environment as argued by Datonjo et al. [27].

The outcome, therefore, was an unacceptable urban environment with disorganised layout, social and economic deprivation and general public health risks, leading to the spread of diseases such as cholera, malaria, typhoid and spinal meningitis amongst others within the period 1830s–1840s [26].

These identified problems necessitated the desire for an ideal environment. Hence, it was from this desire that the concept of a planned and ordered environment emerged [28]. These strategies were adopted for the ordering and control of land use and the implementation of these strategies was legalised by certain statutes. For instance, the first piece of Land use control regulation in Britain was the housing,

Town planning (Act of 1909) and the local authorities with powers to prepare schemes to control the development of new housing estates.

Presently the practice of development control takes various forms. The system is either centralised as in Britain or decentralised as in the Scandinavia. It is either reactive or proactive and regulatory or discretionary in some countries. For instance, American and British land use control methods represent opposite extreme approaches. The American system depends on the establishment of formal and consistent standards and the avoidance of discretionary power in the hands of planning authorities, while the British system involves discretionary development control powers [17, 18].

The variation in development control approaches is influenced by legal and constitutional structures, administration and professional cultures or history. The social and economic ideologies in a country also influence the system. There are, however, countries that have similar planning systems, such as Nigeria and Britain, as the Nigerian town and country planning Acts ordinance 1946 was an offshoot of the 1933 town and country planning Act of the United Kingdom [17, 18]

2.6 Legacies of development control in Nigeria

The rapid rate of population growth in Nigeria and the attendant problems associated with such rapid growth calls for development control measures to avoid environmental, social and economic decay and protect the overall interest of the public [26].

The following development control legacies will be considered in this section see **Table 1**:

2.7 Philosophy of development control

The models (ideas) of development control vary from one society to another. But three models are considered here as outlined by Hassane Cisse et al. [29], as follows:

i. The free market model (traditional/natural model)

ii. The public intervention model (modern model)

iii. The cooperation community model (The emerging model and originated from the western world) [26].

2.8 Development control mechanism

In order to control development, several techniques have been used, see **Table 2**:

3. Methodology

The Jos metropolis, Nigeria was carved out of two local government areas namely; Jos North and Jos South LGAs. The survey instrument used for data collection was a structured questionnaire doctored on the Samsung Galaxy tab N225T. This was to help in obtaining a direct geo-reference and subsequently Geotag and pictures the buildings amongst others. The Survey 123 App with other support applications, such as the Arc GIS Explorer and Measure Height, where satellite images and aerial photographs were used to observe developments within the study area (see **Figure 1**).

Year	Name of legislation	Description of the main features of the legislatives effort made
1863	Township Improvement Ordinance	It represents the first ever legislation on urban planning in Nigeria. It was published in Lagos for the Lagos colony by the colonial government. Its main objectives were to control development and sanitation in Lagos. The enforcement of this ordinance rested with the health Department of the defunct Lagos city council. Because, the focus of the ordinance was on urban sanitation, many planners regard it as a health regulation, rather than an urban and regional planning legislation.
1904	Cantonment Declaration	It was the first real planning legislation in Nigeria. It was derived from attempt to protect Europeans from health hazards so prevalent in the 19th and early 20th Centuries. Thus in 1904, the governor in southern Nigeria was empowered to declare any area a European Reservation within its own Local Health Board. Its main aim was to improve and preserve the health of Europeans in the reservation. The Lagos Municipal Health Board was established in 1908 to administer this ordinance. The regulations were to be administered by a magistrate. It was concerned with Land use and the need to improve health conditions within the cantonment. An interesting part of the provision was that large cities were divided into wards (Ungwa) and the chief of each ward was responsible for its sanitation. The proclamation also introduced the use of incinerators, latrines pits and cemeteries.
1917	The Township Ordinance	Unarguably the first statute on town and country planning in Nigeria was the Town Planning Ordinance, No. 29 of 1917. Its main objective was to establish the broad principles of municipal responsibility, graduated according to the importance of the community and measure of its ability to accept and discharge satisfactorily independent and quasi-independent powers and responsibilities. The ordinance not only served as a legal basis for development of towns for most part of the colonial administration, but also gave guidelines for construction of buildings, control of development and finance of land in the urban areas. It classified cities into three categories: first, second and third class townships (policy of segregation on the basis of enclave for European (reservation), southern Nigerians in the north (Sabon gari) and northern Nigerians in the south (Sabo) (Mabogunje, 1985). Streets in European reservation were to be broad and paralleled. It provided detailed compound size, depth of plot coverage, etc, for development in the reservations. In a way, the ordinance introduced zoning and sub-division regulations into planning practice in Nigeria for the first time. The segregative nature of the ordinance marked the origin of development of slums in Nigerian cities.
1928	Town Planning Ordinance	i. The creation of Lagos Executive Board. ii. It made provision for Slum clearance, Land reclamation, residential and industrial estates and its undertook comprehensive land use planning and development in Lagos. iii. It was in place for 18 years and served as the basis for the framework of the 1946 planning law.
1946	Town and Country Planning Ordinance	Cap 155 of the 1946 edition of the Laws of Nigeria. It was adopted by all the regions in Nigeria under various codes. It covered a wide range of planning operations such as planning schemes, planning authorities, preparation and approval of schemes, execution of schemes, acquisition and disposal of land for scheme preparation, compensation and betterment, the legal procedures for the operations of schemes and financial matters for successful implementation of schemes. Hence, a summary of the functions of planning authorities

Year	Name of legislation	Description of the main features of the legislatives effort made
		under the 1946 ordinance as follows: i. They were autonomous statutory organization as established by the planning laws of the state. ii. Prepare town planning schemes and implement them in stages. iii. Approve layouts and building plans after scrutiny within the framework of building bye-laws, sub-division and zoning regulation. iv. Ensure that owner of landed properties adhere to the building bye-laws and other regulations as contained in the planning scheme. v. Enforce strict compliance with requirements of town and country planning laws of the state. vi. Accepts any rates, money, property or other assistance for the purpose of executing the scheme (S.59) or raise revenue for planning purposes through imposition of levy and collection of rates. vii. They were required to control and manage their own finances and may even buy, sell, or let, hire, lease, exchange or otherwise dispose of any property, movable or immovable.
Source: Compiled by author from various sources		

Table 1.
Development control legacies.

SN	Mechanism	Description of the mechanism
1	Land Use Zoning	It was developed at the beginning of the 20th century in the United States of America as a means of controlling nuisance and protecting landed property values through the regulation of land use activities. The idea of Zoning district arises from the notion that segregation of certain uses from the notion that segregation of certain uses from others reduce the effects of negative externalities which some uses have over others. But it has been severely argued that zoning is not an effective tool for development control as rapid urbanization has rendered it ineffective unlike at the beginning and middle of 20th century when the rate was low. Planners have therefore devised ways of situating different land uses within a district Reps (1964) is of the view that zoning is an unnecessarily prolong land use device. He further stated that the powers of zoning and other development control techniques are grossly inadequate when measured against different development patterns. Zoning is a powerful tool and with recent proliferation of zoning innovations, it is now designed to make it more flexible and responsible to the needs of Land use implementation process.
2	Planning Standards	Planning standards are regulatory mechanisms put in place to ensure that development is conducted in accordance with certain predetermined minimum standards. But planning standards of compatibility and incompatibility are objective, scientific and as a practical matter enforceable. Incompatibility of uses can be measured by factors such as danger to morals, aesthetics, psychological factors, etc. The techniques of using standards for different land use zones, ensures that uses which will adversely affect harmony of the area are not allowed. Some of the relevant planning standard (legislations which serve as guidelines for development control include:- i. Building regulations (bye-laws) ii. Control of advertisement regulations iii. Listed classes order iv. Use classes order.

	Building bye-laws beside others regulates the construction, alteration, maintenance, repair and demolition of building and structures. The bye-laws are implemented through the following:- i. Density and height ii. Setting in accordance with –stipulated floor ratio iii. Standard in regards to minimum space requirement. It is to mainly focus on the degree of compliance and the impacts of theses standard and regulations on the physical environment.

Source: Compiled by author from various sources

Table 2.
Development control mechanisms.

Figure 1.
Area within Jos-Bukuru metropolis, Nigeria.

The second instrument used is the face-to-face interviews where purposive sampling techniques were used. Two heads of planning authorities saddled with the responsibility of development control, three community heads 'mai anguwas' and five house owners (developers) were interviewed, giving a total of 10 (ten) interviews coming from non-streets nomenclature as observed from the figure above. The face-to-face interview questions were structured according to the main questions asked at the inception of the study.

The third instrument used was the review of archived documents on the implementation of master plans and other options used to control development. In addition, there was an observation of buildings, streets, access roads, drainages, utilisation options of electricity, water and some community facilities within the study area to observe non-compliance to building and development regulations to drive home the assertion of nocturnal developments observed in the study area.

The data obtained through the satellite images were analysed using spatial analytic methods to observe the development within the study areas over time. The face-to-face interviews and the archive documents were analysed using the thematic and content analyses.

The next section considers the results obtained from the extensive research activities carried out to meet the aim of the study.

4. Findings and discussion

4.1 Nocturnal development

This section considers land law and planning, land and planning administrative structure in Jos metropolis, Nigeria. The implementation of the development control within the Jos metropolis, Nigeria.

4.1.1 The land law and the planning law adopted in the Jos metropolis, Nigeria

For any development to emerge, there are two basic laws that are directly linked to land operations and the development of buildings and related development are the land and planning laws. These two are statutory control of land use in Nigeria. The 1978 Land use Act is the operative land law in each state under the governor.

He is empowered by the act to grant statutory rights of occupancy to developers conferring on the legal rights over land (Section 5 of the land use Act, [30]).

The right of occupancy is a tool for development control, which requires an individual and corporate organisation to obtain prior consent from the government before they could "Occupy and use: Land."

It has been in the statute books since 1910. The importance of the statutory R of O is that it ensures 'Planned development and zoning programming' as grants are made consistent with planning schemes and policies.

The aim of this chapter is to examine the emerging phenomenon of nocturnal building developments that is developments that are seen springing through the hours of the night only to be seen completed by morning. These 'unplanned developments' are a deviation from the norm. The usage of these laws 'land and planning laws' for physical land use planning in Nigeria is within the competence of state governments, through the planning authorities. These were adopted by the regional governments at various times under different captions. It is the Law in operation within Plateau State until the Nigerian Urban and regional Planning Law No. 88 of March 1992 [31] was promulgated.

The 1992 URP law replaced the planning authorities with planning boards at the state level. Each board is expected to establish a development control department, which grants building permits for all forms of development. In Plateau state, however, there is a planning board with zonal offices, but how effective they are in controlling developments it is against this backdrop that the data collected and analysed to see how developments have presented themselves and how we have the developments also known as nocturnal.

The main aim of the 1978 Land Use Act (Land law) was to ensure that everyone had equal access to land in urban areas for the purpose of house building. But the Government is unable to address the problems that arise, clear land and reform the inherent tenure system that has been prevalent in the country and that has left problems unresolved, even in the face of pressure to acquire or gain access to land for development. The community system vested power of title of land on the families, communities and the village headman/mai ungwa amongst others.

Each regime of government tends to favour its loyal supporters and is usually highly politicised, causing delays in the processing of land titles; certificates of ownership and right of occupancy (C of O and R of O) as argued by Wapwera [1].

DOI: http://dx.doi.org/10.5772/intechopen.110127

Figure 2.
Zinaria Settlement with Haphazard Developments within a difficult Topography.

Figure 3.
Kabong Settlement with Haphazard Developments within a difficult Topography.

This situation has not helped the system making new developers of the settlements in the study, hence the serious development and construction of buildings and structures in the odd hours of the day due to non-permission to build. For further details see **Figures 2** and **3**.

The face-to-face interviews conducted with the developers lead to the revelation that their houses are built without any planning permission; hence, their communities are developing and growing in a completely unplanned manner. This is because they find it difficult to obtain any title documents or a law guiding the development of their houses. Over 80% of the developers agreed that they do not have approved building plans from the relevant planning authority (Jos Metropolitan Development Board) and the land titles documents from (Ministry of Lands Survey and Town Planning) stating the appropriate land used designed for the areas under discussion are not known to them.

4.2 The land and planning administrative structure in Jos Metropolis, Nigeria

The planning administrative structure in Jos Metropolis, Nigeria shows the linkages that help in the execution of the plan and the statutory plan of the given area. This administrative setup is vital in achieving the aim of a well-planned environment. For the purpose of this paper, two physical planning agencies saddled with the responsibility of regulating the activities were identified. The responsibility of granting statutory rights of occupancy (R of O) to private and corporate developers is the function of the Ministry of Lands, Survey and Town Planning (MLSTP). For further details see **Box 1**.

i. application by developer
ii. registration
iii. survey report
iv. land report
v. planning report
vi. land use allocation committee deliberation/recommendation
vii. governor's approval (or otherwise)
viii. issuance of r of o

Box 1.
Processes of obtaining the R of O. Source: MLS & TP Jos (2022).

The procedures do not vary whether for private or corporate developers and whether on isolated sites or within a government scheme (layout). The three technical reports from survey, lands and planning have to be recommended for the granting of an R of O. If any of the reports are negative, the application is refused or rejected and the applicant is duly informed.

The survey report "charts" the application to ensure that the land is not within a railway or major road reservation (i.e., within a town plan, etc). The land report checks that the land is not required for public purpose (Section 50 of the 1978 (Land Use Act)). The nature of existing land use service available, setbacks from streets, access, and conformity with the master plan amongst others are the primary concern of the planning report. Other requirements include tax clearance certification and evidence of ownership (for isolated cases).

i. Application by developer
ii. Registration
iii. Joint site inspection
iv. Planning recommendation
v. Health recommendation
vi. Engineering recommendation, if over two floors or a large development.
vii. Approval (or otherwise) of plan by Chief Town Planning Officer.
viii. Granting of building permit.

Box 2.
Processes of obtaining a permit. Source: Authors Field work and extract from MLS&TP, (2022) and JMDB (2019).

For private companies, they need to make available a certification of incorporation. The main procedures for processing applications for building permits as obtained from Ref. [32] are as Seen in **Box 2**.

Considerations during the building permit stage are on site-specific with details of density, bulk, setbacks, structural soundness, drainage and ventilation amongst others. A detailed examination of procedures for obtaining R of O and building permit shows that the process is cumbersome and un-ending classified by huge bureaucratic bottleneck and red-tapes from one directorate/division/section to the other.

Our face-to-face interviews with the developers led to the revelation that their areas are developing and growing in a completely unplanned and haphazard manner because they find it difficult to describe the location of their houses. However, several of their visitors have observed that locating their houses have always been difficult due to the unplanned nature of their settlement. A total of 80% of the developers agreed that they do not have approved building plans from the relevant planning authority's such as the (Jos Metropolitan Development Board) and the land titles documents from the Ministry of Lands Survey and Town Planning stating the appropriate land used designed for the areas under discussion are not known to them.

This creates a lot of problems for development and land administration and management in Jos metropolis, Nigeria over the years. It is due to the seriousness of this problem that almost every new administration in the state always set a committee to proffer solutions to the cumbersome and unending land administrative processes and subsequent control in most urban areas; 70% of the respondents acknowledged this and look forward to obtaining a positive result. Has that been achieved?

The problem has manifested in the numerous forms of developments with fanciful names with an effort to explain its natural occurring state, for instance, Hilltop

1. Both private and public developers face numerous problems in their efforts to acquire land for development.
2. There was a huge backlog of applications unattended to by the departments of Lands and survey.
3. The governor was pre-occupied with other state matters or to sign certificates of occupancy.
4. There was a serious drain of technical staff to private sector and other public organizations.
5. Haphazard developments abound in the rural, sub-urban and urban areas of the state.
6. Cumbersome procedures and delays in the processing of and granting of statutory R of O
7. Jurisdiction conflict between JMDB, LGAs within the metropolis and MLS & TP in the insurance of certificates of occupancy and development control matters.
8. Duplication of functions by the present structure of having town planning division in JMDB and MLS & T.P as well as the LGAs.
9. Compensation remains the greatest hindrance to urban planning and development not only in Plateau state, but in the country as a whole.
10. Vacant (undeveloped) plots are common features in Jos metropolis. This may not be unconnected with the difficulty of obtaining development permits.
11. Large parcel of land acquired by public bodies remained fallow for long periods of time without being developed.
12. Large chunks of land in the GRA, green belts, government layouts, public open spaces has been excised by past governments and allocated to top government officials and their relations.
13. Record keeping in the M.L.S & T.P was (and still so) chaotic and insufficient.
14. Basic planning and land management equipment such as priority maps are in shreds and outdated.
15. Revenue allocation was totally dismal.
16. The LUAC has not been functioning since 1996(especially in Plateau) and so could not perform its statutory function assigned to it by the Land Use Act of 1978.

No magic can be expected from public land management characterized by these alarming flaws.

Box 3.
Major findings. Source: Authors field work, 2022.

i. Too much power in concentrated on the state governor under the Land Use Act, which has been heavily misused by him.

ii. Lack of based maps for survey, development control and monitoring and evaluation.

iii. The shortage of technical staff in the MLS & T.P and JMDB.

iv. Lack of funds amidst wanton corruption by government officials

v. Conflict and lack of cooperation between the numerous planning agencies in the state.

vi. Corruption by public officials and lack of commitment

vii. Delay in the processing and granting of land titles and building permits which are unaffordable to the poor.

viii. Lack of political commitment.

ix. Land Use Act Committee (LUAC) not functioning

x. No payment of compensation for Public Land Acquisition.

xi. Lack of public participation in the administration of land and formation of policies.

xii. Existence of customary land tenure. The land Use Act has not been able to do away with the existence of customary land tenure as was initially intended (as over 90% of vacant urban land is under customary tenure)

xiii. Campaign against sale of customary lands

Box 4.
legal, administrative and cultural constraints. Source: Authors Field work, 2022.

developments, road-site developments, urban sprawl developments and leap frog developments among the current studies on nocturnal developments.

In time past committees such as the Lot's Report of 1985 and victor Pam's Report on JMDB in 1999, researches from experts in the built environment and related disciplines have yielded some results and are documents as major findings, including the reports from boards of enquiries see **Box 3**.

From the findings indicated in **Box 4**, it is characterised by inadequate manpower, bureaucratic bottleneck, and inadequate materials such as maps, plans, current laws and poor bookkeeping. It demonstrated the theory of institutional deficiency which explains the state of planning and the manifested nocturnal developments which pose a danger to people in the urban areas and specifically to the areas under scrutiny.

4.3 Implementation of the development control within the Jos Metropolis, Nigeria

The process of implementation is facilitated by the usage of Legal tools, administrative and political will to ensure well-planned developments within and outside Jos Metropolis. The process of development control implementation in Nigeria involves

	Stages	Development control processes and
1	Procedures of obtaining planning Permission	Planning permission is synonymous to development permit. It refers to the permission granted by planning authorities to develop a piece of land or building [31]. A relevant development is expected to be approved before commencement of work. An application for a building permit must be followed by information such as plans, designs and righty of occupancy (R of O). The planning authority considers the application with regards to the development plan of the area before recommending approval, refusal or approval; with conditions. The following are reasons why a permit is rejected; 1. The development is not in accordance with an approved plan. 2. The development is likely to have a major (negative) impact on the environment, facilities or inhabitants of the community. Also, according to the law, in addition to the application, a developer must submit a detailed Environment Impact statement (EIS) while applying for: 1. A residential land in excess of 2ha. 2. A factory or an office building in excess of four floors or 500 square metres of lettable space. 3. Permission for a major recreational development. An approved planning permission is expected to remain valid for two years. In a situation where a developer is not satisfied with the decision of the planning authority, he is allowed to make an appeal.
2	Enforcement of development control	Basically, there are three stages in the process of enforcing development control measures. Building Inspectors are charged with the responsibility of investigating cases of contravention in the town and reporting same to the higher town planning officer who may serve contravention notices if need be. A development carried out with or without planning permission or in breach of the condition to an approved planning (building) permit is usually served with stop work notice. The notice specifies: 1. The alleged bridge of planning control 2. The steps to be taken to remedy the bridge. 3. The period (21dyas) within which the steps are to be taken [31]. A developer who fails to comply with the stop-work order is to be served with enforcement order/notice within 21 days. In the case of where a developer fails to comply with the terms of an enforcement notice or disregard a stop-work notice issued, he is guilty of an offence liable to a fine of N50,000 for a corporate body and a fine of N10,000 for an individual [31]. Alternatively, the Control department is empowered to undertake any requisite work (demolition) themselves at the expense of the property owner. It is one of the cardinal objectives of this research to explore the procedures for obtaining planning (building) permission specifically for the construction of shopping malls and the methods of development control enforcement in regards to such development in Jos Metropolis.

Sources: Nigerian Urban and Regional Planning Development Report, [31]

Table 3.
Stages of implementation of development control.

two things: These are granting of planning permits (building permits) and the enforcement of development. Here, there are other forms of control such as control of advertisement, wasteland, land deferment and building of architectural (aesthetic) or historical interest.

The implementation includes the following stages, see **Table 3**.

Furthermore, some of the legal, administrative and cultural constraints affecting urban land management in Jos metropolis, Nigeria see **Box 4**.

From Box 4, it is evident that the control of development by the planning authorities is not effective, leading to recent nocturnal developments coming up in these areas such as Zinari, Azurfa along the Bauchi road along the earmarked slopes, hilltops and river valleys mainly by the urban poor. These settlements face various and serious challenges, such as rugged topography that imposes high cost for the provision of infrastructure, lack of planning and development control, substandard housing, poor access, poor waste disposal systems and poor linkages with other parts of the city. The unique physical and ecological characteristics of the settlements desire environmentally sensitive planning and special intervention by the relevant planning authorities and the government to make them sustainable.

These are some reasons that led to the nocturnal developments observed in most settlements due to high demand for housing land. The general spatial evolvement of developments within the last two decades has expanded rapidly in all directions, see locations such as Jenta Adamu, Mangoro, Kabong, Azurfa and Zinaria along Bauchi Road, Jos North LGA.

From the Azurfa and Zinaria settlements along Bauchi Road in Jos North LGA. The Greater Jos Urban Master (1975-2000) prepared by Doxiades had designated hilly areas as country parks and green areas as well as Buffer Zones to cool off the inner core of the city of Jos, but due to rural–urban migration and the natural increase from the inhabitants of the neighbouring settlements, the area has taken over by developments which are substandard.

However, due to the pressure from rural–urban migration, natural increase as well as the consistent crisis in Jos and its environs that has created isolated segregation of inhabitants across religious differences has given rise to developments of residential housing and anaemic urban planning and inadequate development control strategy, uncontrolled slums, hilltops development and now the nocturnal development. The Jos crises, which started in 2001, further exacerbated the process of sporadic developments in areas with their original land uses around Jenta, Kabong, Azurfa and Zinariya areas. These results are further affirmed by the works of Dung-Gwom and Bashir [2].

The latter two areas being close to the University of Jos have sprung up to provide affordable housing to the rapidly growing staff and students population of the University and spill over population from the nearby high-density areas of Anguwan Rogo and Anguwa Rimi.

Furthermore, recent studies by the World Bank in conjunction with some ministries in the Built Environment on the issue of housing-property enumeration indicated that most of the developments have emerged overnight as could be seen manifested and making further enquiries shows no corresponding information or details in the areas, permit to develop properties, non-possession of C of O and R of O as well as poor planning standard within the study area.

From the foregoing, the non-availability of information about developed properties within and outside the metropolis indicates that there is a pending danger to both individuals and the government, as infrastructure would be provided based on the available people within each area as data and information for such areas are not available for proper planning leading to sustainability. For details of the developments under discussion see the photo gallery obtained from various locations in Jos Metropolis, Nigeria and the subsequent findings obtained from the studies. These are presented as **Figures 4–9** (horizontally presented).

Figure 4.
Nocturnal development along difficult topography. Source: Authors fieldwork, 2022.

Figure 5.
Nocturnal development lacking in accessibility, absences of drainage. Source: Authors fieldwork, 2022.

Figure 6.
Nocturnal development along various locations within the metropolis.

Figure 7.
Nocturnal without sewage disposal and draining.

Nocturnal development along difficult topography, that is, rocky/rugged areas Rukuba Road in Jos North LGA, Jos Metropolis, Nigeria, making the houses to be operating without basic amenities.

These developments also show that the location is also a difficult one, lacking in accessibility, absences of drainage and houses are always work in progress.

Figure 8.
Nocturnal without sewage disposal & Children's' Playground.

Figure 9.
Nocturnal developments without basic infrastructure i.e without sewage disposal, drainages & Children Play ground.

See the photo gallery showing the physical planning characteristics of the nocturnal developments within the Jos metropolis, Nigeria.

The general characteristic of the areas with nocturnal developments as observed from this study shows the following:

- More around areas with difficult terrains, that is, bad topography that is generally rocky and rugged,
- No playgrounds for children and social facilities within such communities,
- Difficulty in making basic infrastructure available, that is, roads, pipe born water, gas and electricity amongst others,
- Buildings are erected along with telecommunication masks of either MTN, AIRTEL and GLO amongst others, which is risky as a result of vibration and emissions of waves,
- Absence of proper drainages for easy sewage disposal, hence the streets are littered with effluence from houses making the dirty and unkempt,
- Housing developments are generally in the state of work in progress,
- No provision for waste dumping facilities, hence heaps of waste are dumped on the streets and
- All the photographs, aerial and skirts do not depict a single plan as the developments all over are done haphazardly.

5. Conclusions and recommendations

At the onset of this chapter, it sought to examine nocturnal developments that emerged within the Jos metropolis, Nigeria without any permission and planning standard with a view to make recommendations towards addressing the problem of building without permission. To achieve the aim, a review of related literature and the usage of appropriate methods and design for data collection and subsequent analysis were employed to examine the time of development and the process followed to put the development in the locations within the study area. The result revealed that the developments have been carried out haphazardly, which was affected by unproved legislations and the plethora of planning authorities covering the planning region.

The findings revealed that these areas are developing and growing in a completely unplanned and haphazard manner. A total of 80% of the developers agreed that they do not have approved building plans from the relevant planning authority (Jos Metropolitan Development Board) and the land titles documents from the Ministry of Lands Survey and Town Planning stating the appropriate land used designed for the areas under discussion. With all these inadequacies and the pressure of increase in the population of the areas and urban, metropolis also has forced them to build their houses and raise structures and when the planning authorities come to the site cannot destroy such structures considering the fact that it is an investment which usually cost a fortune.

Based on these results and findings from the study the following sets of recommendations were made:

- The time of development for most of the buildings and related structures is usually overnight which goes to say they do not sleep, over gathering and assembling their building materials, hence planning scouts and watchmen or even vigilantes should be engaged to keep watch over most planned and designated areas. Moreover, cameras could be planted all over the place for easy monitoring of any activities. Satellites could also be deployed for monitoring areas with a high tendency to develop overnight.

- The process followed to put the developments has been frustrated by the processing of documents in the planning authorities over the years as it is cumbersome and un-ending, making it difficult to obtain a land document. The minimum duration to obtain a land document of either R of O or C of O in these organisations is usually ten (10) years, due to long or chain of steps involved in the processing of the document and subsequent approval and granting of development permits. The process could be shortened to enhance and facilitate people to register faster and obtained the documents.

From this chapter, nocturnal developments are encouraged by a number of factors that goes from the behaviour of the developers and the planning authorities saddled with the responsibility of development control within the areas. It shows a gross inadequacy in the development and management of the urban metropolis, even though it was unable to bring about controlled urban development and management.

Furthermore, in-depth research is required to compare different methods of organisational-based models of urban development and its management in different countries that would facilitate knowledge transfer amongst developing countries.

Finally, confirmation of the theory of Institutional deficiency and the theory of planning practice, this is confirmed in the elements of the institutional deficiency theory are evident and bringing about the non-implementation of the Greater Jos Urban Master Plan due to the following: inadequate man-power, high level of corruption, Bureaucratic Bottleneck and Red Tape, delay and negligence of responsibility amongst others. The theory of planning practice is also made manifest in all the planning authorities within the built environment, as the areas have continued to grow and develop haphazardly without any form of development control.

Acknowledgements

The authors wish to sincerely thank the physical planning authorities saddled with the responsibility of urban development and its management in Jos metropolis, Nigeria for allowing them access to their employees' views and opinions during the questionnaire administration. Thank you very much. The authors take total responsibility for any typographical errors observed in this article.

Conflict of interests

The authors declare that they have no conflicts of interest.

Author details

Samuel Danjuma Wapwera
Faculty of Environmental Sciences, Department of Urban and Regional Planning, University of Jos, Jos City, Nigeria

*Address all correspondence to: wapwera@gmail.com

References

[1] Wapwera SD. Spatial Planning Framework for Urban Development and Management in Jos Metropolis. Nigeria; 2014. Submitted in Partial Fulfillment of the Requirement of the Degree of Doctor of Philosophy as the University of Salford, Greater Manchester, United Kingdom

[2] Dung-Gwom JY, Bashir SM. Planning, housing and environmental challenges of hilltop settlements in Zinariya and Azurfa areas of Jos, Nigeria. Urban Studies and Public Administration. 2022; **5**(2):6-35

[3] UN-HABITAT. Urban trends: Urban sprawl now a global problem; State of the worlds cities, Bridging the urban divide. 2010/2011

[4] Wapwera SD, Parsa A, Egbu C. Financing low income housing in Nigeria. Journal of Financial Management of Property and Construction. 2011;**16**(3):283-301 q Emerald Group Publishing Limited 1366-4387. DOI: 10.1108/13664381111179242

[5] Ratcliffe J, Stubbs M, Shepherd M. Urban Planning and Real Estate Development. London & New York: Taylor & Francis Group; 2002

[6] Sustainable Development Goals 33 (SDGs, 2015-2030). Available from: https://www.undp.org/sustainable-development-goals/no-poverty?gclid=Cj0KCQiAx6ugBhCcARIsAGNmMbiGhWfM43PsyZeF4EnuFpfvZ4R3KRiByju4bRgp6JnrKsgAIpJ788EaAtaZEALw_wcB

[7] Dung-Gwom JY, Jugu AS. Characteristics and planning challenges of hilltops settlements in Jos Metropolis, Nigeria. UPLanD—Journal of Urban Planning, Landscape & Environmental Design. 2017;**2**(2):129-149

[8] Dung-Gwom JY. The Nature of Peri-Urban Developments in Jos, Nigeria. In: A Paper Presented at the XXXVI World Congress on Housing Science. National Housing Programme – New Vision. Kolkata, India; 2008

[9] Amanda B. Ten Geographic Facts about Vancouver. Vancouver, British Columbia; 2009

[10] Jiriko K. Urban Master Planning Paradigm in Nigeria: What Future? Kaduna: Mba Prints+Graphics; 2008

[11] CAP. Physical planning and urban development in West Africa region. In: A Paper Presented during the World Planners Congress Agenda Setting Workshop. Abuja, Nigeria; 2005

[12] Litman T. Evaluating Public Transits Benefit & Cost. Victoria Transport Policy Institute; 2011

[13] Adezandeh EA, Akintunde JA, Akintunde EA. Analysis of urban growth agents in Jos Metropolis, Nigeria. International Journal or Remote Sensing and GIS. 2015;**4**(2):41-50. www.rpublishing.org

[14] Mahalingam A, Levitt RE. Institutional theory as a framework for analysinig conflict on global projects. Journal of Construction Engineering and Management. ASCE. 2007;7:517-528

[15] Greater Jos Master Plan (2008–2025). The proposed greater Jos Master plan prepared by Fola Konsults, 2009.

[16] Illesanmi FA. Development Control Tools and Practice. Yola, Nigeria: Modibbo Adama University of Technology; 2016

[17] RTPI. 2011. Remarkable Group.co.uk/localism-blog/rtpi [Online]. Available

from: http://www.consultationonline [Accessed: August 31, 2011]

[18] Rydin Y. The Purpose of Planning Creating Sustainable Towns and Cities. Bristol, UK: The Policy Press, University of Bristol; 2011. Available from: https://sustainabledevelopment.un.org/partnerships.

[19] Sudjic D. The Language of Cities. Published by Penguin Books; 2017

[20] Reed Business Information, Limited. Quadrant House, The Quadrant Brighton Road Sutton, Surrey United Kingdom SM2 5AS. Available from: https://trid.trb.org/view/377560

[21] Wapwera SD, Akujuru VA, Angau AM. An analysis of self-help housing conditions in Jos South LGA of Plateau State, Nigeria. Swift Journal of Geography and Regional Planning. 2016; **2**(1):001-008

[22] Ola CS. Town and Country Planning and Environment Laws in Nigeria. Jericho, Ibadan, Nigeria: University Press Ltd; 1984

[23] Ogundele FO, Ayo O, Odewumi SG, Aigbe GO. Challenges and prospects of physical development. A case study of Festac Town, Lagos, Nigeria. African Journal of Political Science and International Relations. 2011;**5**(4): 174-178

[24] Osinbanjo Y. Problems and prospects of development control. In: Paper presented at the 2004 Annual Delegates Conference of Nigerian Bar Association Holding at the international conference Centre. Abuja; 2004

[25] Dung-Gwom JY. The Implementation of Procedures for Controlling Development in Jos. Nigeria: Department of Planning and Landscape, University of Manchester, UK; 1991

[26] Aluko O. Development control in Lagos State: An assessment of public compliance to space standards for urban development. 2011. DOI: 10.4314/afrrev.v5i5.14

[27] Datonjo L, Marcus D, John BD, Eebee AI. The challenge of development control in Nigeria capital cities: A case of some selected cities in Niger Delta. Developing Country Studies. 2016;**6**(2): 1-8

[28] Tini NH. An Appraisal of the Effectiveness of Development Control in Mubi Town Adamawa State. Nigeria: University of Jos; 2001

[29] Hassane Cisse NR, Menon M, Segger MC, Nmhielle VO. Planning laws, development controls, and social equity. The World Bank Legal Review. 2014;**5**. Fostering Development through Opportunity Inclusion and Equity

[30] Land Use Act (Degree No.6) 1978: Law of the Federation of Nigeria

[31] Nigeria Urban and Regional Planning (NURP) Decree NO. 88 of 1992

[32] Jos Metropolis Development Board. Development control standards and regulations for Development in the Jos-Bukuru metropolitan area. In: JMDB, editor. Town Planning. Jos Plateau State, Nigeria; 2006

Chapter 8

The Evolution of Land-Use Changes in the Alto Tâmega Region, Portugal: From 1990 to 2018 – A Vision of Sustainable Planning

Sérgio Lousada, José Manuel Naranjo Gómez and Luís Loures

Abstract

Considering the complex dynamics, patterns, and particularities that the Alto Tâmega region present—e.g., the fragility, shown to achieve sustainable development and growth—a study that analyzes the Land-Use of this region is seen as pivotal to identifying barriers and opportunities for long-term sustainable development, get a vision of sustainable planning. Using GIS (Geographic Information Systems), the present chapter enables us to identify the dynamics and patterns of the evolution of the Land-Use Changes in the Alto Tâmega Region from 1990 to 2018 (years 1990, 2000, 2012, and 2018 using CORINE (Coordination of Information on the Environment) data). Land-Use Changes studies are reliable tools to evaluate the human activities and footprint of proposed strategies and policies in a territory. This study permits us to reinforce that the Land-Use Changes in the Alto Tâmega Region have undergone multiple changes—marked by increasing and decreasing periods. Also, can be considered a surveying baseline for the comparative analysis of similar works for different Land-Use Changes trends in Europe or worldwide. This chapter also enables us to understand that the main actors should design development policies to protect, preserve and conserve these incomparable landscapes, environments, ecosystems, and the region as a whole.

Keywords: Alto Tâmega region, GIS tools, land-use changes, regional studies, sustainable planning, territorial planning and management

1. Introduction

1.1 Sustainable development

Institutions and citizens have been paying more attention to issues of sustainable development in the last few years [1, 2]. In 1987, Mrs. Brundtland – at the time, the Prime Minister of Norway – described in the United Nations World Commission on

Environment and Development's report "Our Common Future" the concept of sustainable development as "meeting the needs of the present without satisfying the needs of future generations". The required capacity constitutes the development of a hazard [3–5]. Thus, considering the concept of sustainable development, it's possible to adopt the scientific and technological spheres of progress in order to enhance and improve sustainable development management. That could be accomplished, for instance, by increasing mankind's knowledge regarding the laws of nature, by developing new ways to explore natural resources, and by improving the levels of efficiency in terms of resources' utilization [5].

The concept of sustainable development (SD) has been on the global agenda for fifty years and is now widely accepted among nations, organizations, and individuals alike [6–9]. According to the Brundtland Commission, SD is defined as a development that meets the needs of the present generation without compromising the ability of future generations to meet their needs [5, 9]. Theoretical perspectives of SD have been broadly discussed and applied, including several different requirements for what should constitute a sustainable organization [9, 10]. However, despite decades of research, there is a need for further studies on the integration of SD with core business measurements and processes [6, 9]. Moreover, SD's definition reflects a scenario where output is produced without generating a natural resources depletion of any kind [2, 8]. Additionally, one can mention the Sustainable Development Goals (SDGs), a blueprint with the aim to guide society into a more sustainable development [2].

SDGs can be characterized as a list of objectives, based on the long-term Outline Perspective Plan (OPP), that consider both development and sustainability. Indeed, SDGs were created in 2015 by the United Nations General Assembly, highlighting 17 interconnected main goals that would be applied to all nations. These goals will continue from 2016 to 2030. This series of goals has a total of 17 goals and 169 indicators attached to it [1–3, 8, 11].

Those 17 goals, basically, stand for: eradicating poverty and hunger; promoting health cares, well-being, education, access to potable water, to sanitation, and to an affordable and clean source of energy; fighting gender or other forms of inequality; providing dignified jobs to people; stimulating a sustainable economic growth; developing and promoting innovation within industry, while supporting the arise of infrastructure; reinforcing sustainability within cities and communities; promoting a responsible production process and conscious consumption; boosting climate action; protecting and preserving underwater and land life diversity; establishing strong institutions aiming to promote peace and justice; and establishing partnerships to enable the accomplishment of these goals by 2030, in a clear attempt to reach a higher level of human equality [2, 12, 13].

Current issues for urban studies are focused on implementing sustainable principles and addressing spatial demands related to the needs of modern society, maintaining environmental quality, protecting biodiversity, and managing the increasingly urgent issues of climate change adaptation and mitigation [14–16]. Urbanization is a process that creates and leaves an impact on the economic and social development of developing countries [2]. These issues globally depend on the unsolved conflict between anthropic systems and environmental components [17]. The need for tools to support policy-making oriented to sustainable planning arises at several levels of governance. In this regard, the fast progress of Geographic Information Systems (GIS) during the last few decades provided researchers with powerful tools with which to conduct spatial analyses and modelling [18, 19].

Due to the need to examine the value of several urban indicators for each selected single field unit and find potential relationships between them, it was decided to conduct the research in the geographic information system (GIS) environment. Conducting urban research and analyses with the use of spatial information systems has been growing since the 1960s [20–24]. The use of GIS in urban and spatial planning is becoming more and more popular as they are undoubtedly tools that facilitate the work of urban planners and architects [21–24]. GIS systems enable the collection, processing, analysis, visualization and sharing of spatial data and their primary function is to support decision-making [25, 26]. No wonder that after 2000 they have been widely used in the work of town planners, planners and more and more often, also architects. GIS systems have become the basic tool in spatial research, the results of which may be helpful in the preparation of planning studies, i.e., local spatial development plans or studies of the conditions and directions of spatial development [24]. In the literature, we can find many examples of the use of GIS in such analyses, including determining the maximum allowable height of newly designed buildings, assessment of the city sprawl based on land use analysis using GIS and remote sensing, identification of areas for new investments, analysis of building layouts, land absorption analysis with the use of GIS, the study of the impact of building morphology on energy consumption, construction of a spatial database as a tool for participatory activities and many others [24, 27–29]. As a consequence of that, this article resorts to maps in order to geo-visualize the data, instead of simply exhibiting the obtained results in tables. GIS systems has also as one of its main advantages the fact that it provides spatial-based solutions [2, 20, 22, 26], meaning that these systems are able to detect variations through the integration of data from multiple sources. An example of that is the land use information that can be obtained for a specific region after combining hydrological data, soil, topographic maps, and GIS's images [30–33].

1.2 Land-use changes

The CORINE Land Cover (CLC) multi-temporal European datasets, delivered by the European Spatial Agency and Copernicus land monitoring services, were developed with a main purpose: to be a cognitive reference framework regarding the spread of land consumption, allowing comparisons to be established between different European nations [34–37].

In fact, Information on the Environment Coordination CORINE Land Cover (CLC), which is a combined effort of all European Community countries initiated in 1985, collects and interprets geospatial data aiming to: (a) obtain and coordinate interdisciplinary data regarding the environment's state; (b) establish a particular focus on EU countries' areas that are seen as priorities; (c) coordinate the organization and management of data, both locally and internationally; and (d) assure that data is compatible [34].

The CLC database is a tool for performing complex spatial analyses based on diverse land-use kinds. CORINE land cover classes (CLC) have three levels in their hierarchical organization. The first covers five main types of land-use and land cover: artificial areas, agricultural areas, forest and semi-natural areas, wetlands, and water bodies. The next level has fifteen departments. The third level includes forty-four departments that note that the methodological scope of the three individual-level three classes is strictly defined [38, 39].

In fact, the Geographic Information System (GIS) groups and makes available data on land changes. For that, it resorts to high-resolution land cover assessments and change evaluations, with a particular focus on urban areas [34, 40, 41]. In addition to that, these systems also allow an analysis on the variations of human activities and land cover [41, 42]. Moreover, Urban Atlas (UA) allows the differentiation of coverage classes by classifying high-resolution satellite pictures (SPOT 2.5 m, ALOS 2.5 m, RapidEye 5 m). Considering that the lowest mapping unit is 0.25 hectares, the development of land cover maps is limited to 305 European larger cities – where populations exceed 100,000 people – and they present an estimated accuracy of 5 meters. The fact is that the UA only has 20 classes, regarding land cover, a considerably lower number when compared to the CLC [34, 42].

Humans have made major modifications to the earth's surface over time in order to generate food via farming methods. Over half of the earth's surface has been transformed in the last few years, and over one-third of the earth's surface is believed to be agricultural [33, 43]. Humans are simultaneously confronting environmental problems on multiple fronts, such as climate change, loss of biodiversity, soil degradation, water pollution, and loss of ecosystem services, and each of these is caused either directly or indirectly by land-use changes [44–50]. It is estimated that approximately 60% of global land-use changes are directly associated with human land-use activities and 40% with indirect drivers, such as climate change [45]. The expansion of built-up land in urban and rural areas, agricultural intensification, energy, and material consumption are the primary drivers of land-use change [51].

Urbanization is a phenomenon that has followed mankind throughout History, being considered as a relevant feature in every society. One can classify urbanization as the transformation process of rural areas into urban areas, or, simply, as the agglomeration of built-up areas and different ways of life [36, 42]. Urbanization started to grow at exponential rates in the nineteenth century, as a consequence of the industrialization process, and, after that, has become a usual phenomenon within cities in expansion [42]. Moreover, urbanization directly affects approximately 80% of the European population since that's equivalent to the parcel of people currently living in urban areas [47, 52]. Additionally, multiple researchers point out that, by 2030, 75% of European landscapes will be urban in nature [45]. Nowadays, urban areas' continuous expansion can be observed mostly in large agglomerations' surrounding areas, which can be justified by multiple processes, such as suburbanization and urban sprawl [53]. The processes of landscape urbanization have a fundamental impact on its structure and physiognomy and depend on the intensity of changes and technological progress. The above processes change the structure of the landscape. In this paper, landscape is understood as a combination of different types of land cover [42].

Land use and land cover change (LULC) has become a fundamental and essential component in current strategies that concern monitoring environmental change and managing natural resources [54], is a global subject of study. Most studies have focused on metropolitan areas, and other environmental targets of interest, such as the tropics, karsts, coastal zones, ecosystem services, climate change [45, 51, 55], etc. The land surface is being considerably affected due mostly to the growth registered in terms of anthropogenic activities in the biosphere, which might ultimately end up affecting global systems' efficiency levels [54]. In fact, LULC and the resources associated to it have been utilized with the purpose of satisfying humans' social, material, cultural, and even spiritual necessities, which may be tracked down as the origin of significant changes [56, 57]. Quick changes regarding LULC in developing countries have created a depletion of important resources, such as water, soil, and vegetation [54, 58]. Furthermore, because of their quickness, extent, and intensity, global-scale implications – in terms of natural resources and greenhouse gas emissions, for instance – are also related with these land use changes [54, 59]. However, these implications might start to be observed on the local, regional, and national environments due to the massive and quick changes that have been taking place [54, 60].

In order to have a deeper knowledge on land dynamics, it becomes necessary to resort to LULC changes [54]. Researcher from different disciplines conducted multiple empirical studies that concluded that these changes in terms of LULC are important to numerous applications, like hydrology, forestry, agriculture, environmental studies, geology, and ecology [61]. Indeed, some researchers believe that LULC changes could end up playing a decisive role in terms of climate change and ecosystem's imbalance [61, 62]. As a result of that, mostly in developing countries, various resources – land, forest, water, among many others – are significantly declining [63]. Precise information regarding LULC changes, despite being difficult to get [54], is key to determine the root causes, as well as the consequences of these changes. Conducting an accurate analysis regarding the causes behind LULC changes is also fundamental to understand current changes and to produce realistic forecasts [54, 64].

Given the increasing number of disasters over recent years, one of the most efficient and accessible methods for reducing the pressure posed by natural or technological risks is reducing the vulnerability level of communities exposed to a particular hazard [65–68]. At all levels of government, there is a demand for instruments to enhance policy-making aimed at long-term planning [34]. Thus, territorial planning strategy can be considered as a key instrument to attribute wealth pre-conditions to inhabitants; consequently, this will enhance the levels of prosperity of a certain region's future generations by lowering social and spatial inequalities, ultimately leading to sustainable development of that same region [34].

The aim of current study is to map and explain the Land-Use Changes in the Alto Tâmega Region between 1990 and 2018. In this regard, we emphasize that the current study will contribute to science by enabling the collection of big data connected to Land-Use Changes, as well as an overview of how they have evolved in the Alto Tâmega Region over the last three decades.

Finally, we are able to discuss the spatial patterns observed, and to give some principles and recommendations for future regional planning and management strategies and policies to be developed and implemented throughout the Alto Tâmega Region.

2. The Alto Tâmega: a brief overview

The Alto Tâmega Region (NUTS III) (**Figure 1**), comprising the municipalities of Boticas, Chaves, Montalegre, Ribeira de Pena, Valpaços and Vila Pouca de Aguiar, covers an area of approximately 2922 km^2 and recorded, according to data from the 2021 Census of the National Institute of Statistics (INE), a resident population of 84,330 inhabitants, corresponding to roughly 2.5% of the population of the northern region of Portugal, with an average population density of 28.00 inhabitants per km^2 [69]. It is the 12th largest sub-region, the 24th most populous sub-region and the 19th densest sub-region in the country. It is made up of six municipalities and 118 parishes, with the city of Chaves being the administrative city and the main urban center of the sub-region. With 17,207 inhabitants in its urban area and 37,592 inhabitants in the

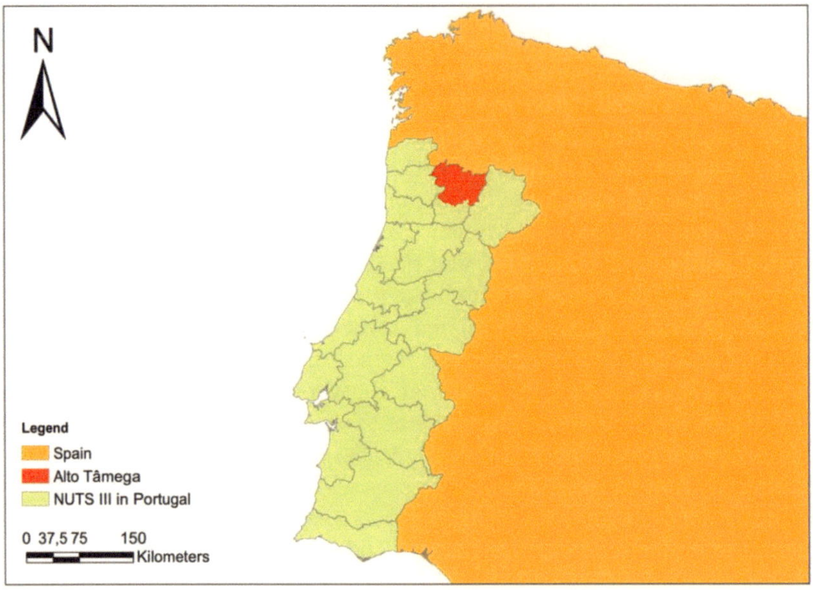

Figure 1.
Delimitation of the study area - Alto Tâmega Region (Source: Authors by ESRI ArcGIS, 2020).

entire municipality, it is the largest city and the largest municipality in Alto Tâmega Region, limited to the north by Galicia (Spain), to the east with the Terras de Trás-os-Montes, south with Douro and west with Cávado and Ave [70].

This territory is one of the twenty-three national Intermunicipal Communities (CIM), bordering four other CIMs – Cávado, Ave, Douro and Terras de Trás-os-Montes – and to the north with Spain, which places the region in a privileged position, in view of possibilities for promoting cross-border relations, especially with Orense [71].

The region under analysis corresponds to the territory covered by the Municipalities of Boticas, Chaves, Montalegre, Ribeira de Pena, Valpaços and Vila Pouca de Aguiar, according to data from the 2021 Census, Alto Tâmega Region registered 84,330 inhabitants, 9813 inhabitants less compared to the 2011 Census, where 94,143 inhabitants were registered. All six municipalities recorded a decrease in inhabitants by −10.4% (**Table 1**) [72, 73].

Municipalities	**Area (km^2)**	**Population (2021 census)**	**Population (2011 census)**	**Variation (%)**
Boticas	322.00	5.002	5.750	−13.0
Chaves	591.20	37.623	41.243	−8.8
Montalegre	805.50	9.279	10.537	−11.9
Ribeira de Pena	217.50	5.887	6.544	−10.0
Valpaços	548.70	14.714	16.882	−12.8
Vila Pouca de Aguiar	437.10	11.825	13.187	−10.3
Total	**2922.00**	**84.330**	**94.143**	**−10.4**

Table 1.
CORINE land cover nomenclature (Source: Authors).

Figure 2.
Alto Tâmega region - Municipality of Boticas (Source: Authors).

Next, a brief description of each of the municipalities is processed, starting with the municipality of Boticas. Founded on November 6, 1836, the municipality of Boticas is a land of unique historical, cultural and natural legacies, hills and mountains of indescribable natural beauty, secular mills, forts, dolmen ruins and Roman milestones (**Figure 2**). From here are the statues of the Calaico-Lusitano Warrior, ex-libris of portuguese archaeology. Barroso's "Enough of Oxen" are a legacy of communitarianism still present. Barrosã Meat, an autochthonous breed of this region and delicacy of kings in the past, stands out as a product of excellence, as well as Honey of Barroso and Honey Brandy. The gastronomic offer is completed with the famous Barroso-style stew, sausages, ham and "Wine of the Dead". The uniqueness of this territory and its inhabitants, constitute one of the most attractive posters in the region and led the United Nations Food and Agriculture Organization to distinguish it and declare it as a World Agricultural Heritage [70–73].

In relation to the Municipality of Chaves, the highlights are heritage, traditions, gastronomy, genuine flavors and ancestral knowledge that make Chaves a healthy land, with soul, which keeps its identity with extraordinary persistence (**Figure 3**). With its squares and spacious streets, parks and colorful gardens, imposing houses, buildings with facades that testify to ancient civilizations, but whose modernity enriches. Chaves offers multiple proposals for tours that allow you to discover impressive landscapes of the river and the mountains to discover a little more of the millenary history of the region. The thermal heritage, legacy of the Romans, is still a pole of attraction for Chaves and Vidago. The hotel and restaurant offer is vast and of excellent quality, a pretext for discovering this territory of well-being [70–73].

Montalegre is a border municipality, in the district of Vila Real, with approximately 800 km^2 of extension (**Figure 4**). Declared World Agricultural Heritage and

Figure 3.
Alto Tâmega Region - Municipality of Chaves (Source: Authors).

Figure 4.
Alto Tâmega region - Municipality of Montalegre (Source: Authors).

Biosphere Reserve, it has an unsurpassed tourist heritage. Ethnography, history, folklore, gastronomy and landscape are undoubtedly leftovers. Another of Barroso's great features is its natural formations, which, spared the action of generations, endow this territory with a majestic landscape and ethnic wealth, which this people preserve. The Peneda-Gerês National Park (PNPG), the secrets of Mourela, the fresh air of Larouco, the ham and smoked meat, the veal, the rye bread, the customs and customs, but above all the hospitality, honesty and human warmth that you experience, are strong reasons to visit the "Marvelous Kingdom" [70–73].

In the municipality of Ribeira de Pena you will find a rich and varied gastronomy, a cultural heritage enriched by the passage of the writer Camilo Castelo Branco and the Museum Network (**Figure** 5). The Tâmega, Póio, Louredo and Beça rivers invite you to take a dip and practice sports activities such as rafting or canyoning. The municipality stands out for its landscapes that mix between a majestic flora and so different between the green valleys and the altitudes, which extends from the Alvão mountain range to the Barroso region. Ribeira de Pena clearly assumes itself as a nature tourism destination, largely through the hotel and leisure facilities that were born in the county. It is here that we find one of the biggest Slides in the world, the "Fantasticable", which is the ex-libris of Pena Aventura Park. Also, in Ribeira de Pena you will be able to enjoy the environmental park and a network of walking paths that will guide you throughout the discovery of the territory [70–73].

The municipality of Valpaços is more than its 550 km^2 of beautiful landscapes, more than its cultivated land with excellent products, more than its 25 magnificent parishes full of heritage that are worth discovering, great stories that make us holders of a unique and wonderful gastronomy, rich and variety of fauna and flora, which enhances the practice of leisure activities, attracting different audiences, from various

Figure 5.
Alto Tâmega Region - Municipality of Ribeira de Pena (Source: Authors).

Figure 6.
Alto Tâmega Region - Municipality of Valpaços (Source: Authors).

parts of the country, at different times of the year (**Figure 6**). But it is also synonymous with modernity, with various facilities available to the visitor, as well as tourist projects of superior quality. Be sure to visit the Interactive Tourism Store, Wine House, Rabaçal Ecovia, the river beaches, among many other unique places that you can only find in Valpaços [70–73].

The municipality of Vila Pouca de Aguiar is located in Alto Tâmega, in Trás-os-Montes, in one of the highest relief regions in Northern Portugal, marked by mountains and plateaus with embedded valleys and rivers of excellent quality (**Figure 7**). Here, we find well-preserved ecosystems, coexisting with family farming and common forest management. Rare species persist in nature, such as the Iberian wolf, the blue peat butterfly, the "Veronica micrantha" and the bats (in Tresminas). Known for the abundance of megalithic monuments, and for the exploitation of gold in Roman times, Vila Pouca de Aguiar allows us to relate the human presence and the exploitation of natural resources, throughout history. Gold, Granite or Medicinal waters were (and are ...) treasures that allowed the settlement of peoples and the development of society and local culture [70–73].

Briefly, the Alto Tâmega Region is characterized by the agricultural activity and agro-industry play an important role in the economic landscape of this low-density territory. The endogenous resources of outstanding quality are one of the differentiating factors of these municipalities, with emphasis on the endogenous products of recognized quality, many of which are guaranteed a Protected Designation of Origin (DOP) and Protected Geographical Indication (IGP), among which stand out the meat, honey, olive oil, chestnuts, potatoes, folar, smoked products and sausages, among others [70–73].

Figure 7.
Alto Tâmega Region - Municipality of Vila Pouca de Aguiar (Source: Authors).

The tourist activity constitutes another of the strategic bets of Alto Tâmega Region, based on an offer of thermal tourism and rural areas tourism (TER) that complements the natural beauty of the region, contextualized by large areas of natural interest and that provide a privileged framework in terms of biodiversity and landscape richness [70–73].

In addition, cultural heritage also plays an important role in the tourist, economic and social panorama of Alto Tâmega Region. In this case, the national monuments with great tourist interest, distributed throughout the six municipalities, as well as diverse cultural manifestations such as handicrafts and ethnographic traditions, should be highlighted [70–73].

The climate of the Alto Tâmega Region is generally considered to be comfortable and cool, however in winter the climate is cool, with frequent snowfall [74].

As for the soils, most of them have some aptitude for pastures and forestry, with fewer of those with good agricultural aptitude, these, in general, concentrated in the plains [74].

According to the 2021 Census, we can draw some conclusions regarding the Alto Tâmega Region [70–74]:

- The percentage of young residents stands at 9.2%, below the average for the North region with 12.5% and below the national average of 13.5%;
- The percentage of seniors residents is 30.7%, above the regional average in the North with 21.2% and above the national average with 22.3%;
- The percentage of foreign residents stands at 1.5%, below the regional average of the North with 2.5% and well below the national average with 6.4%;
- The sectors with the most workers are commerce, with 17.6% of all employed workers, followed by construction with 11.7%, manufacturing with 10.6% and accommodation and catering with 7.5%;
- The unemployment rate in 2020 was 6.0, 0.2% below the regional average in the North, which was 6.2 and 0.2% above the national average, which was in the 5.8%;
- Purchasing power stood at 70.4%, below the regional average of 93% in the North, with Portugal at 100%. Only the municipality of Chaves surpassed the average purchasing power of the Alto Tâmega Region with 79.1%;
- The average monthly salary in the region in 2019 was €954.20, below the average of €1100.40 recorded in the North region and below the national average of €1206.30.

3. Methodology

The data used was two layers of information. These can be utilized to duplicate this work in another work area because they are public and open. The analyzed area is the Alto Tâmega Region, in Portugal.

Data on land use were first gathered. The European Space Agency (EEA) provides a geodatabase employing polygonal graphic characteristics that suggest land uses throughout the European Union for the years 1990, 2000, 2006, 2012, and 2018 through the CORINE Land Cover (Coordination of Information-CLC) project [48, 75]. Regarding the use of remote sensing data, the information was supplied by means of shapefiles. So, these files were managed by using ArcGis 10.5. A project was generated and subsequentially the shapefiles were added as layers, that is to say, vectorial information.

The scale used is 1:100,000 in the Geodesic Reference System corresponding to the European Terrestrial Reference System 1989 (ETRS89) and the Mapping System is Universal Transverse Mercator (UTM), with the minimum cartographic unit (MCU) being equal to 25 hectares. The accuracy obtained has been increasing over the years, since in 1990 it was less than 50 meters, in 2000, 2006 and 2012 it was less than 25 meters, and finally, in 2018 it is less than 10 meters. Also, the information contained in these polygons is hierarchical in three levels of information (**Table 2**).

Level 1	Level 2	Level 3
1. Artificial surfaces	1.1. Urban fabric	1.1.1. Continous urban fabric
		1.1.2. Discontinuous urban fabric
	1.2. Industrial, commercial and transport	1.2.1.Industrial or commercial units
		1.2.2.Road and rail networks and associated land
		1.2.3.Port areas
		1.2.4.Airports
	1.3. Mine, dump and construction sites	1.3.1.Mineral extraction sites
		1.3.2.Dump sites
		1.3.3.Construction sites
	1.4. Artificial, non-agricultural vegetated areas	1.4.1.Green urban areas
		1.4.2.Sport and leisure facilities
2. Agricultural areas	2.1. Arable land	2.1.1. Non-irrigated arable land
		2.1.2. Permanently irrigated land
		2.1.3. Rice fields
	2.2. Permanent crops	2.2.1.Vineyards
		2.2.2.Fruit trees and berry plantations
		2.2.3.Olive groves
	2.3. Pastures	2.3.1.Pastures
	2.4. Heterogeneous agricultural areas	2.4.1.Annual crops associated with permanent crops
		2.4.2.Complex cultivation
		2.4.3.Land occupied by agriculture
3. Forests and semi-natural areas	3.1. Forests	3.1.1. Broad-leaved forest
		3.1.2.Coniferous forest
		3.1.3.Mixed forest

Level 1	Level 2	Level 3
	3.2.Shrub and/or herbaceous vegetation association	3.2.1. Natural grassland
		3.2.2.Moors and heathland
		3.2.3.Scierophyllous vegetation
		3.2.4.Transitional woodland shrub
	3.3. Open spaces with little or no vegetation	3.3.1.Beaches, dunes, and plains
		3.3.2.Bare rock
		3.3.3.Sparsely vegetated areas
		3.3.4.Burnt areas
		3.3.5.Glaciers and perpetual snow
4. Wetlands	4.1. Inland wetlands	4.1.1.Inland marshes
		4.1.2.Peatbogs
	4.2. Coastal wetlands	4.2.1.Salt marshes
		4.2.2.Salines
		4.2.3.Intertidal flats
5. Water bodies	5.1. Inland waters	5.1.1.Water courses
		5.1.2.Water bodies
	5.2. Marine waters	5.2.1.Coastal lagoons
		5.2.2.Estuaries
		5.2.3.Sea and ocean

**For detailed information about the CLC Codes, the authors recommend the following source: www.eea.europa.eu/publications/COR0-landcover, accessed on 10 November 2022.*

Table 2.
CORINE Land Cover nomenclature (Source: [75]).*

The second layer of information corresponds to the administrative delimitation of the Alto Tâmega Region. From the National Geographic Information System of Portugal (SNIG), as shown in **Figure 8**.

ArcGIS 10.5 Geographic Information Systems (GIS) management software was then used to process both levels of data. When Lambert-2001's ETRS-LAEA was first adopted as the official coordinate reference system, all layers of information were converted to it by means of designation of projection in the project [75], due to the fact that they are the inputs, the projection of equivalent regions within the territory, on which ETRS-LAEA is based. In this way, it acts as a standard for homogeneous units across all of Europe. Thus, the representation of analytical and statistical data uses this coordinate system.

Subsequently, the layer relating to the administrative divisions of the country, Portugal has carried out a selection query through alphanumeric information and the Alto Tâmega Region was selected. Subsequently, this single region was kept in a single layer of information. This layer of information was the limit of the scope of action of this work. The clip tool was then used, with Alto Tâmega's boundary as the reference layer. This procedure was used for each of the years studied (1990, 2000, 2006, 2012 and 2018). In this way, land uses were obtained, but only those that were included in

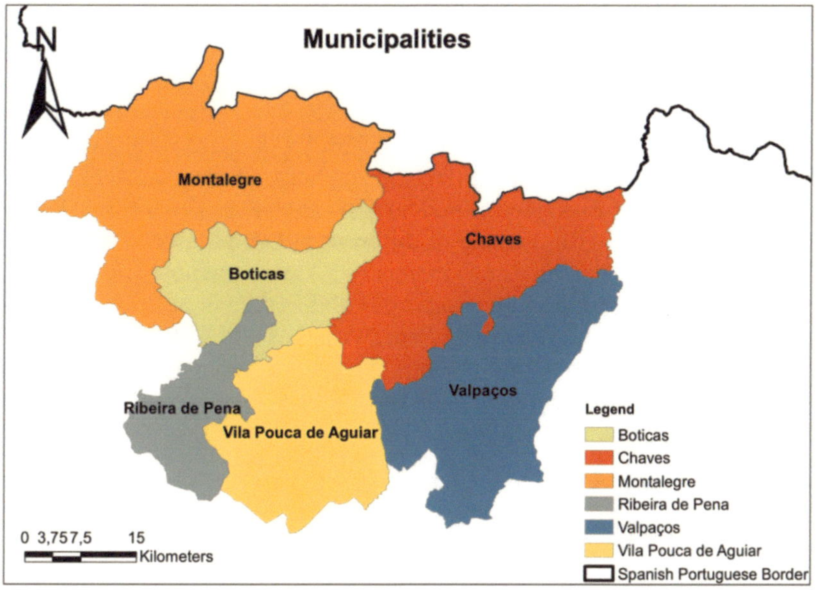

Figure 8.
Delimitation of the study area - Alto Tâmega Region - Municipalities (Source: Authors by ESRI ArcGIS, 2020).

the region. Subsequently, geometric measurement of the area of each of the polygons was performed in hectares. This resulted in the number of hectares of each polygon representative of land uses according to the CLC nomenclature.

The alphanumeric data stored in each of the tables for the years under analysis was then exported by means of the command export and imported into a database that was maintained by the Microsoft Access database management program, a component of the Microsoft Office 365 suite of applications.

Using Structured Query Language (SQL), selection queries were created to the database to select in accordance with the CLC nomenclature, and then another grouping query was appended to the original query. Finally, for the years 1990, 2000, 2006, 2012, and 2018, the number of hectares for each land use was determined.

However, to take into account not only numerical but also geographical results, thematic maps were also obtained for each year in order to take into consideration both the numerical and geographic outcomes. This made it possible to pinpoint both the areas with the highest variety in land uses and those with predominant land uses.

A system has been designed to facilitate understanding of the technique utilized and the criteria for selecting case studies (**Figure 9**).

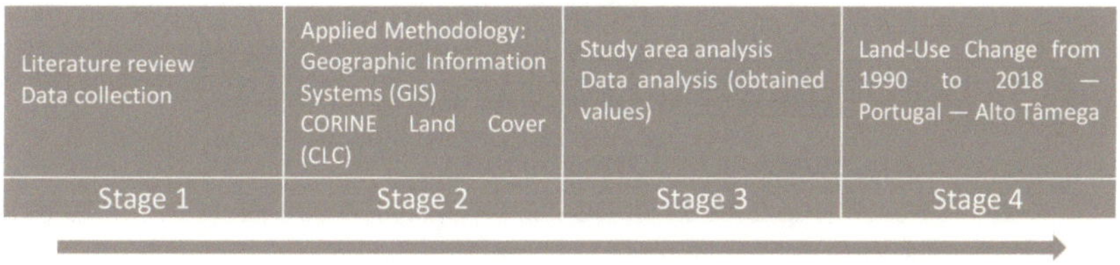

Figure 9.
Summary of the methodology's selection criteria and case study selection process (Source: Authors).

4. Results

The results come from the analysis of the land-use changes for the Alto Tâmega region in the years 1990, 2000, 2006, 2012 and 2018. The results will be exposed through the tables, and thematic cartography. This typology of results exposed allows for extracting the most relevant information and characterizing the evolution of land use based on the 44 land uses determined by CLC in level 3. The information is organized and presented with the help of different plots and tables obtained using ArcGIS 10.5 Geographic Information Systems (GIS) management software, and CLC, in percentage, according to the following order of description of the Municipalities of Boticas, Chaves, Montalegre, Ribeira de Pena, Valpaços and Vila Pouca de Aguiar.

As an example, in **Figure 10**, we present the results shown in **Table 3**, namely the Evolution of land uses according to level 3 of the CLC nomenclature in the Alto

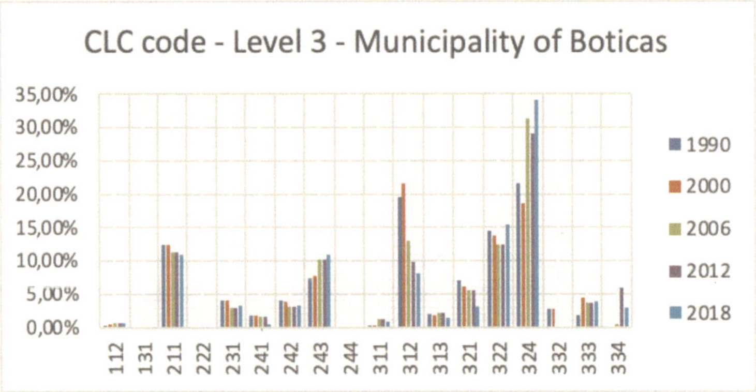

Figure 10.
Evolution of land uses according to level 3 of CLC nomenclature in the Alto Tâmega Region – Municipality of Boticas in 1990, 2000, 2006, 2012, and 2018 (Source: Authors by ESRI ArcGIS, 2020).

CLC Code\Year	1990	2000	2006	2012	2018
112	0.36%	0.46%	0.71%	0.71%	0.71%
131	0.00%	0.10%	0.09%	0.09%	0.09%
211	**12.41%**	**12.42%**	**11.36%**	**11.36%**	**11.05%**
222	0.08%	0.08%	0.08%	0.08%	0.00%
231	4.03%	4.03%	2.90%	2.90%	3.25%
241	1.80%	1.80%	1.58%	1.58%	0.58%
242	3.97%	3.88%	3.20%	3.20%	3.35%
243	**7.46%**	**7.81%**	**10.12%**	**10.12%**	**11.05%**
244	0.19%	0.00%	0.00%	0.00%	0.00%
311	0.38%	0.38%	1.31%	1.29%	0.95%
312	**19.47%**	**21.70%**	**12.99%**	**9.90%**	**8.12%**
313	2.08%	1.77%	2.14%	2.12%	1.49%
321	7.08%	6.08%	5.53%	5.53%	3.12%

CLC Code\Year	1990	2000	2006	2012	2018
322	**14.56%**	**13.74%**	**12.47%**	**12.49%**	**15.44%**
324	**21.65%**	**18.56%**	**31.41%**	**29.00%**	**34.11%**
332	2.66%	2.66%	0.00%	0.00%	0.00%
333	1.80%	4.36%	3.67%	3.67%	3.82%
334	0.02%	0.16%	0.44%	5.94%	2.88%

**Values in bold corresponding to the higher value founded.*

Table 3.
Percentage of land uses according to level 3 of CLC nomenclature in the Alto Tâmega Region – Municipality of Boticas (Source: authors).

Tâmega Region - Boticas Council in 1990, 2000, 2006, 2012 and 2018. This will be the same procedure for the remaining municipalities analyzed in this study.

In order to know what are the differences in area extension for every land use, the differences in percentage areas between years are calculated.

From the information in **Table 4**, it can be seen that the four greatest differences occur for land uses (according to the land uses identified in **Table 3** - Values in bold

CLC Code\year range	2000–1990	2006–2000	2012–2006	2018–2012
112	0.10%	0.25%	0.00%	0.00%
131	0.10%	−0.01%	0.00%	0.00%
211	0.02%	−1.06%	0.00%	−0.31%
222	0.00%	0.00%	0.00%	−0.08%
231	0.00%	−1.13%	0.00%	0.35%
241	0.00%	−0.22%	0.00%	−1.00%
242	−0.09%	−0.68%	0.00%	0.14%
243	**0.35%**	**2.31%**	0.00%	0.92%
244	−0.19%	0.00%	0.00%	0.00%
311	0.00%	0.92%	−0.02%	−0.34%
312	**2.23%**	**−8.70%**	**−3.09%**	**−1.78%**
313	−0.31%	0.37%	−0.01%	−0.63%
321	−1.00%	−0.55%	0.00%	−2.41%
322	−0.82%	−1.27%	**0.02%**	**2.95%**
324	**−3.09%**	**12.85%**	−2.40%	5.10%
332	0.00%	−2.66%	0.00%	0.00%
333	2.56%	−0.69%	0.00%	0.14%
334	0.14%	0.28%	5.51%	−3.07%

**Values in bold corresponding to Land-Use Changes - significant changes.*

Table 4.
Percentage difference of land uses according to level 3 of CLC nomenclature in the Alto Tâmega Region – Municipality of Boticas (Source: authors).

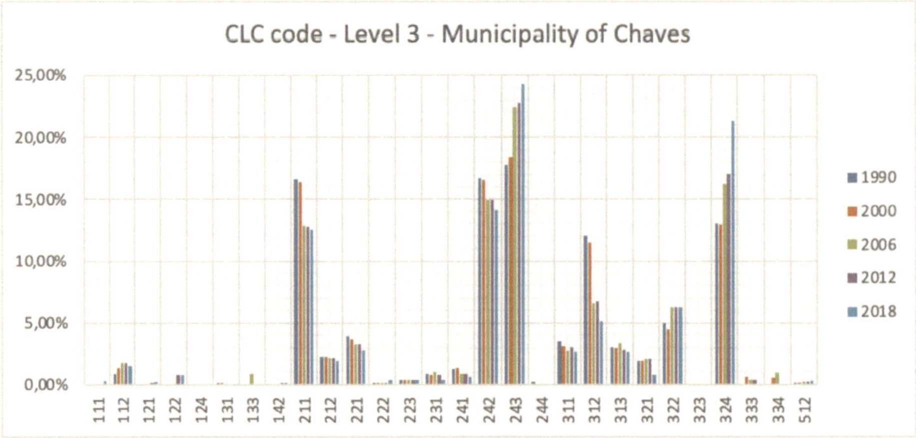

Figure 11.
Evolution of land uses according to level 3 of CLC nomenclature in the Alto Tâmega Region – Municipality of Chaves in 1990, 2000, 2006, 2012, and 2018 (Source: Authors by ESRI ArcGIS, 2020).

corresponding to the higher value founded) 243 between 2000 and 2006 (growing), 312 between 2000 and 2006 (decreasing), 322 between 2012 and 2018 (growing), and 324 between 2000 and 2006 (growing) (**Figure 11**).

In order to know what are the differences in area extension for every land use, the differences in percentage areas between years are calculated (**Table 5**).

From the information in **Table 6**, it can be seen that the four greatest differences occur for land uses (according to the land uses identified in **Table 5** - Values in bold corresponding to the higher value founded) 211 between 2000 and 2006 (decreasing),

CLC Code\Year	1990	2000	2006	2012	2018
111	0.08%	0.08%	0.08%	0.08%	0.33%
112	0.84%	1.38%	1.78%	1.78%	1.50%
121	0.00%	0.00%	0.07%	0.15%	0.27%
122	0.00%	0.00%	0.00%	0.77%	0.77%
124	0.00%	0.00%	0.05%	0.05%	0.05%
131	0.14%	0.14%	0.10%	0.10%	0.10%
133	0.00%	0.00%	0.89%	0.10%	0.10%
142	0.07%	0.07%	0.07%	0.11%	0.16%
211	**16.63%**	**16.43%**	**12.88%**	**12.81%**	**12.54%**
212	2.21%	2.26%	2.13%	2.13%	1.91%
221	3.92%	3.66%	3.28%	3.29%	2.83%
222	0.13%	0.13%	0.17%	0.17%	0.41%
223	0.37%	0.37%	0.36%	0.36%	0.42%
231	0.86%	0.77%	1.06%	0.75%	0.39%

CLC Code\Year	1990	2000	2006	2012	2018
241	1.25%	1.38%	0.91%	0.91%	0.64%
242	**16.72%**	**16.57%**	**14.92%**	**14.93%**	**14.18%**
243	**17.81%**	**18.38%**	**22.42%**	**22.77%**	**24.28%**
244	0.26%	0.00%	0.00%	0.00%	0.00%
311	3.52%	3.15%	2.74%	3.05%	2.64%
312	**12.06%**	**11.46%**	**6.59%**	**6.71%**	**5.12%**
313	3.04%	2.99%	3.39%	2.83%	2.61%
321	1.93%	1.91%	2.07%	2.06%	0.80%
322	4.97%	4.52%	6.25%	6.30%	6.30%
323	0.05%	0.00%	0.00%	0.00%	0.00%
324	**13.00%**	**12.94%**	**16.24%**	**17.08%**	**21.31%**
333	0.00%	0.65%	0.39%	0.39%	0.06%
334	0.00%	0.57%	0.92%	0.06%	0.00%
512	0.13%	0.18%	0.24%	0.24%	0.28%

**Values in bold corresponding to the higher value founded.*

Table 5.
Percentage difference of land uses according to level 3 of CLC nomenclature in the Alto Tâmega Region – Municipality of Chaves (Source: authors).

CLC Code\Year Range	2000–1990	2006–2000	2012–2006	2018–2012
111	0.00%	0.00%	0.00%	0.25%
112	0.54%	0.40%	0.00%	−0.29%
121	0.00%	0.07%	0.08%	0.12%
122	0.00%	0.00%	0.77%	0.00%
124	0.00%	0.05%	0.00%	0.00%
131	0.00%	−0.05%	0.00%	0.00%
133	0.00%	0.89%	−0.78%	−0.01%
142	0.00%	−0.01%	0.05%	0.04%
211	**−0.20%**	**−3.56%**	−0.07%	−0.27%
212	0.06%	−0.13%	0.00%	−0.22%
221	−0.26%	−0.38%	0.01%	−0.46%
222	0.00%	0.05%	0.00%	0.24%
223	0.00%	−0.01%	0.00%	0.05%
231	−0.09%	0.29%	−0.31%	−0.36%
241	0.13%	−0.47%	0.00%	−0.26%
242	−0.15%	−1.64%	0.01%	−0.75%
243	**0.58%**	**4.04%**	0.35%	1.50%
244	−0.26%	0.00%	0.00%	0.00%

CLC Code\Year Range	2000–1990	2006–2000	2012–2006	2018–2012
311	−0.37%	−0.41%	0.31%	−0.40%
312	**−0.60%**	**−4.87%**	0.12%	−1.59%
313	−0.05%	0.40%	−0.56%	−0.22%
321	−0.03%	0.17%	−0.02%	−1.25%
322	−0.45%	1.73%	0.05%	0.00%
323	−0.05%	0.00%	0.00%	0.00%
324	**−0.06%**	**3.30%**	**0.84%**	**4.23%**
333	0.65%	−0.26%	0.00%	−0.33%
334	0.57%	0.35%	−0.86%	−0.06%
512	0.05%	0.05%	0.00%	0.04%

**Values in bold corresponding to Land-Use Changes - significant changes.*

Table 6.
Percentage of land uses according to level 3 of CLC nomenclature in the Alto Tâmega Region – Municipality of Chaves (Source: authors).

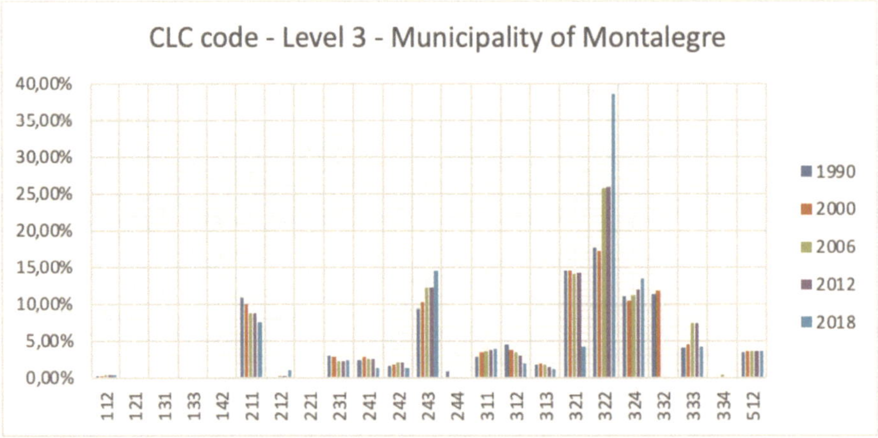

Figure 12.
Evolution of land uses according to level 3 of CLC nomenclature in the Alto Tâmega Region – Municipality of Montalegre in 1990, 2000, 2006, 2012, and 2018 (Source: Authors by ESRI ArcGIS, 2020).

243 between 2000 and 2006 (growing), 312 between 2000 and 2006 (decreasing), and 324 between 2000 and 2006 (growing) (**Figure 12**).

In order to know what are the differences in area extension for every land use, the differences in percentage areas between years are calculated (**Figure 13**) (**Table** 7).

From the information in **Table 8**, it can be seen that the four greatest differences occur for land uses (according to the land uses identified in **Table** 7 - Values in bold corresponding to the higher value founded) 243 between 2012 and 2018 (growing), 321 between 2012 and 2018 (decreasing), 322 between 2000 and 2006 and 2012 and 2018 (growing), and 332 between 2000 and 2006 (decreasing) (**Figure 14**).

In order to know what are the differences in area extension for every land use, the differences in percentage areas between years are calculated (**Table 9**).

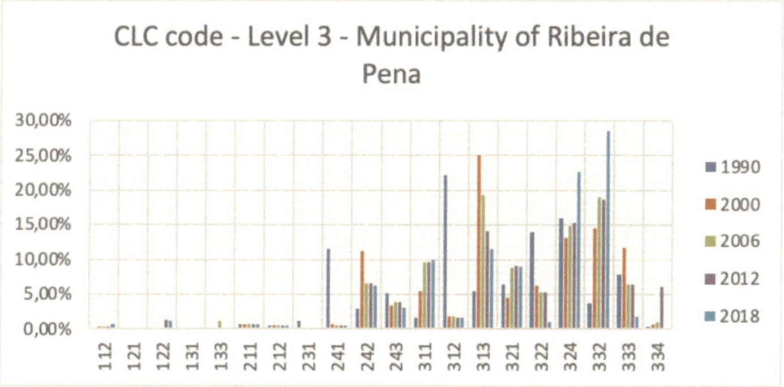

Figure 13.
Evolution of land uses according to level 3 of CLC nomenclature in the Alto Tâmega Region – Municipality of Ribeira de Pena in 1990, 2000, 2006, 2012, and 2018 (Source: Authors by ESRI ArcGIS, 2020).

CLC Code\Year	1990	2000	2006	2012	2018
112	0.22%	0.34%	0.37%	0.37%	0.43%
121	0.00%	0.00%	0.03%	0.03%	0.03%
131	0.06%	0.06%	0.06%	0.06%	0.06%
133	0.00%	0.05%	0.00%	0.00%	0.00%
142	0.00%	0.00%	0.05%	0.05%	0.05%
211	**10.89%**	**10.03%**	**8.75%**	**8.75%**	**7.58%**
212	0.00%	0.00%	0.22%	0.22%	1.04%
221	0.13%	0.13%	0.00%	0.00%	0.00%
231	2.99%	2.82%	2.25%	2.25%	2.34%
241	2.42%	2.91%	2.50%	2.50%	1.29%
242	1.68%	1.87%	2.06%	2.06%	1.37%
243	**9.44%**	**10.31%**	**12.29%**	**12.30%**	**14.46%**
244	0.84%	0.15%	0.00%	0.00%	0.00%
311	2.82%	3.43%	3.65%	3.83%	3.96%
312	4.48%	3.77%	3.48%	2.99%	1.92%
313	1.86%	1.89%	1.75%	1.45%	1.16%
321	**14.55%**	**14.54%**	**14.14%**	**14.17%**	**4.29%**
322	**17.63%**	**17.22%**	**25.75%**	**25.86%**	**38.68%**
324	**10.99%**	**10.51%**	**11.13%**	**11.94%**	**13.39%**
332	**11.40%**	**11.77%**	0.00%	0.00%	0.00%
333	4.05%	4.50%	7.42%	7.45%	4.29%
334	0.07%	0.11%	0.41%	0.05%	0.00%
512	3.48%	3.58%	3.66%	3.66%	3.66%

**Values in bold corresponding to the higher value founded.*

Table 7.
Percentage difference of land uses according to level 3 of CLC nomenclature in the Alto Tâmega Region – Municipality of Montalegre (Source: authors).

CLC Code\Year Range	2000–1990	2006–2000	2012–2006	2018–2012
112	0.12%	0.02%	0.00%	0.06%
121	0.00%	0.03%	0.00%	0.00%
131	0.00%	0.00%	0.00%	0.00%
133	0.05%	−0.05%	0.00%	0.00%
142	0.00%	0.05%	0.00%	0.00%
211	−0.86%	−1.28%	0.00%	−1.18%
212	0.00%	0.22%	0.00%	0.82%
221	0.00%	−0.13%	0.00%	0.00%
231	−0.16%	−0.57%	0.00%	0.09%
241	0.49%	−0.41%	0.00%	−1.21%
242	0.19%	0.19%	0.00%	−0.69%
243	0.87%	1.99%	**0.01%**	**2.15%**
244	−0.68%	−0.15%	0.00%	0.00%
311	0.61%	0.22%	0.18%	0.13%
312	−0.71%	−0.29%	−0.49%	−1.07%
313	0.03%	−0.14%	−0.30%	−0.29%
321	−0.01%	−0.40%	**0.03%**	**−9.88%**
322	**−0.41%**	**8.53%**	**0.11%**	**12.82%**
324	−0.48%	0.62%	0.81%	1.46%
332	**0.36%**	**−11.77%**	0.00%	0.00%
333	0.45%	2.92%	0.03%	−3.16%
334	0.04%	0.30%	−0.36%	−0.05%
512	0.10%	0.08%	0.00%	0.01%

**Values in bold corresponding to Land-Use Changes - significant changes.*

Table 8.
Percentage of land uses according to level 3 of CLC nomenclature in the Alto Tâmega Region – Municipality of Montalegre (Source: authors).

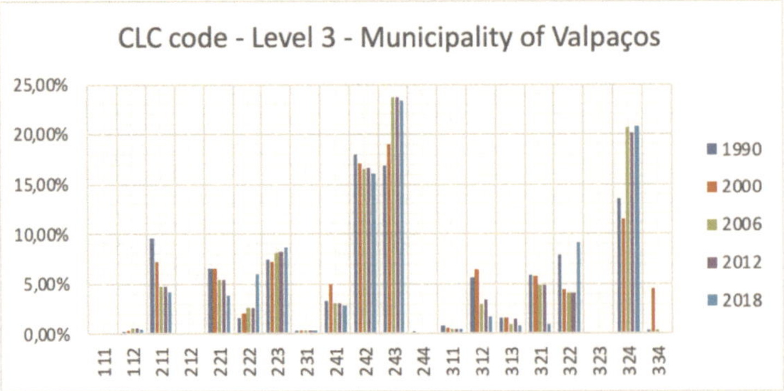

Figure 14.
Evolution of land uses according to level 3 of CLC nomenclature in the Alto Tâmega Region – Municipality of Valpaços in 1990, 2000, 2006, 2012, and 2018 (Source: Authors by ESRI ArcGIS, 2020).

From the information in **Table 10**, it can be seen that the four greatest differences occur for land uses (according to the land uses identified in **Table 9** - Values in bold corresponding to the higher value founded) 241 between 1990 and

CLC Code\SWYear	1990	2000	2006	2012	2018
112	0.16%	0.31%	0.46%	0.46%	0.65%
121	0.00%	0.00%	0.00%	0.00%	0.27%
122	0.00%	0.00%	0.10%	1.32%	1.24%
131	0.00%	0.00%	0.00%	0.00%	0.06%
133	0.00%	0.00%	1.24%	0.00%	0.13%
211	0.75%	0.75%	0.75%	0.75%	0.77%
212	0.49%	0.49%	0.47%	0.47%	0.61%
231	1.22%	0.00%	0.00%	0.00%	0.00%
241	**11.57%**	0.72%	0.54%	0.54%	0.54%
242	2.98%	**11.15%**	6.54%	6.54%	6.28%
243	5.20%	3.47%	3.83%	3.83%	3.06%
311	1.59%	5.54%	9.61%	9.61%	9.95%
312	**22.23%**	1.78%	1.78%	1.59%	1.73%
313	5.53%	**25.05%**	**19.24%**	**14.05%**	**11.56%**
321	6.49%	4.52%	8.74%	9.12%	9.02%
322	**13.89%**	6.28%	5.31%	5.31%	1.03%
324	**15.94%**	**13.17%**	**14.90%**	**15.27%**	**22.71%**
332	3.77%	**14.47%**	**19.06%**	**18.60%**	**28.57%**
333	7.81%	**11.62%**	6.46%	6.46%	1.81%
334	0.39%	0.68%	0.98%	6.09%	0.00%

**Values in bold corresponding to the higher value founded.*

Table 9.
Percentage difference of land uses according to level 3 of CLC nomenclature in the Alto Tâmega Region – Municipality of Ribeira de Pena (Source: authors).

CLC Code\Year Range	2000–1990	2006–2000	2012–2006	2018–2012
112	0.15%	0.15%	0.00%	0.19%
121	0.00%	0.00%	0.00%	0.27%
122	0.00%	0.10%	1.22%	−0.07%
131	0.00%	0.00%	0.00%	0.06%
133	0.00%	1.24%	−1.24%	0.13%
211	0.00%	0.00%	0.00%	0.02%
212	0.00%	−0.02%	0.00%	0.14%
231	−1.22%	0.00%	0.00%	0.00%
241	**−10.84%**	**−0.19%**	0.00%	0.00%

CLC Code\Year Range	2000–1990	2006–2000	2012–2006	2018–2012
242	**8.17%**	**−4.60%**	0.00%	−0.26%
243	−1.73%	0.36%	0.00%	−0.77%
311	3.95%	4.06%	0.00%	0.34%
312	**−20.45%**	**0.01%**	−0.20%	0.15%
313	**19.53%**	**−5.81%**	**−5.19%**	**−2.49%**
321	−1.97%	4.22%	0.38%	−0.10%
322	**−7.61%**	**−0.97%**	**0.00%**	**−4.28%**
324	**−2.77%**	**1.73%**	**0.38%**	**7.44%**
332	**10.71%**	**4.59%**	**−0.46%**	**9.96%**
333	**3.80%**	**−5.16%**	**0.00%**	**−4.65%**
334	0.29%	0.30%	**5.11%**	**−6.09%**

**Values in bold corresponding to Land-Use Changes - significant changes.*

Table 10.
Percentage of land uses according to level 3 of CLC nomenclature in the Alto Tâmega Region – Municipality of Ribeira de Pena (Source: authors).

2000 (decreasing), 242 between 1990 and 2000 (growing) and 2000 and 2006 (decreasing), 312 between 1990 and 2000 (decreasing), 313 between 1990 and 2000 (growing) and 2000 and 2006, 2006 and 2012, 2012 and 2018 (decreasing), 322 between 1990 and 2000 and 2012 and 2018 (decreasing), 324 between 2012 and 2018 (growing), 332 between 1990 and 2000, 2000 and 2006, 2012 and 2018 (growing), and 333 between 1990 and 2000 (growing), 2000 and 2006, 2012 and 2018 (decreasing).

In order to know what are the differences in area extension for every land use, the differences in percentage areas between years are calculated (**Figure 15**) (**Table 11**).

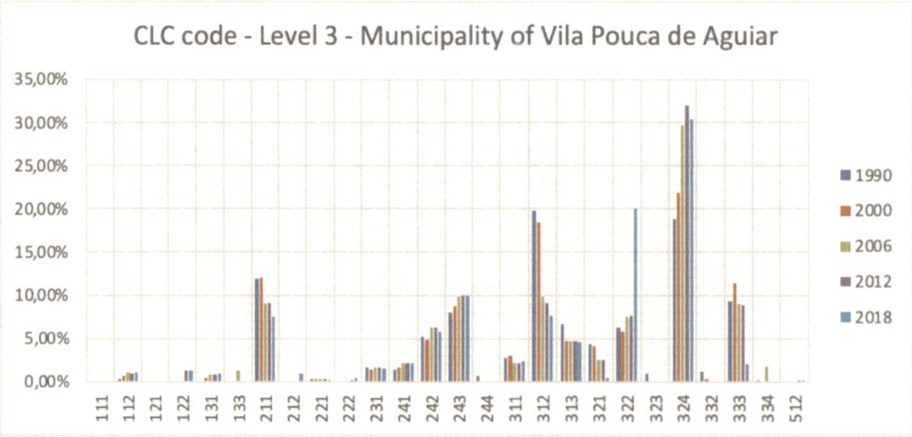

Figure 15.
Evolution of land uses according to level 3 of CLC nomenclature in the Alto Tâmega Region – Municipality of Vila Pouca de Aguiar in 1990, 2000, 2006, 2012, and 2018 (Source: Authors by ESRI ArcGIS, 2020).

From the information in **Table 12**, it can be seen that the four greatest differences occur for land uses (according to the land uses identified in **Table 11** - Values in bold corresponding to the higher value founded) 243 between 2000 and 2006 (growing), and 324 between 2000 and 2006 (growing).

CLC Code\Year	1990	2000	2006	2012	2018
111	0.00%	0.00%	0.00%	0.00%	0.20%
112	0.28%	0.39%	0.64%	0.66%	0.48%
211	9.59%	7.23%	4.79%	4.79%	4.17%
212	0.00%	0.00%	0.00%	0.00%	0.10%
221	6.52%	6.58%	5.41%	5.42%	3.91%
222	1.61%	2.06%	2.63%	2.64%	5.96%
223	7.42%	7.25%	8.16%	8.25%	8.67%
231	0.44%	0.44%	0.42%	0.42%	0.33%
241	3.32%	4.98%	3.12%	3.12%	2.86%
242	**17.95%**	**17.08%**	**16.58%**	**16.65%**	**16.07%**
243	**16.87%**	**18.96%**	**23.76%**	**23.75%**	**23.39%**
244	0.23%	0.00%	0.00%	0.00%	0.00%
311	0.81%	0.66%	0.46%	0.50%	0.50%
312	5.69%	6.44%	2.94%	3.37%	1.69%
313	1.57%	1.61%	0.98%	1.48%	0.86%
321	5.86%	5.82%	4.88%	4.86%	0.91%
322	7.86%	4.46%	4.10%	4.03%	9.15%
323	0.18%	0.00%	0.00%	0.00%	0.00%
324	**13.45%**	**11.46%**	**20.70%**	**20.07%**	**20.75%**
334	0.35%	4.57%	0.43%	0.00%	0.00%

**Values in bold corresponding to the higher value founded.*

Table 11.
Percentage difference of land uses according to level 3 of CLC nomenclature in the Alto Tâmega Region – Municipality of Valpaços (Source: authors).

CLC Code\Year Range	2000–1990	2006–2000	2012–2006	2018–2012
111	0.00%	0.00%	0.00%	0.20%
112	0.12%	0.25%	0.02%	−0.18%
211	−2.36%	**−2.44%**	**0.00%**	−0.62%
212	0.00%	0.00%	0.00%	0.10%
221	**0.06%**	**−1.17%**	**0.01%**	**−1.51%**
222	0.46%	0.57%	**0.00%**	**3.32%**
223	**−0.18%**	**0.91%**	0.09%	0.43%
231	0.00%	−0.01%	0.00%	−0.09%

CLC Code\Year Range	2000–1990	2006–2000	2012–2006	2018–2012
241	**1.66%**	**−1.86%**	0.00%	−0.26%
242	−0.87%	−0.50%	0.07%	−0.57%
243	**2.10%**	**4.80%**	**−0.01%**	−0.36%
244	−0.23%	0.00%	0.00%	0.00%
311	−0.15%	−0.20%	0.04%	0.00%
312	0.75%	−3.50%	0.42%	−1.68%
313	0.03%	−0.63%	0.50%	−0.61%
321	−0.04%	−0.94%	−0.02%	−3.95%
322	**−3.39%**	**−0.37%**	**−0.06%**	**5.11%**
323	−0.18%	0.00%	0.00%	0.00%
324	**−1.99%**	**9.24%**	**−0.63%**	**0.68%**
334	**4.23%**	**−4.15%**	**−0.43%**	0.00%

**Values in bold corresponding to Land-Use Changes - significant changes.*

Table 12.
Percentage of land uses according to level 3 of CLC nomenclature in the Alto Tâmega Region – Municipality of Valpaços (Source: authors).

In order to know what are the differences in area extension for every land use, the differences in percentage areas between years are calculated (**Table 13**).

From the information in **Table 14**, it can be seen that the four greatest differences occur for land uses (according to the land uses identified in **Table 13** - Values in bold corresponding to the higher value founded) 312 between 2000 and 2006 (decreasing),

CLC Code\Year	1990	2000	2006	2012	2018
111	0.00%	0.00%	0.00%	0.00%	0.15%
112	0.32%	0.69%	1.02%	0.95%	1.04%
121	0.00%	0.00%	0.00%	0.00%	0.00%
122	0.00%	0.00%	0.07%	1.36%	1.33%
131	0.12%	0.46%	0.85%	0.86%	0.95%
133	0.00%	0.00%	1.31%	0.00%	0.00%
211	**11.90%**	**12.01%**	9.01%	9.03%	7.45%
212	0.00%	0.00%	0.00%	0.00%	0.96%
221	0.28%	0.28%	0.29%	0.29%	0.26%
222	0.00%	0.00%	0.10%	0.21%	0.48%
231	1.69%	1.48%	1.68%	1.68%	1.55%
241	1.44%	1.67%	2.12%	2.17%	2.20%
242	5.18%	4.83%	6.28%	6.28%	5.86%
243	8.01%	8.76%	9.87%	9.91%	9.93%
244	0.72%	0.00%	0.00%	0.00%	0.00%

CLC Code\Year	1990	2000	2006	2012	2018
311	2.73%	3.05%	2.15%	2.19%	2.42%
312	**19.84%**	**18.40%**	9.81%	9.08%	7.57%
313	6.70%	4.69%	4.71%	4.75%	4.60%
321	4.39%	4.15%	2.52%	2.52%	0.48%
322	6.28%	5.85%	7.53%	7.60%	**20.01%**
323	0.88%	0.00%	0.00%	0.00%	0.00%
324	**18.87%**	**21.90%**	**29.71%**	**31.95%**	**30.41%**
332	1.22%	0.28%	0.15%	0.15%	0.15%
333	9.27%	**11.41%**	8.99%	8.86%	2.02%
334	0.15%	0.08%	1.84%	0.00%	0.00%
512	0.00%	0.00%	0.00%	0.17%	0.17%

**Values in bold corresponding to the higher value founded.*

Table 13.
Percentage difference of land uses according to level 3 of CLC nomenclature in the Alto Tâmega Region – Municipality of Vila Pouca de Aguiar (Source: authors).

CLC Code\Year Range	2000–1990	2006–2000	2012–2006	2018–2012
111	0.00%	0.00%	0.00%	0.15%
112	0.37%	0.33%	−0.07%	0.09%
121	0.00%	0.00%	0.00%	0.00%
122	0.00%	0.07%	1.29%	−0.03%
131	0.35%	0.38%	0.01%	0.09%
133	0.00%	1.31%	−1.31%	0.00%
211	**0.11%**	**−3.00%**	**0.02%**	**−1.58%**
212	0.00%	0.00%	0.00%	0.96%
221	0.00%	0.01%	0.00%	−0.03%
222	0.00%	0.10%	0.11%	0.28%
231	−0.21%	0.20%	0.00%	−0.13%
241	0.23%	0.45%	0.05%	0.04%
242	−0.35%	1.44%	0.00%	−0.41%
243	0.75%	1.11%	0.04%	0.01%
244	−0.72%	0.00%	0.00%	0.00%
311	0.32%	−0.90%	0.04%	0.24%
312	**−1.44%**	**−8.59%**	**−0.73%**	−1.51%
313	**−2.01%**	**0.02%**	0.04%	−0.15%
321	**−0.25%**	**−1.63%**	**0.00%**	**−2.04%**
322	**−0.42%**	**1.68%**	**0.07%**	**12.40%**
323	−0.88%	0.00%	0.00%	0.00%

CLC Code\Year Range	2000–1990	2006–2000	2012–2006	2018–2012
324	**3.04%**	**7.80%**	**2.25%**	**−1.54%**
332	−0.94%	−0.14%	0.00%	0.00%
333	**2.14%**	**−2.42%**	**−0.14%**	**−6.84%**
334	**−0.07%**	**1.77%**	**−1.84%**	**0.00%**
512	0.00%	0.00%	0.17%	0.00%

**Values in bold corresponding to Land-Use Changes - significant changes.*

Table 14.
Percentage of land uses according to level 3 of CLC nomenclature in the Alto Tâmega Region – Municipality of Vila Pouca de Aguiar (Source: authors).

322 between 2012 and 2018 (growing), 324 between 2000 and 2006 (growing) and 333 between 2012 and 2018 (decreasing).

In addition, using ArcGIS 10.5 Geographic Information Systems (GIS) management software, it was possible to more accurately represent the location of each area (thematic cartography) – i.e., according to their respective CLC nomenclature and temporal variance, **Figures 16–20**.

The results showed a gradual Land-Use Changes related to "Agricultural areas" and "Forests and semi-natural areas". According to **Tables 4, 6, 8, 10, 12** and **14**, it seems that the increase in certain kinds of land uses such as "243 the land occupied by agriculture" and "322 moors and heathland" is compensated by the decrease of other land uses like "312 coniferous forest", "332 bare rock" and "333 sparsely vegetated áreas". Nonetheless, it is advisable to execute more exhaustive study research to know it.

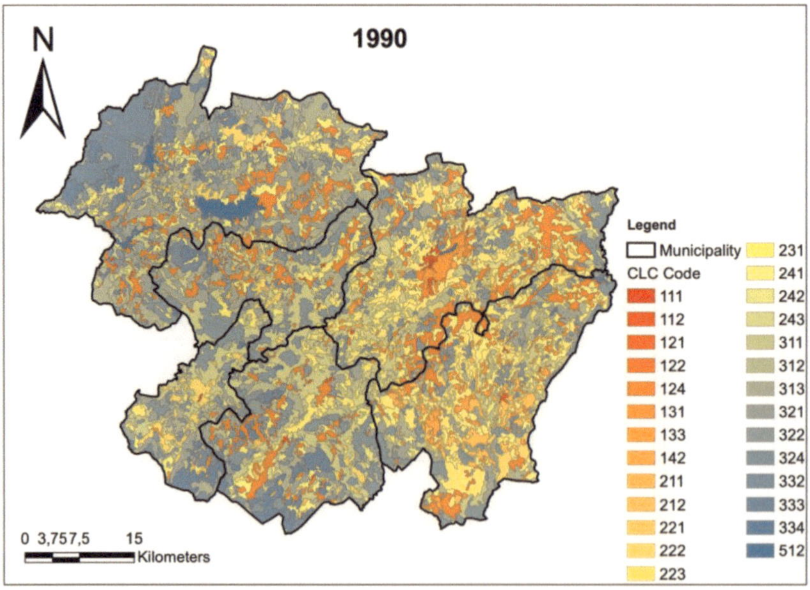

Figure 16.
Land uses according to level 3 of CLC nomenclature in the Alto Tâmega Region in 1990 (Source: Authors by ESRI ArcGIS, 2020).

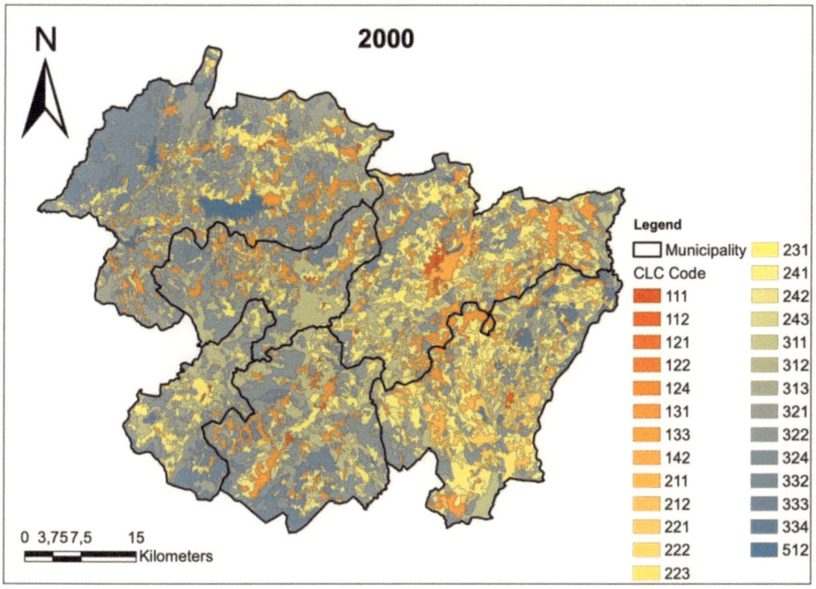

Figure 17.
Land uses according to level 3 of CLC nomenclature in the Alto Tâmega Region in 2000 (Source: Authors by ESRI ArcGIS, 2020).

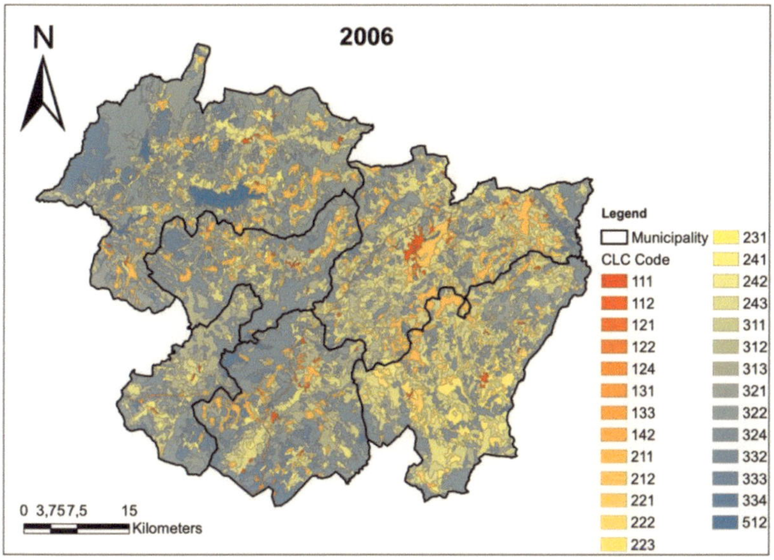

Figure 18.
Land uses according to level 3 of CLC nomenclature in the Alto Tâmega Region in 2006 (Source: Authors by ESRI ArcGIS, 2020).

In summary, most land-use changes for the Alto Tâmega region in the years 1990, 2000, 2006, 2012 and 2018, in the Municipalities of Boticas, Chaves, Montalegre, Ribeira de Pena, Valpaços and Vila Pouca de Aguiar, according to the methodology presented (**Table 2**), namely CLC in level 1, correspond to areas related to "Agricultural areas" and "Forests and semi-natural areas".

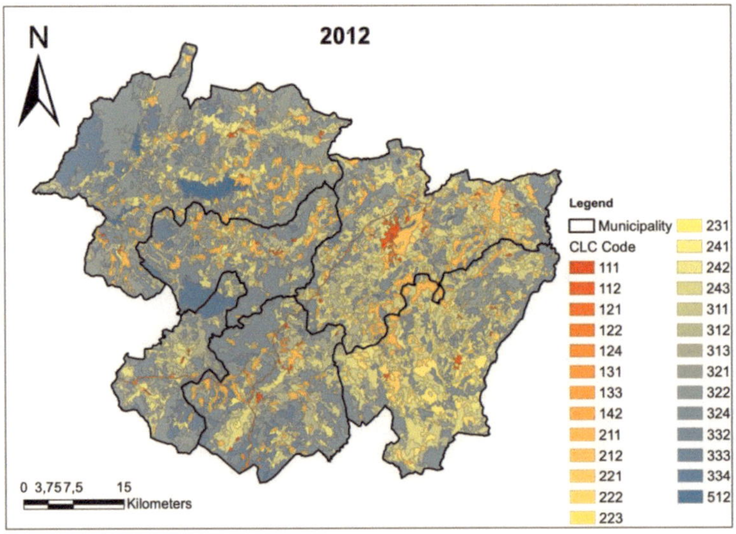

Figure 19.
Land uses according to level 3 of CLC nomenclature in the Alto Tâmega Region in 2012 (Source: Authors by ESRI ArcGIS, 2020).

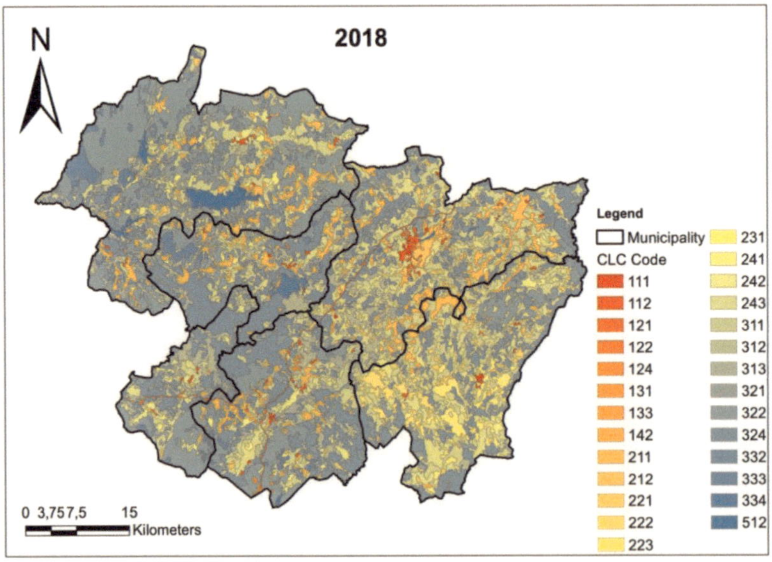

Figure 20.
Land uses according to level 3 of CLC nomenclature in the Alto Tâmega Regio in 2018 (Source: Authors by ESRI ArcGIS, 2020).

5. Discussion and conclusions

In this section, we will address the results that come from the analysis of the land-use changes for the Alto Tâmega region in the years 1990, 2000, 2006, 2012 and 2018.

Therefore, the results presented through the tables and thematic cartography, in the previous section are related to the characterization of the evolution of land use based on the 44 uses of the soil determined by CLC. So, as we are analyzing the Land-Use Changes, we will give more importance to the CORINE Land Cover nomenclature associates, not neglecting the rest.

As described earlier in the penultimate paragraph of the discussion section, and can be validated by the observation of the thematic cartography (**Figures 16–20**). Corroborating what has already been portrayed concerning the Alto Tâmega Region, namely climatology, the rural depopulation, are factors that contribute to the increase of the Land-Use Changes related to "Agricultural areas" and "Forests and semi-natural areas" [76–79].

This temporal evolution – in the Municipalities of Boticas, Chaves, Montalegre, Ribeira de Pena, Valpaços and Vila Pouca de Aguiar – has been influenced by the land tenure regime and, as expected, by the land management carried out [80].

In addition, the demographic aspects linked to these territorial units have contributed, directly or indirectly, to these agricultural and forestry changes. The valorization of the agricultural and forestry heritage of the Alto Tâmega Region should be preserved and protected in a sustainable way. So, we continue to have densely populated areas in each of the municipalities (centers), however and due to the demographic reduction, the sustainable growth of agricultural and forestry areas, do not have a relevant weight, possibly due to the lack of economic incentives towards the future owner's of these areas [80, 81].

The research of Land-Use Changes is critical for understanding regional trends and developments [82, 83]. It was feasible to discern changes in all CLC levels in the Alto Tâmega Region from 1990 to 2018 throughout this examination.

Thus, it was credible to establish that these Land-Use Changes of the Alto Tâmega Region suffered some changes, characterized by increasing and decreasing periods. Some of those decreasing values are disturbing and should have special attention by the government authorities to provide preservation and conservation of these unique Alto Tâmega landscapes and environments.

The changes in the Land-Use could be understood as a direct manifestation of human activity over natural environments [84, 85]. Therefore, the natural factors and features—i.e., geomorphology, slope, relief, soil, and vegetation, among many others— are critical for the proper organization and distribution of the territory and their consequent land uses [84]. The lack of knowledge aligned with the existence of planning conducts to the destruction of the natural resources causing a relevant (negative) impact on the local communities [86].

Therefore, the study of the Land-Use Changes is seen as pivotal to understanding the dynamics and tendencies of these territories as well as to provide clues for the main actors to where the efforts toward sustainable development and growth should be placed.

In the final remarks, the Land-Use Changes could be understood as another tool for the knowledge of the territory—assessing the past and envisioning the future.

6. Limitations of the study and future research directions

Despite all the material discussed in this chapter, regarding the dynamics, trends, and specificities related with the variations occurred within Alto Tâmega Region land-use changes, it becomes clear that there's a need for more research on this topic, in order to find new variables that will, eventually, lead to relevant findings.

Due to frequent changes, mainly in terms of policies and people's behavior in this region, it's fundamental to accurately monitor and analyse possible variations in terms of land-use. Furthermore, it's important that the data obtained from the monitoring and analysis aforementioned is also applied in sustainable development strategies.

Still, if this issue wasn't previously identified, some land-use in this region wouldn't be considered in this study because of CLC's minimum cartographic unit that was used – 25 hectares. Therefore, resorting to more recent versions of CLC that have a higher level of resolution can mostly avoid this type of problems.

Moreover, future research in this region should be able to combine cartography with aspects such as its protected natural spaces, their multiple figures, and land-use changes throughout time. Additionally, although this study's focus on barriers and opportunities regarding sustainable development in this region, to obtain more specific limitations and chances, considering all the specificities of Alto Tâmega Region, other aspects, methodologies, and prospects should be also taken into account.

Acknowledgements

This work was supported by national funds through the Fundação para a Ciência e a Tecnologia, I.P. (Portuguese Foundation for Science and Technology) by the project UIDB/05064/2020 (VALORIZA – Research Centre for Endogenous Resource Valorization).

Conflict of interest

"The authors declare no conflict of interest."

Author details

Sérgio Lousada[1,2,3,4*], José Manuel Naranjo Gómez[2,4,5] and Luís Loures[2,6,7]

1 Faculty of Exact Sciences and Engineering (FCEE), Department of Civil Engineering and Geology (DECG), University of Madeira (UMa), Funchal, Portugal

2 VALORIZA - Research Centre for Endogenous Resource Valorization, Portalegre, Portugal

3 RISCO - Civil Engineering Department of University of Aveiro, Aveiro, Portugal

4 CITUR - Madeira - Centre for Tourism Research, Development and Innovation, Madeira, Portugal

5 School of Agricultural Engineering, University of Extremadura, Badajoz, Spain

6 Polytechnic Institute of Portalegre (IPP), Portalegre, Portugal

7 Research Centre for Tourism, Sustainability and Well-being (CinTurs), University of Algarve, Faro, Portugal

*Address all correspondence to: slousada@staff.uma.pt

References

[1] Assumma V, Bottero M, Caprioli C, Datola G, Mondini G. Evaluation of ecosystem services in mining basins: An application in the Piedmont region (Italy). Sustainability. 2022;**14**:872. DOI: 10.3390/su14020872

[2] Yaakub NF, Masron T, Marzuki A, Soda R. GIS-based spatial correlation analysis: Sustainable development and two generations of demographic changes. Sustainability. 2022;**14**:1490. DOI: 10.3390/su14031490

[3] Zhang S, Xu M, Yang Y, Song Z. Technological innovation, production efficiency, and sustainable development: A case study from Shenzhen in China. Sustainability. 2021;**13**:10827. DOI: 10.3390/su131910827

[4] Ciegis R, Ramanauskiene J, Martinkus B. The economic conditions of enterprise functioning the concept of sustainable development and its use for sustainability scenarios. n.d.

[5] Yumashev A, Ślusarczyk B, Kondrashev S, Mikhaylov A. Global indicators of sustainable development: Evaluation of the influence of the human development index on consumption and quality of energy. Energies. 2020;**13**: 2768. DOI: 10.3390/en13112768

[6] Ranängen H, Cöster M, Isaksson R, Garvare R. From global goals and planetary boundaries to public governance—A framework for prioritizing organizational sustainability activities. Sustainability. 2018;**10**:2741. DOI: 10.3390/su10082741

[7] Purvis B, Mao Y, Robinson D. Three pillars of sustainability: in search of conceptual origins. Sustainability Science. 2019;**14**:681-695. DOI: 10.1007/s11625-018-0627-5

[8] Mensah J. Sustainable development: Meaning, history, principles, pillars, and implications for human action: Literature review. Cogent Social Sciences. 2019;**5**:1653531. DOI: 10.1080/23311886.2019.1653531

[9] Westin L, Hallencreutz J, Parmler J. Sustainable development as a driver for customer experience. Sustainability. 2022;**14**:3505. DOI: 10.3390/su14063505

[10] Dahlsrud A. How corporate social responsibility is defined: An analysis of 37 definitions. Corporate Social Responsibility and Environmental Management. 2008;**15**:1-13. DOI: 10.1002/csr.132

[11] Botticini F, Auzins A, Lacoere P, Lewis O, Tiboni M. Land take and value capture: Towards more efficient land use. Sustainability. 2022;**14**:778. DOI: 10.3390/su14020778

[12] Bottero MC, Polo Pérez I, Taddia G, Lo RS. A geodatabase for supporting planning and management of mining activities: the case of Piedmont Region. Environment and Earth Science. 2020; **79**:83. DOI: 10.1007/s12665-020-8815-x

[13] World Population Prospects - Population Division - United Nations. n.d. Available from: https://population.un.org/wpp/ [Accessed October 17, 2022]

[14] Pilogallo A, Scorza F. Regulation and Maintenance Ecosystem Services (ReMES): A Spatial Assessment in the Basilicata Region. Southern Italy: Cham: Springer; 2021. pp. 703-716. DOI: 10.1007/978-3-030-87007-2_50

[15] Pilogallo A, Scorza F. Mapping regulation ecosystem services specialization in Italy. Journal of Urban

Planning and Development. 2022;**148**: 04021072. DOI: 10.1061/(ASCE) UP.1943-5444.0000801

[16] Pilogallo A, Scorza F. Ecosystem services multifunctionality: An analytical framework to support sustainable spatial planning in Italy. Sustainability. 2022;**14**: 3346. DOI: 10.3390/su14063346

[17] Nelson E, Sander H, Hawthorne P, Conte M, Ennaanay D, Wolny S, et al. Projecting global land-use change and its effect on ecosystem service provision and biodiversity with simple models. PLoS One. 2010;**5**:e14327. DOI: 10.1371/journal.pone.0014327

[18] Juknelienė D, Kazanavičiūtė V, Valčiukienė J, Atkocevičienė V, Mozgeris G. Spatiotemporal patterns of land-use changes in Lithuania. Land. 2021;**10**:619. DOI: 10.3390/land10060619

[19] Paul A. Longley, Michael F. Goodchild, David J. Maguire, David W. Rhind. Geographic Information Science and Systems, 4th Edition — New York: Wiley; WileyCom. 2015. Available from: https://www.wiley.com/en-us/Geographic+Information+Science+and+Systems%2C+4th+Edition-p-9781119031307 [Accessed October 17, 2022]

[20] Stylianidis E, Karanikolas N, Kaimaris D. A GIS for urban sustainability indicators in spatial planning. International Journal of Sustainable Development and Planning. 2012;**7**:1-13. DOI: 10.2495/SDP-V7-N1-1-13

[21] Felke TP. Geographic information systems: Potential uses in social work education and practice. Journal of Evidence-Based Social Work. 2006;**3**: 103-113. DOI: 10.1300/J394v03n03_08

[22] Andreev DV. The use of GIS technology in modern conditions. IOP Conference Series: Earth and Environmental Science. 2020;**421**: 042001. DOI: 10.1088/1755-1315/421/4/042001

[23] Burrough PA. Principles of geographical information systems for land resources assessment. Geocarto International. 1986;**1**:54-54. DOI: 10.1080/10106048609354060

[24] Jachimowicz AM. Morphology of Warsaw city structure using urban indexes and GIS tools. Land. 2022;**11**:135. DOI: 10.3390/land11010135

[25] Gillespie MAK, Baude M, Biesmeijer J, Boatman N, Budge GE, Crowe A, et al. A method for the objective selection of landscape-scale study regions and sites at the national level. Methods in Ecology and Evolution. 2017;**8**:1468-1476. DOI: 10.1111/2041-210X.12779

[26] Bielecka E. GIS spatial analysis modeling for land use change. A bibliometric analysis of the intellectual base and trends. Geosciences. 2020;**10**: 421. DOI: 10.3390/geosciences10110421

[27] Mondal B, Das DN, Bhatta B. Integrating cellular automata and Markov techniques to generate urban development potential surface: A study on Kolkata agglomeration. Geocarto International. 2017;**32**:401-419. DOI: 10.1080/10106049.2016.1155656

[28] Musiaka Ł, Nalej M. Application of GIS tools in the measurement analysis of urban spatial layouts using the square grid method. ISPRS International Journal of Geo-Information. 2021;**10**:558. DOI: 10.3390/ijgi10080558

[29] Murshed SM, Duval A, Koch A, Rode P. Impact of urban morphology on energy consumption of vertical mobility in Asian cities—A comparative analysis with 3D city models. Urban Science.

2019;**3**:4. DOI: 10.3390/urbansci3010004

[30] Lousada S, Gonçalves L, Atmaca A. Hydraulic planning in insular urban territories: The Case of Madeira Island—São Vicente. Water. 2022;**14**:1-19. DOI: 10.3390/w14010112

[31] Lousada S, Cabezas J, Castanho RA, Gómez JM. Hydraulic planning in insular urban territories: The case of Madeira Island-Ribeira Brava. Water. 2021;**13**. DOI: 10.3390/w13212951

[32] Bildirici ME, Lousada S, Yılmaz Genç S. Terrorism, Freshwater, and environmental pollution: Evidence of Afghanistan, Burkina Faso, Iraq, Arab Republic of Egypt, Cameroon, Mali, Mozambique, Niger, Nigeria, Somalia, Syrian Arab Republic, and Pakistan. Water. 2022;14. DOI: 10.3390/w14172684.

[33] Baig MF, Mustafa MRU, Baig I, Takaijudin HB, Zeshan MT. Assessment of land use land cover changes and future predictions using CA-ANN simulation for Selangor, Malaysia. Water. 2022;**14**:402. DOI: 10.3390/w14030402

[34] Naranjo Gómez JM, Lousada S, Garrido Velarde J, Castanho RA, Loures L. Land-use changes in the canary archipelago using the CORINE data: A retrospective analysis. Land. 2020;**9**:232. DOI: 10.3390/land9070232

[35] Bielecka E, Jenerowicz A. Intellectual structure of CORINE land cover research applications in web of science: A Europe-Wide review. Remote Sensing. 2019;**11**:2017. DOI: 10.3390/rs11172017

[36] Bielecka E, Jenerowicz A, Pokonieczny K, Borkowska S. Land Cover Changes and Flows in the Polish Baltic Coastal Zone: A Qualitative and Quantitative Approach. Remote Sensing. 2020;**12**:2088. DOI: 10.3390/rs12132088

[37] Statuto D, Cillis G, Picuno P. Analysis of the effects of agricultural land use change on rural environment and landscape through historical cartography and GIS tools. Journal of Agricultural Engineering. 2016;**47**:28-39. DOI: 10.4081/jae.2016.468

[38] CIEP – Corine. n.d. Available from: http://clc.gios.gov.pl/index.php/o-clc/program-clc [Accessed October 17, 2022]

[39] Martínez-Fernández J, Ruiz-Benito P, Bonet A, Gómez C. Methodological variations in the production of CORINE land cover and consequences for long-term land cover change studies. The case of Spain. International Journal of Remote Sensing. 2019;**40**:8914-8932. DOI: 10.1080/01431161.2019.1624864

[40] Melchiorri M, Florczyk AJ, Freire S, Schiavina M, Pesaresi M, Kemper T. Unveiling 25 years of planetary urbanization with remote sensing: Perspectives from the global human settlement layer. Remote Sensing. 2018;**10**:768. DOI: 10.3390/rs10050768

[41] Washaya P, Balz T. SAR coherence change detection of urban areas affected by disasters using sentinel-1 imagery. International Archives of the Photogrammetry, Remote Sensing and Spatial Information Sciences. 2018;**XLII–3**:1857-1861. DOI: 10.5194/isprs-archives-XLII-3-1857-2018

[42] Myga-Piątek U, Żemła-Siesicka A, Pukowiec-Kurda K, Sobala M, Nita J. Is there urban landscape in metropolitan areas? An unobvious answer based on Corine land cover analyses. Land. 2021;**10**:51. DOI: 10.3390/land10010051

[43] Houghton RA. The Worldwide extent of land-use change: In the last few

centuries, and particularly in the last several decades, effects of land-use change have become global. Bioscience. 1994;**44**:305-313. DOI: 10.2307/1312380

[44] Middle-range theories of land system change — Elsevier Enhanced Reader. n.d. DOI: 10.1016/j.gloenvcha.2018.08.006.

[45] Foley JA, DeFries R, Asner GP, Barford C, Bonan G, Carpenter SR, et al. Global consequences of land use. Science. 2005;**309**:570-574. DOI: 10.1126/science.1111772

[46] The emergence of land change science for global environmental change and sustainability. n.d. DOI: 10.1073/pnas.0704119104.

[47] Song X-P, Hansen MC, Stehman SV, Potapov PV, Tyukavina A, Vermote EF, et al. Global land change from 1982 to 2016. Nature. 2018;**560**:639-643. DOI: 10.1038/s41586-018-0411-9

[48] Peters MK, Hemp A, Appelhans T, Becker JN, Behler C, Classen A, et al. Climate–land-use interactions shape tropical mountain biodiversity and ecosystem functions. Nature. 2019; **568**:88-92. DOI: 10.1038/s41586-019-1048-z

[49] The superior effect of nature based solutions in land management for enhancing ecosystem services — Elsevier Enhanced Reader. n.d. DOI: 10.1016/j.scitotenv.2017.08.077

[50] Changes in the global value of ecosystem services — Elsevier Enhanced Reader. n.d. DOI: 10.1016/j.gloenvcha.2014.04.002

[51] Lai Y, Huang G, Chen S, Lin S, Lin W, Lyu J. Land use dynamics and optimization from 2000 to 2020 in East Guangdong Province, China. Sustainability. 2021;**13**:3473. DOI: 10.3390/su13063473

[52] Sustainability — Free full-text — Cartography and landscape of Agrarian colonisation villages in Spain during the 20th Century: Graphic recording and reconstruction of the productive territory, the case of El Chaparral (Granada) n.d. Available from: https://www.mdpi.com/2071-1050/14/7/4324 [Accessed September 29, 2022]

[53] Multi-scale analysis of urban sprawl in Europe: Towards a European de-sprawling strategy — Elsevier Enhanced Reader. n.d. DOI: 10.1016/j.landusepol.2015.08.001

[54] Heredia-R M, Torres B, Cabrera-Torres F, Torres E, Díaz-Ambrona CGH, Pappalardo SE. Land use and land cover changes in the diversity and life zone for uncontacted indigenous people: Deforestation hotspots in the Yasuní biosphere reserve, Ecuadorian Amazon. Forests. 2021;**12**:1539. DOI: 10.3390/f12111539

[55] Future scenarios based on a CA-Markov land use and land cover simulation model for a tropical humid basin in the Cerrado/Atlantic forest ecotone of Brazil — Elsevier Enhanced Reader. n.d. DOI: 10.1016/j.landusepol.2020.105141

[56] Scenarios of land use and land cover change for NW Amazonia: Impact on forest intactness — Elsevier Enhanced Reader. n.d. DOI: 10.1016/j.gecco.2019.e00567

[57] Haindongo PN, Kalumba AM, Orimoloye IR. Local people's perceptions about land use cover change (LULCC) for sustainable human wellbeing in Namibia. GeoJournal. 2022;**87**:1727-1741. DOI: 10.1007/s10708-020-10337-7

[58] Akinyemi FO. Vegetation Trends, Drought Severity and Land Use-Land Cover Change during the Growing Season in Semi-Arid Contexts. Remote Sensing. 2021;**13**:836. DOI: 10.3390/rs13050836

[59] Searchinger TD, Wirsenius S, Beringer T, Dumas P. Assessing the efficiency of changes in land use for mitigating climate change. Nature. 2018; **564**:249-253. DOI: 10.1038/s41586-018-0757-z

[60] Llerena-Montoya S, Velastegui-Montoya A, Zhirzhan-Azanza B, Herrera-Matamoros V, Adami M, de Lima A, et al. Multitemporal analysis of land use and land cover within an oil block in the Ecuadorian Amazon. ISPRS International Journal of Geo-Engineering. 2021;**10**:191. DOI: 10.3390/ijgi10030191

[61] dos Santos JYG, Montenegro SMGL, da Silva RM, Santos CAG, Quinn NW, Dantas APX, et al. Modeling the impacts of future LULC and climate change on runoff and sediment yield in a strategic basin in the Caatinga/Atlantic forest ecotone of Brazil. Catena 2021;203:105308. DOI: 10.1016/j.catena.2021.105308.

[62] Dosdogru F, Kalin L, Wang R, Yen H. Potential impacts of land use/cover and climate changes on ecologically relevant flows. Journal of Hydrology. 2020;**584**:124654. DOI: 10.1016/j.jhydrol.2020.124654

[63] Global change pressures on soils from land use and management - Smith - 2016 - Global Change Biology - Wiley Online Library. n.d. Available from: https://onlinelibrary.wiley.com/doi/10.1111/gcb.13068 [accessed October 18, 2022]

[64] Genç SY, Behradfar A, Castanho RA, Kırıkkaleli D, Gómez JMN, Loures L. Land use changes in turkish territories: Patterns, directions and socio-economic impacts on territorial management. Current World Environment: An International Research Journal of Environmental Sciences. 2021;**16**: 105-122. DOI: 10.12944/CWE.16.1.11

[65] Kundzewicz ZW, Kanae S, Seneviratne SI, Handmer J, Nicholls N, Peduzzi P, et al. Flood risk and climate change: Global and regional perspectives. Hydrological Sciences Journal. 2014;**59**:1-28. DOI: 10.1080/02626667.2013.857411

[66] Armaş I. Social vulnerability and seismic risk perception. Case study: The historic center of the Bucharest Municipality/Romania. Natural Hazards. 2008;**47**:397-410. DOI: 10.1007/s11069-008-9229-3

[67] Barichivich J, Gloor E, Peylin P, Brienen RJW, Schöngart J, Espinoza JC, et al. Recent intensification of Amazon flooding extremes driven by strengthened Walker circulation. Science Advances. 2018;**4**:eaat8785. DOI: 10.1126/sciadv.aat8785

[68] Vieira I, Barreto V, Figueira C, Lousada S, Prada S. The use of detention basins to reduce flash flood hazard in small and steep volcanic watersheds – A simulation from Madeira Island. Journal of Flood Risk Management. 2018;**11**: S930-S942. DOI: 10.1111/jfr3.12285

[69] INE - Plataforma de divulgação dos Censos 2021 – Resultados Provisórios. n.d. Available from: https://www.ine.pt/scripts/db_censos_2021.html [accessed September 28, 2022]

[70] Caracterização – CIM Alto Tâmega. n.d. Available from: https://cimat.pt/caracterizacao/ [Accessed September 28, 2022]

[71] A Região – Visit Alto Tâmega. n.d. Available from: https://visitaltotamega.com/a-regiao/ [Accessed October 19, 2022]

[72] Rodrigues E, João F, Mendes L. Desenvolvimento Turístico na CIM Alto Tâmega "Ecomuseu do Barroso". Seia: Polytechnic Institute of Guarda – ESTH. 2018.

[73] Tâmega A. Wikipédia Enciclopédia Livre. 2022

[74] Caracterização – AMAT. n.d. Available from: https://www.amat.pt/caracterizacao/ [Accessed October 20, 2022]

[75] CORINE Land Cover — European Environment Agency. n.d. Available from: https://www.eea.europa.eu/publications/COR0-landcover [Accessed April 28, 2022]

[76] López-Rodríguez G, Rodríguez-Vicente V, Marey-Pérez MF. Study of Forest Productivity in the Occurrence of Forest Fires in Galicia (Spain). Sustainability. 2021;**13**:8472. DOI: 10.3390/su13158472

[77] Chas-Amil M-L, García-Martínez E, Touza J. Iberian Peninsula October 2017 wildfires: Burned area and population exposure in Galicia (NW of Spain). International Journal of Disaster Risk Reduction. 2020;**48**:101623. DOI: 10.1016/j.ijdrr.2020.101623

[78] de Diego J, Rúa A, Fernández M. Designing a model to display the relation between social vulnerability and anthropogenic risk of Wildfires in Galicia, Spain. Urban Science. 2019;**3**:32. DOI: 10.3390/urbansci3010032

[79] Alexandre Castanho R, Lousada S, Manuel Naranjo Gómez J, Escórcio P, Cabezas J, Fernández-Pozo L, et al. . In: Carlos Loures L, editor. Dynamics of the land use changes and the associated barriers and opportunities for sustainable development on peripheral and insular territories: The Madeira Island (Portugal), Land Use - Assess. Past Envisioning Future. London: IntechOpen; 2019. DOI: 10.5772/intechopen.80827

[80] Marey Pérez M, Rodríguez Vicente V, Crecente MR. Using GIS to measure changes in the temporal and spatial dynamics of forestland: experiences from north-west Spain. Forestry: An International Journal of Forest Research. 2006;**79**:409-423. DOI: 10.1093/forestry/cpl027

[81] Marey-Pérez M, Díaz-Varela E, Calvo-González A. Does higher owner participation increase conflicts over common land? An analysis of communal forests in Galicia (Spain). iForest - Biogeosciences and Forestry. 2015;**8**: 533-543. DOI: 10.3832/ifor1060-008

[82] Gómez N, Manuel J. Impacts on the social cohesion of mainland Spain's future motorway and high-speed rail networks. Sustainability. 2016;**8**:624. DOI: 10.3390/su8070624

[83] Vulevic A, Castanho RA, Naranjo Gómez JM, Loures L, Cabezas J, Fernández-Pozo L, et al. Accessibility dynamics and regional cross-border cooperation (CBC) perspectives in the Portuguese—Spanish borderland. Sustainability. 2020;**12**:1978. DOI: 10.3390/su12051978

[84] Gao P, Niu X, Wang B, Zheng YL. Land use changes and its driving forces in hilly ecological restoration area based on gis and rs of northern china. Scientific Reports. 2015;**5**:11038. DOI: 10.1038/srep11038

[85] Bertrand N, Vanpeene-Bruhier S. Periurban landscapes in mountain areas. Journal of Alpine Research - Revue de Géographie Alpine. 2007:69-80. DOI: 10.4000/rga.363

[86] Loures L. Land Use: Assessing the Past, Envisioning the Future. BoD – Books on Demand. London: IntechOpen Limited; 2019

Chapter 9

Challenges of Urban Plan Implementation in Small Towns of West Showa Zone, Oromia Regional State, Ethiopia

Kassahun Gashu and Gobena Feyisa

Abstract

The purpose of this study was to examine challenges of urban plan implementation in small towns of Ethiopia, in the case of Babich town West Sowa Zone, Oromia Regional State. Mixed research approaches and cross-sectional design were used. Primary data were collected by using questionnaire, FGD, and KII from residents of the town. Secondary data were collected from published and unpublished documents. Quantitative data were analyzed by using descriptive statistics such as percentage using tables, while qualitative data from FGD and KII were analyzed using thematic analysis method. The result indicated that poor plan implementation was emanated from the establishment of town before plan, socioeconomic issues such as less awareness of the society about the importance of the town plan, lack of budget, lack of skilled man power in the town, and environmental issues such as conflict with surrounding kabeles on paying compensation for land plot. The basic plan of Babich town is not properly implemented. It lacks plan implementing strategy, monitoring, and evaluation. There is irregular and irrational development and also incompatible land uses in the town. The concerned bodies have to preparing implementation guidelines, performing scheduled monitoring and evaluation, providing capacity building by giving training for town plan implementer.

Keywords: basic plan, challenges of implementation, Ethiopia, small towns, urban plan

1. Introduction

Urban planning is a design to order the use of land and other physical resources for public interest with objective of increasing the quality of life and well-being of people living in cities [1, 2], or it is a design and regulation of the uses of space that focuses on the physical form, economic functions, and social impacts of the urban environment and on the location of different activities within it [3].

In most countries, urban or spatial planning refers to the planning of the physical structure of development or land use planning [4]. Nadeem and Hameed [5] defined urban planning as the planning of the physical structure of land use. It is also named as city planning or town planning. It is a technical and political process concerned with the development and use of land, protection and use of the environment, public welfare, and the design of the urban environment, including air, water, and the infrastructure passing into and out of urban areas such as transportation, communications, and distribution networks. It is the process of developing urban, suburban, and rural areas orderly [6].

Rapid urban explosion is agreed to be the most complex and important socioeconomic and environmental phenomenon that has emerged between the twentieth and twenty-first centuries, and urban planning emerged as a scholarly discipline in the 1900s in Great Britain for the first time as academic planning program [7].

Modern-day urban planning in Africa can be traced back to colonial roots [8]. The long tradition of urban planning practice in Eastern Africa indicates that there is an understanding of physical land use planning, which mainly comprised of master planning [9], planning and building standard and regulation, and a system of development control. Master plans, sometimes named as "end-state" plans or "blueprint" plans, refer to the physical plans that depict on a map the future scenario of the town when the plan is fully implemented [10]. Master plan is a tool to guide and manage the growth of cities in a planned manner [5].

Urban planning and implementation experience in Ethiopia has a short history. Review of urban planning practice in the country indicated that less than a quarter of the recognized urban centers have no plans to guide their spatial development. Even those that have plans have difficulty to implement the proposals [10]. The first urban planning practice that was traditional in nature is known to be started in Ethiopia at Entoto during the reign of Menelik II [11].

Regarding its importance, Workineh [12] pointed out that the main importance of urban planning is division of a city into specialized functional quarters; development of commanding central sites for palaces, temples, and civic buildings; and advanced systems of fortification, water supply, and drainage. But urban planning implementation is found at low stage of development in developing countries including Ethiopia [8]. The major cause of poor plan implementation in the country includes lack of accountability, lack of skills, commitment, technical and material capacity of the municipalities, lack of transparency, and lack of awareness at community level.

2. Urban plan and plan implementation challenges

There are many challenges of plan implementation. In line with this, a study conducted by Slaev and Nedovic-Budic [13] in Sofia's master plan in Bulgaria reviled that large inconsistencies exist between the plan's overall goals and some of its measures and implementation tools. Among the various challenges the most common are high crime level, unemployment, health issues, poverty, overstretched public facilities/infrastructures, high population density, and poor planning and implementation policies.

Urban plan is a means of smart, liveable, and inclusive development of urban areas for better urban life. It promotes planned development through allocation of plots for various land uses functions in an orderly and rational manner. This helps to make sure efficient utilization of scarce land resource. Because allocation of plots for various land use function is one among the ultimate goals of urban plan preparation and implementation [14].

Studies [15] within the challenges of urban plan implementation indicated that the process of plan implementation effort in Ethiopia is observed to be at its infancy. However, a study conducted by Simie [11] and Jillo [16] indicated that though urban plan is often viewed as positive, it lacks proper implementation. The main cause of poor plan implementation in the country includes lack of accountability, lack of skills (converting map in to the ground), commitment, technical and material capacity of the municipalities, lack of transparency, and lack of awareness at community level. In the same line, Dube [4] conducted study on urban planning and land management challenges in Arba Minch town acknowledged that archaic land information management system, informal land acquisition, corruption, land speculation, and land-related conflicts became the challenges of land management and plan implementation in the town.

Concerning implementation efforts of urban plans within the Oromia region, most urban centers have faced difficulty in implementing the proposed land use plan. Consistent with the implementation report of the OUPI, most of the prepared structure plans for urban centers are failed to transfer graphics into the ground. As a result, most urban plans are left aside or outdated before their implementation [17]. The above-mentioned problems of plan implementation also are true in Babich town. Like most towns of Ethiopia Babich is found at very infancy stage of urbanization. Babich town got the primary basic plan preparation in 2009. The second revised basic plan was done in 2019, when the town is that the capital of newly emerged woreda called Liban-Jawi [18].

Even though some empirical studies are available on urban planning and land management, there is no sufficient study regarding the challenges of urban plan implementation in small towns in Ethiopia in general and in West Showa Zone, Oromia Regional State in particular. Thus, this study contributes to the existing gaps in the study area.

Many small towns in Ethiopia grew without having available urban plan. Like many other small towns in Ethiopia, Babichi town faced difficulty in implementing the basic plan because the town was established before without the basic plan. On the opposite hand, the basic plan by itself has problem, because it fails to include design for social service institutions, which was previously occupied by residential area, This was happened due to unprofessional personnel in municipality who lacks appropriate skill to transfer graphics into the ground. A study by Mulugeta [10] about existing structural plan implementation situation in Injibara town indicates that the main factors that have had impact on the preparation and implementation of urban plans are lack of qualified personnel to prepare and implement urban plans and lack of budget to pay compensation for nearby farmers, which will result in illegal housing practice. Therefore, the main objective of this study was to investigate the challenges of urban plan implementation in small towns of Ethiopia in the case of Babich town, West Shawa Zone, Oromia Regional State, Ethiopia.

3. Study area

Babich town is found in West Showa Zone, Liban-Jawi district. It is a seat of the district located at 48 km west from Ambo capital, of West Showa Zone and 176 km

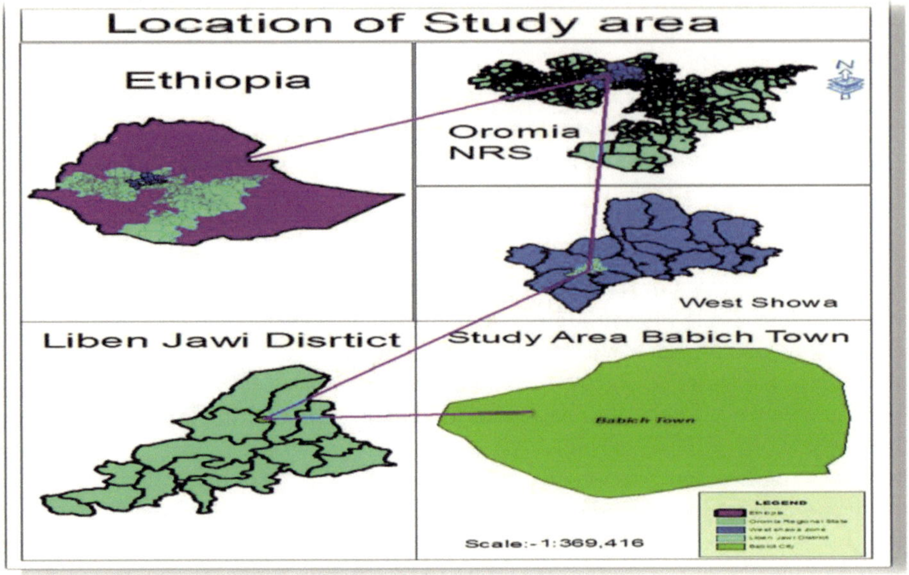

Figure 1.
Study area map (source, own GIS application).

from Addis Ababa to the West (**Figure 1**). It is found between $8^0 58'29''$ latitude, North and $37^0 30'45''$logitude, East. The average altitude of the town is 2261 m asl.

Babich town has a total area is about 580.72 hectares. It is bounded by Mata-Arba, Lemmichoand Dhoke rivers. These rivers are serving as the boundary of the town and source of potable water, vegetable, and fruit production [18]. It has Woina-Dega agro-ecology. The maximum and minimum mean rainfall ranges from 1500 to 2000 mm and from 500 to 900 mm, respectively. The town has the highest annual mean temperature of 28°C, while the lowest is 16°C [19].

The total population of Babich town is about 12,348, and of this 6170 is males and 6178 is females [18]. Income and livelihood of peoples in Babich town is based on various activities. Daily labor and peaty trade had the largest source of income for the livelihood of the community. Government employment and agricultural activities had also significant contributions in the livelihood of the people. In the town, there are small-scale manufacturing's which are privately owned and operated traditionally and labor insensitive [18].

4. Methodology

4.1 Data and data sources

This research used mixed approach, which is a mixture of both quantitative and qualitative methods. The research used both primary and secondary sources of data. Primary data were collected using questionnaire, structured and semi-structured interviews, focus group discussion (FGD), key informant interviews (KIIs), and observation, while secondary data were collected from both published and unpublished books, articles, journals, legal documents, government reports, and related studies.

The structured and semi-structured survey questionnaire was developed using two points (Yes or No). Two FGDs were done from two kebeles of the town (one from each kebele). Each FGDs was having six discussants who were residents in each kebele (such as elders, youth) government workers from municipality, environmental protection, and land administration). Moreover, KIIs were done from mayor of the town, each kebel administrator, and professional planners in the municipality.

4.2 Sampling technique and sample size

This research was done in Babich town, which was picked purposively i) due to prior knowledge by the researchers and ii) prevalence of less urban plan implementation in the town.

The study town and the sample kebeles were selected purposively based on i) the researchers' prior knowledge and ii) prevalence of less urban plan implementation in the town. There are a total of 12,348 household heads in both urban kebeles of Babich town. Of these, 7162 (58%) were found in urban kebel 01 and 5186 (52%) in urban kebele 02 [18]. The total sample size from the selected urban kebeles can be determined using Yamane [20] formula, which is specified as follows:

$$n = \frac{N}{1 + N(e)^2} \tag{1}$$

where n = designates the sample size the research used; N = designates total number of respondents; e = designates the maximum variability or margin of error 5%(0.05); 1 = designates the probability of the event occurring.

Hence, using the above formula, the total sample size computed was 387. In order to include representative samples from the two urban kebeles, proportional sampling technique using the following formula was used:

$$ni = \left[\frac{Ni}{N}\right] xn. \tag{2}$$

where ni is the sample size of each urban kebele; Ni is the number of each urban kebele household heads, N is the total number of household heads in the sample town, and n is the total sample size.

Therefore, 224 and 162 were sample from urban kebele 01and 02, respectively, in the study area. Sample respondent household heads were chosen using simple random sampling technique.

4.3 Data analysis

The study employed both qualitative and quantitative methods of data analysis. Descriptive statistics (mean, percentage) was used to summarize data obtained through questionnaires, while thematic analysis was used to analyze qualitative data collected from FGDs and KIIs.

5. Results and discussion

5.1 Respondents' profile

The socio-demographic condition of sample respondents is presented in **Table 1**. In terms of gender 52% are males, while 48% are females. The average age of sample respondents is 42 years, 85% are married, and the majority of them (40%) are traders.

5.2 Challenges of urban plan implementation

According to Adams [21], a planning course of action is not always crowned with victory. There are often unexpected happenings, which cannot be solved sufficiently. Furthermore, other negative factors can lead to failures to implement the appropriate plan. **Figure 2** indicates the challenges that influence urban plan implementation.

These challenges could be of various types but the most common and dominants in decreasing order are the absence of fixed demarcation (359, 93%), lack of enough municipal man power/capacity (259, 67%), lack of monitoring and evaluation (244,

Variables		Number
Gender	Male	202 (52.2%)
	Female	185 (47.8%)
Age	Average (yrs)	42
Marital status	Single	42 (10.9%)
	Married	328 (84.7%)
	Widowed	8 (2.1%)
	Divorced	9 (2.3%)
Education	Illiterate	13 (3.4%)
	Can read and write	147 (37.9%)
	Primary school	86 (22.2%)
	Secondary school	59 (15.2%)
	Diploma	46 (11.9%)
	Degree and above	36 (9.3%)
Occupation	Daily laborer	78 (20.2%)
	Farmer	62 (16%)
	Trader	154 (39.8%)
	Driver	13 (3.3%)
	Government employee	76 (19.6%)
	NGO	4 (1%)
	Total	387 (100%)

Source: Survey Data, 2021.

Table 1.
Socio-demographic characteristics of respondents (N = 387).

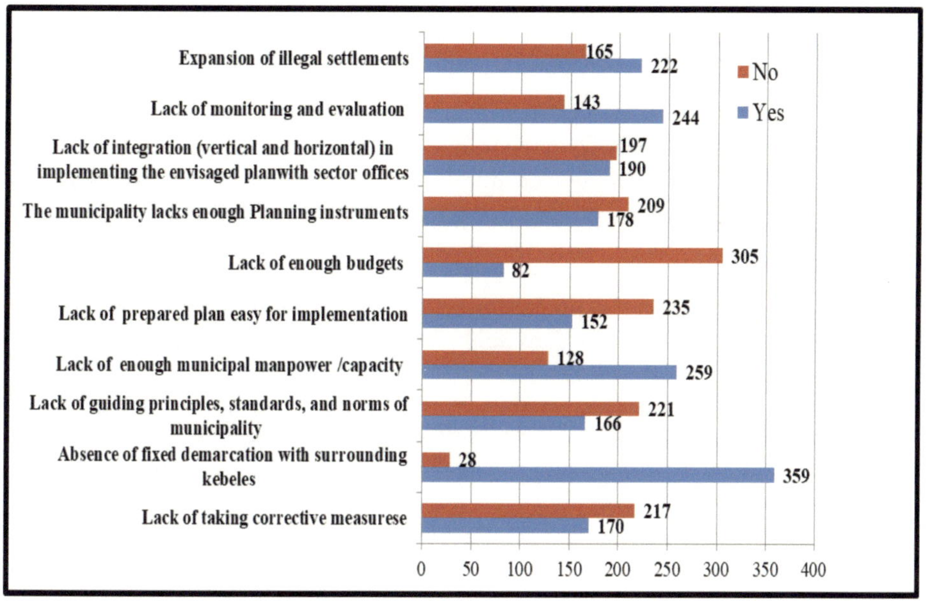

Figure 2.
Challenges of urban plan implementation, source: Survey data.

63%), expansion of illegal settlement (222, 53%), lack of integration (vertical and horizontal) in implementing the envisaged plan with sector offices (190, 49%), lack of enough planning instruments (178, 46%), lack of taking corrective measures (170, 44%).

On the other hand, lack of budget, lack of guiding principles, standards, norms, and lack of easy plan for implementation are challenges that have relatively least influence on urban plan implementation as compared to those listed above. A study by Habtamu (2011) vividly stressed out on the negligence of plan implementation bodies and the role of invisible hand in influencing plan implementation process. In principle, any land uses should not be changed without the consent of the plan preparing body. Besides any land use change should be supervised and monitored by responsible office.

In addition to the above-listed challenges, the main challenges of urban plan implementation mostly happen in small towns like Babich as stated by FGD discussants and KII interviews include financial problems, lack of skilled man power, absence of trained and skillful representatives to become influential in plan implementation, lack of effective monitoring and evaluation, lack of equipment's and violations of plan, problems of good governance, continuous practice of illegal land tenure, lack of community participation during basic plan preparation as well as basic plan implementation, and lack of transparency. The details of these main challenges are presented below.

a. **Financial problems**: Financial problem is one of the important hindrances of urban plan implementation. This is because of improper consideration to implementation issues during the budget allocation period and releasing allocated budget for implementing bodies on their request is raised as the problematic [22]. The budget allocated for the municipality of the Babich town is not sufficient. If there is no sufficient amount of budget, it is difficult to provide basic infrastructural services for the people. Therefore, problems such as waste disposals, canals, green areas, parks, and infrastructural services are not properly provided in the town.

b. **Good governance problem**: In the municipality, there is lack of good governance problem, which includes accountability, transparency, responsiveness, rule of low, efficiency and effectiveness, equity, and fairness [23]. The municipality should be highly committed, but in Babich town there is intervention of politicians and they are not devoted to implement the plan. So there is no smooth relation among politicians, professionals, and the community and this in turn leads to lagging the town development and hampering plan implementation in the town [24].

c. **Lack effective monitoring and evaluation problem**: The FGD and KII were widely discussed on monitoring and evaluation of the plan implementation in Babich town was very weak. The monitoring and evaluation system for the proper implementation of the urban plan in the town has been poor, false, and inconsistent and failed to provide corrective measures by the concerned bodies [25]. The failure of timely evaluation and monitoring has resulted in a loss of proper urban plan implementation at right time for right goal [26].

d. **Illegal land tenure/transfer problem**: It is the act of transferring land based up on the wish of land owner in informal method [27]. This implies transferring plots of land illegally without considering the legal procedure. This could be due to the absence of unity or coordination among different stakeholders. Farmers in the peri-urban area and urban dwellers practiced illegal land tenure through two important ways: i) change of farm land to residential land without appropriate planning and ii) illegal sell plots of land on cash based on the seasonal price. Therefore, illegal land transfer results in hampering of basic plan implementation and causes continuous conflict in the study area (**Figure 2**).

e. **Plan violations problem:** It shows the failure of all stakeholders to act according to the prepared base plan for town [25]. There is plan violation in different direction of the study area. The violations were made based on the desires of personalities or certain group choices. In this case, the land proposed for one purpose was changed to another without any legal background. The practical example for this problem in the town is Ehud-Gabia market was changed to bus station, residential place, and kindergarten (KG) school in the southern margin of the town. This indicates that there is land use change without legal background.

f. **Absence of stakeholders' involvement in plan preparation and implementation.**

According to OUPI [17] plan preparation and implementation should be consultative and participatory where sector organization and institutions have been concerned and play a role. Sectoral, physical, and socioeconomic development program should be integrated at both planning and implementation stages. However, there is no adequate participation of stakeholders' involvement in basic plan preparation and implementation in Babich town. In line with this, FGD discussants indicated that there is not stakeholder participation in any of the basic plan preparation activities of Babich town. This means the prepared basic plan of the Babich town has failed to include the needs and priorities of residents of the town. So according to this point of view it is possible to generalize that the challenges of plan implementation range from political to economical; from financial to institutional; and from physical to sociocultural factors.

Problems caused by absence of plan implementation in Babich town:

- **Lagging the development of town:** Urban planning is the method of shaping and guiding the physical growth of the town by creating the plan-based building for the beautifulness of the town (see **Figure 3**). Urban planning is also very important to use the urban scarce of land wisely without wasting and it plays a vital role to ensure the growth and development of urban centers to make towns competitive centers of integrated and sustainable development that are adequately responsive to the needs of their inhabitants [28].

- **Social conflicts with in the town and with nearby kebeles:** Urban expansion has both pros and cons on the surrounding communities. The pro is urban expansion provided infrastructures, such as pure water, electricity, road, shop, mill, accessible to the community, whereas the con is the issue of the amount of compensation paid to the community is not enough with the urban economy [29].

Figure 3.
Illegal settlements in the peri-urban area of Babich town. Source: Photo by authors.

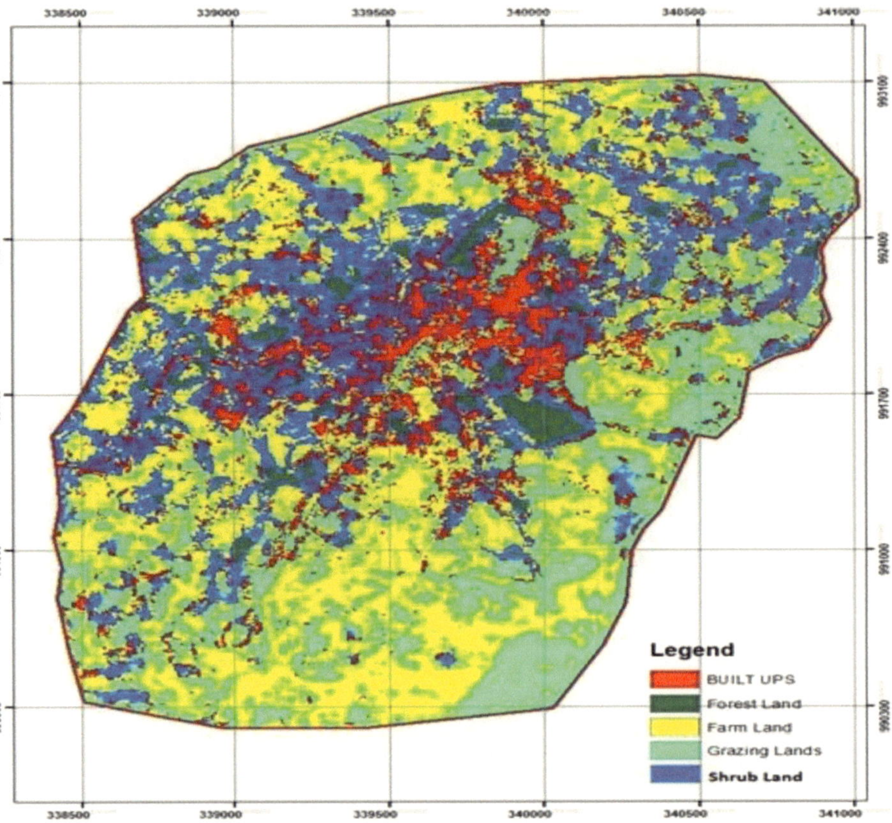

Figure 4.
Land use map of Babich town, source: Babich town municipality.

- **Failure of different sectors to carry out their activities**: There are different offices, for example, land management office, launched to perform different tasks assigned to it in Babich town. The role of the office is to manage the land in the town, but it does not provide proposed role effectively and efficiently. The land proposed for solid waste disposal site in the town plan is not implemented. This situation has impact on high way/asphalt/and disturbing the attractiveness of the town for dweller and polluting town (see **Figure 4**).

5.3 Importance of urban planning

The goal of urban planning and its effective implementation are to create urban areas, which is economically vibrant, environmentally sustainable, socially viable and livable, and spatially integrated and linked (**Figure 5**). On top of this, the current planning approach has its own core values and key development principles. The key development principles of structural plan are strategic, comprehensive and the needs for urban-rural harmony; dynamic, flexible, and sustainable; and participatory and equitable. This vividly shows that urban planning has been existed to serve human being long years [30] (**Figure 6**).

Figure 5.
Current situation of the Babich town. Source: Photo by authors.

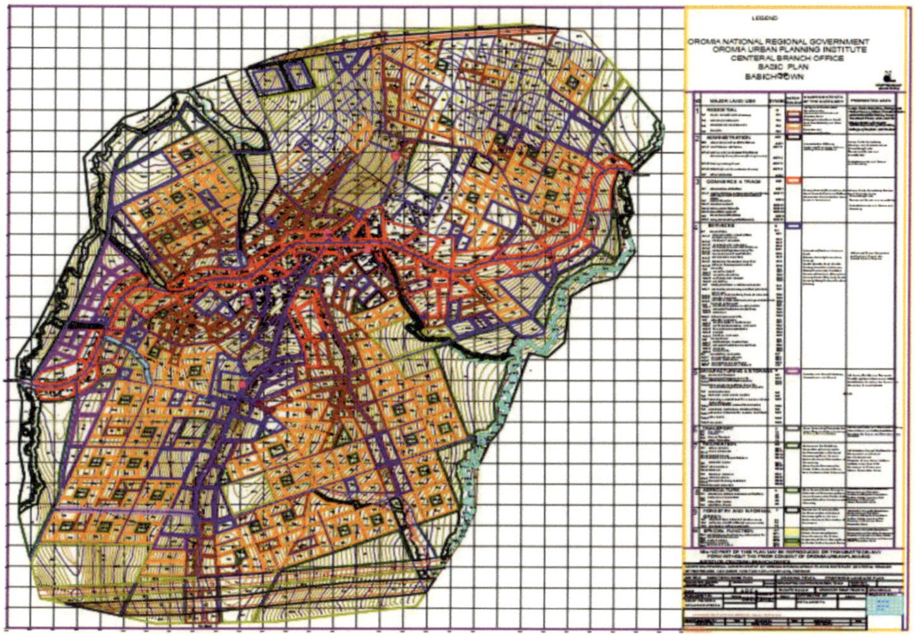

Figure 6.
Basic plan of Babich town, source: Babich town municipality.

Urban planning is believed to have been evolved and considered to be as old as the history of human civilization. However, the standard, development, and quality of planning show a discrepancy from time to time and from place to place. The relatively long tradition of planning practice in Eastern Africa indicates that there is an understanding of physical land use planning, which mainly comprised of master planning, planning, and building standard and regulation and a system of development control [8].The then pre-twentieth-century rise and fall of political capital period was taken as a historic landmark that shaped the current development of the urban system of Ethiopia in general and Addis Ababa in particular [31]. There is a continuous change in urban planning process such as developing a theoretical underpinning to the study of urban physical structure, changing in the involvement of professionals involved in urban planning, planning method, the way urban plan has respond to the rapid change.

6. Conclusion

Challenges of urban plan implementation in Babich town were started from the pre-plan establishment of the town. In order to do these activities, enough budget and communal participation about the advantage of plan for urban development and wisely use of plots of land is very important; however, in Babich town it is at low stage. The socioeconomic and administration factors are also another important barrier of plan implication in the study area. All the administrative bodies from region to the town municipality did not perform their responsibilities according to the mandate empowered on them. There is no proper monitoring and evaluation by regional UPI. Politicians' intervention is serious challenge plan implementation in the study town.

The community participation in plan preparation and plan implementation process, in plan changing process and decision making about their town issue, is completely ignored. Illegal land tenure is also another big problem. The illegal land tenure is practiced in the town in the areas proposed for greenery and parking. Urban plan implementation problem may also happen due to the lack of detail plans. Preparation of detail plan is vital for every kind of plan implementation. Any work of plan implementation has to be supported with detail plans; otherwise, it will be problematic to implement it. Plan should be supported by essential effort to implement the proposed plan.

Acronyms

FGD	Focus Group Discussion
KII	Key Informant Interview
UPI	Urban Plan Institute
OUPI	Oromia Urban Plan Institute

DOI: http://dx.doi.org/10.5772/intechopen.109122

Author details

Kassahun Gashu* and Gobena Feyisa
Department of Geography and Environmental studies, University of Gondar, Ethiopia

*Address all correspondence to: kg19me@gmail.com

References

[1] Beltrao G, Kessler E. Urban Planning and Land Management for Promoting Inclusive Cities. India: Technical assistance consultant's report for ministry of housing and urban poverty alleviation; 2013

[2] Mouratidis K. Urban planning and quality of life: A review of pathways linking the built environment to subjective well-being. Cities. 2021;**115**:103229

[3] Handy SL, Boarnet MG, Ewing R, Killingsworth RE. How the built environment affects physical activity: Views from urban planning. American Journal of Preventive Medicine. 2002;**23**(2):64-73

[4] Dube EE. Urban planning & land management challenges in emerging towns of Ethiopia: The case of Arba Minch. Journal of Urban and Environmental Engineering. 2013;**7**(2):340-348

[5] Nadeem O, Hameed R. Evaluation of environmental impact assessment system in Pakistan. Environmental Impact Assessment Review. 2008;**28**(8):562-571

[6] Woldeamanuel MG. Urban Issues in Rapidly Growing Cities: Planning for Development in Addis Ababa. Oxon: Routledge; 2020

[7] Momoh J, Opaluwah A, Haithem H. Challenges Facing Nigeria's Urban Environment: The Abuja Case. Journal of Sustainable Development in Africa. 2018;**20**(2):34-49

[8] Admasu TG, Jenberu AA. Urban planning implementation challenges in Arba Minch Town, Southern Ethiopia. Urban Forum. 2020;**31**(4):549-572

[9] Rebira G. Practices and Challenges of Urban Planning in Ethiopia: A Case of Nekemte Town [Unpublished MA Thesis]. Addis Ababa: Addis Ababa University; 2017

[10] Mulugeta Z. Assessment of the majore challenges/ factores for implementation of structural plan implementation in Injibara town, Amhara regional state, Ethiopia. International Journal of Advanced Multidisciplinary Research. 2017;**4**(4):4-11

[11] Simie HL. Challenges of Urban Plan Implementation in Small Towns of Ethiopia: The Case of Gelan Town [Unpublished MA Thesis]. Addis Ababa: Addis Ababa University; 2012

[12] Workineh NA. Land-use deviation from the structural plan in Injibara Town, Amhara National Regional State, Ethiopia. Ethiopian Renaissance Journal of Social Sciences and the Humanities. 2021;**8**(1):104-124

[13] Slaev AD, Nedovic-Budic Z. The challenges of implementing sustainable development: The case of Sofia's master plan. Sustainability. 2016;**9**(1):15

[14] Kitie BF. Challenges of Urban Plan Implementation in Oromia Region: The case of Sebeta town: Three Kebeles in Focus. 2012

[15] Kalekiristos B. Urban development planning and implementation problem in Ethiopia and future prospects. In: Paper Presented on Proceedings of the 2nd national Conference on Urban Development Planning and Implementation; Towards Paving the Way for Partnership. Nazareth: UPI; 2002

[16] Jillo YD. Assessment of Urban Plan and Design Implementation and Management In Ethiopian Secondary

Towns: The Case of Dilla [MA Thesis]. Addis Aababa: Addis Ababa University; 2012

[17] OUPI. Report on Monitoring and Evaluation of Structure Plan in the Region. Ethiopia: Addis Ababa; 2011

[18] Babich town municipality. Annual Report. West showa zone, Ethiopia: Babich Town; 2022

[19] EMA. Annual Report for West Showa Zone. Ethiopia: Addis Ababa; 2020

[20] Yamane T. Statistics, An Introductory Analysis, 2nd Ed. New York: Harper and Row; 1967

[21] Adams D. Urban Planning and the Development Process. London: Routledge; 2012

[22] Mishra AK. Urban master plans in Rajasthan, India: The case of Alwar. Management Research and Practice. 2012;**4**(1):31

[23] Bekele WB, Ago FY. Good governance practices and challenges in local government of Ethiopia: The case of Bonga town administration. Research in Educational Policy and Management. 2020;**2**(2):97-128

[24] Šaparnienė D, Valukonytė I. Implementation of good governance principles in local self-government: The case of Šiauliai city. Socialiniai tyrimai. 2012;**3**:98-112

[25] Watson V. 'The planned city sweeps the poor away...': Urban planning and 21st century urbanisation. Progress in Planning. 2009;**72**(3):151-193

[26] Deininger K, Selod H, Burns A. The Land Governance Assessment Framework: Identifying and Monitoring Good Practice in the Land Sector. Nairobi: World Bank Publications; 2012

[27] Whittal J. A new conceptual model for the continuum of land rights. South African Journal of Geomatics. 2014;**3**(1):13-32

[28] Parolek DG, Parolek K, Crawford PC. Form Based Codes: A Guide for Planners, Urban Designers, Municipalities, and Developers. New Jersey: John Wiley & Sons; 2008

[29] Dutta V. Land use dynamics and peri-urban growth characteristics: Reflections on master plan and urban suitability from a sprawling north Indian city. Environment and Urbanization ASIA. 2012;**3**(2):277-301

[30] Mohamed A, Worku H, Lika T. Urban and regional planning approaches for sustainable governance: The case of Addis Ababa and the surrounding area changing landscape. City and Environment Interactions. 2020;**8**:100050

[31] Mulugeta Z. Assessment of the major challenges/factors for implementation of structural plan implementation in Injibara town, Amhara Regional State, Ethiopia. Journal of Civil, Structural, Environmental Water resources and Infrastructure Engineering Research (JCSEWIER). 2016;**6**(1):7-18

Section 3

Challenges and Future Directions

Chapter 10

Role of Biodiversity – Opportunities, Threats, and Strategic Interventions for a Resilient Indian City

Souporni Paul, Suchandra Bardhan and Sankeerthana Ananthula

Abstract

Cities are at the core of current environmental problems due to their ever-increasing demand for land surfaces and their impact on natural resources. Urban expansion compromises the functioning capacity of the ecosystems and creates complex effects on local and regional biodiversity. As cities grow, vital habitats are altered, destroyed, or fragmented into patches not big enough to support complex ecological communities, which is presently the most prominent factor contributing to the current global extinction and one of the pressing environmental issues. Hence, biodiversity conservation is integral to sustainable development and a significant concern of this millennium. Accurate assessment of urban biodiversity and implementation of strategies to arrest its loss at the local level is one of the most discussed topics in contemporary environmental research and international policies. The present study attempts to understand and analyse the urban biodiversity of Kolkata - a high-density megacity in eastern India with multiple environmental issues. It addresses the global agenda of biodiversity loss through a detailed assessment of the biodiversity status in the Indian city of Kolkata, followed by formulating relevant biodiversity strategies.

Keywords: biodiversity, landscape ecology, urban resilience, city biodiversity index, urban biodiversity strategies, habitat restoration, biodiversity conservation

1. Introduction

We live in the Urban Age and the Anthropocene [1]. Since the mid-twentieth century, rising population and extensive urbanisation have surpassed the earth's carrying capacity beyond its ability to recover [2]. These anthropogenic impacts contribute to two significant global implications: climate change and loss of biodiversity.

Biodiversity is the "variety of life" on earth [3] and is crucial to the survival of humanity. Article 2 of the Convention on Biological Diversity (CBD) refers to biodiversity as the "variability among living organisms from all sources, including terrestrial, marine, and other aquatic ecosystems and the ecological complexes of which they are part; this includes diversity within species, between species and ecosystems". A robust biological diversity is nature's resilience and adaptability to unprecedented natural calamities [4]. A large adaptive capacity and adaptive management of natural areas and native species are essential conditions for a climate-proof ecosystem. In sustainability and climate change research, much attention is being given to issues like air pollution, greenhouse gas emissions, fossil fuels, etc. However, urban biodiversity and cities' preparedness for climate change have not been explored. Especially in India, one of the world's most biodiverse countries, maintaining a stable ecosystem and rich biodiversity in cities is an essential and fundamental aspect of urban development.

The Convention on Biological Diversity (CBD), for the first time, recognised the importance of biodiversity for the well-being, resilience, and sustenance of human civilisation at the United Nations Conference on Environment and Development, UNCED (also known as "Rio Earth Summit"), held at Rio de Janeiro in June 1992 [5]. Accurate assessment of urban biodiversity and implementation of biodiversity plans to arrest its loss at the local level is one of the most discussed topics in contemporary environmental research and international policies. Despite the immense significance and contributions of biodiversity towards community resilience and human health, wealth, and sustenance, its global loss is considered one of the significant environmental challenges of this century [6]. Urbanisation, accompanied by changes in land-use patterns, habitat fragmentation, degradation of terrestrial and aquatic ecosystems, urban heat island effect, and air pollution, is a significant driver of both climate change and biodiversity loss (ibid). Therefore, there is an urgent need to draw our focus on protecting our biodiversity for a sustainable urban environment and resilient community structure. Biodiversity conservation is an integral component of the Sustainable Development Goals 2030 in many ways, as it provides the foundation for a "sustainable" and "resilient" society. At present, climate resilience efforts involve social, economic, technological, and political strategies at every scale of society. From local community action to global treaties, addressing climate resilience is becoming a priority, but the outcomes are still debatable.

With 2% of the world's surface area, India is home to 7–8% of verified species, four out of 34 global biodiversity hotspots across 10 different bio-zones, and several largest and fastest-growing cities [7]. Considering the magnitude of India's multiple crises, foremost being its population, inequality, illiteracy, lack of healthcare and sanitation, and pollution, climate change and biodiversity issues remain ignored. With limited awareness and support, it becomes an enormous challenge for city administrators to address climate change concerns or manage urban biodiversity and natural resources.

The current research has been taken up for the Indian city of Kolkata (formerly Calcutta). Located on the eastern bank of River Bhagirathi-Hooghly in the lower Gangetic delta, Kolkata is India's third-largest city and the capital of West Bengal. Once bestowed with rich biological diversity and efficient ecological systems within the city, it is considered an 'environmentally subsidised' city [8]. The natural features have not only articulated its urban growth but also influenced its infrastructure, sustenance, and liveability by offering essential ecosystem services, preserving natural habitat, and fulfilling the recreational need of its residents [9]. Currently, with a population density of around 24,252/Sq. Km., it experiences ever-increasing pressure on its urban biodiversity and natural resources due to its growing footprint and continuous infrastructural augmentation to meet the apparent developmental needs [10, 11]. It suffers from urban environmental challenges of deforestation, pollution, land-use changes, biodiversity loss, and over-exploitation of natural systems resulting in deterioration (**Figure 1**).

DOI: http://dx.doi.org/10.5772/intechopen.109528

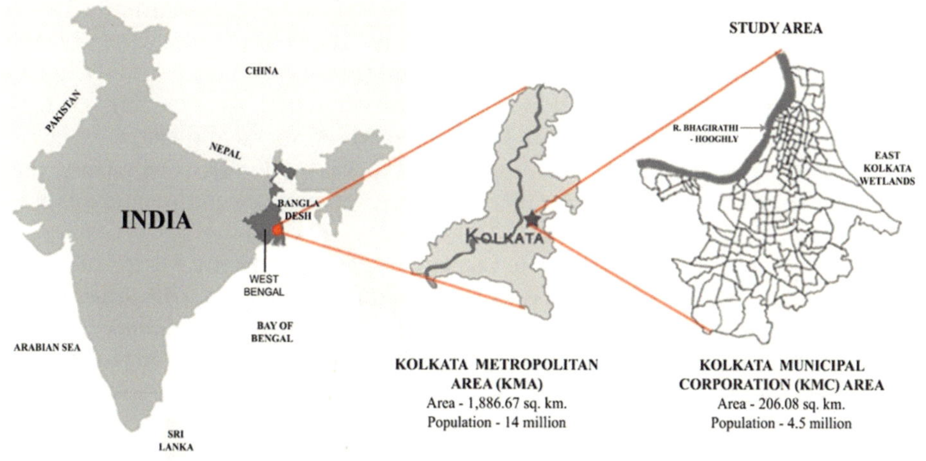

Figure 1.
Location and profile of Kolkata city.

2. Background and research proposal

Kolkata is one of the eight megacities most vulnerable to disaster-related mortality, in line with the recent United Nations (UN) report on climate change. According to the UN Intergovernmental Panel on Climate Change (IPCC) report, it is also one of the 20 largest coastal cities with potentially the highest flood losses by 2050, owing to the additional risk of subsidence due to sea-level rise and flooding. Cyclones in India are most frequent in the state of West Bengal in districts adjoining Kolkata, as per a recent report prepared by the Indian Meteorological Department (IMD). A UN report said that Kolkata had already lost a substantial portion of its green cover due to Cyclone Amphan, which struck India's east coast in May 2020. According to the IPCC report, the city is also highly vulnerable to extreme heat waves. Future assumptions about climate change impact by JICA include a rise in temperature by 1.20 to 1.80°C, an increase in precipitation by about 16%, and a sea level rise of 0.27 m by 2050. Thus, multiple reports underline the city's vulnerability to multiple climatic risks. At the same time, there is a lack of resilience in combating those risks. Kolkata demonstrates limited resilience plans to address the effects of climate change, such as the provision of shelters for the informal settlements and urban poor. Past studies conclude that the drainage and sewage network is inefficient enough to prevent the annual waterlogging the city faces annually in events of high rainfall or high tide of the river Bhagirathi-Hooghly [12–14]. Hence, urban infrastructure systems need to be designed for greater resource efficiency and disaster resilience by integrating biodiversity and other natural assets for improved liveability and sustainability.

Detailed studies show that Kolkata is steadily declining in terms of its urban biodiversity and open spaces. Studies of urban growth by [10] and Paul and Bardhan [13, 14] display the severe loss of valuable natural areas over the past decades. Rapid encroachment of vacant lands and water bodies in the eastern periphery drastically influenced land use and resource management. In 2015, collaborative research stated that the open areas in the city dropped from 25% of the total area in 1990 to a meagre 10% by 2015, while residential and commercial land covered 79% of the total area [15]. Paul and Bardhan [13, 14] showed that presently the urban-blue-and-green-spaces (UBGS) of Kolkata occupy only 11.51% of the total land area with an individual share of UBGS is 5.08 sq. m/person (**Figure 2**), which is far below the minimum requirement set by national and global standards. Kolkata faces an acute shortage and unequal distribution of UBGS concerning social and environmental injustice and is critically insufficient to support urban biodiversity. Ironically, the prominent Master Plans of the city refer to climate change mitigation and adaptation but have not mentioned the word "biodiversity" to date.

The biodiversity of Kolkata has been accounted for sporadically. Historical accounts by A.P. Benthall provide a detailed account of as many as 276 species of trees found till 1944, including 61 endemic, 69 naturalised, 41 truly indigenous, 20 natives to other parts of India, and six introduced species. After Benthall, many naturalists and professional scientists have also recorded butterflies, fishes, herpetofauna, birds and small mammals available in the city. The most recent accounts by [16] and Ghosh [17] reveal that at least 21 per cent of the species have disappeared from the cityscape within 64 years, of which 12 were indigenous. Ghosh [17] reported about 84 species of butterflies and at least 70–75 bird species available throughout the year in and around Kolkata. However, their numbers are decreasing due to a decrease in greenery and wetlands, a decrease in trees appropriate for nesting, accumulation of garbage, growth of high-rise buildings, increased paved areas and vehicular traffic causing noise and air pollution. The People's Biodiversity Register (PBR) is the most comprehensive accounting of biodiversity, released in 2012 and 2022. The PBR 2012

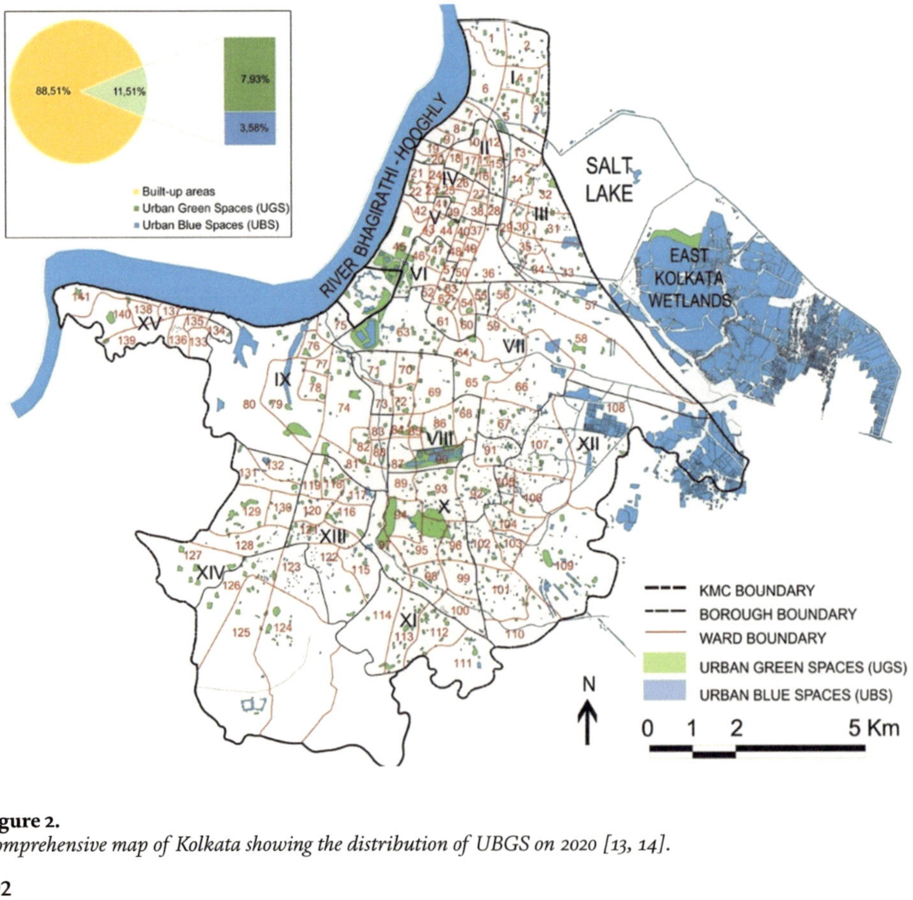

Figure 2.
Comprehensive map of Kolkata showing the distribution of UBGS on 2020 [13, 14].

recorded 676 plants, 30 mammals, 40 reptiles, 12 amphibia, and 34 fish species. The latest PBR released in May 2022 has documented 138 species of trees, 26 types of Chinese vegetables, 33 species of medicinal plants and about 100 other plant species and around 290 animal species, including about 70 species of butterflies, 47 varieties of fish, 84 varieties of birds and 22 varieties of mammals.

Recent application of the City Biodiversity Index (CBI) by Paul and Bardhan [18, 19] gives an overview of the present status of biodiversity in Kolkata, including their governance and management. The assessments show that the valuable urban-blue-and-green-spaces (UBGS) are continually decreasing, affecting the overall environmental conditions and, eventually, the city's native biodiversity. Similarly, there is good institutional capacity but a significant lack of coordination, broad-level policy planning, adequate budget and research. The studies revealed the root causes, underlying causes, and immediate causes leading to biodiversity loss in the city. They helped thoroughly understand the status, trends, and threats to biodiversity. With an increasing population, extending city limits, and decreasing natural resources, Kolkata needs a comprehensive conservation and management policy for its urban biodiversity for community resilience against climate change impacts.

3. Results for biodiversity conservation and community resilience for climate change preparedness

Urbanisation is a significant global driver of land-use conversion and deforestation, leading to biodiversity loss. Thus, cities can remain healthy by integrating biodiversity augmentation measures in city planning and management policies. The article focuses on community resilience as the ability to address climate-related hazards and their impacts with the help of robust biodiversity and efficient ecological systems within the city. It aims to mainstream biodiversity into the local policies with species management, large-scale preservation of the city's natural areas, institutional capacity, and active participation by the community. Conservation strategies are rooted protection, preservation, and enrichment of a city's native species and natural resources, while management strategies focus on governance, mandates, awareness, and community participation (**Figure 3**).

Aligned with the Sustainable Development Goals 2030, the framework consists of the following:

- Short-term resiliency planning encourages awareness about species and their habitats, awareness of the ecosystem benefits and establishes a connection with nature

- Long-term resiliency planning focuses on the ability to adapt and thrive despite changing climate, environmental, social, and economic conditions with biodiversity planning and management

The context is challenging since Kolkata is a divergent urban area with multiple complex ecological systems and an extensive network of people, infrastructures, and services that are strongly interconnected. The complexity is enhanced due to the diversity of stakeholders and areas involved in the administration and planning processes (**Table 1**).

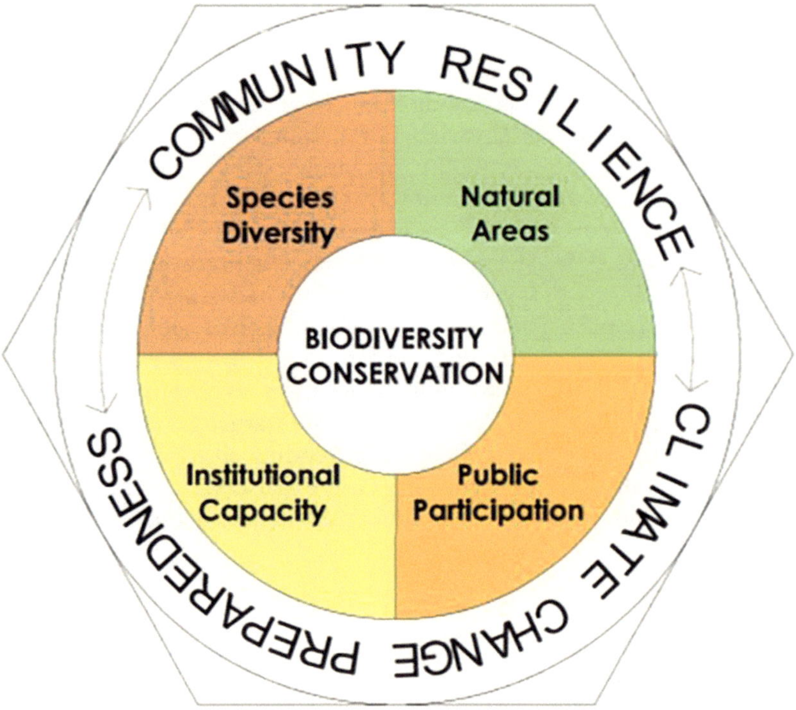

Figure 3.
Conceptual framework of the research (Source: Authors).

Urban biodiversity results from the city's inherent natural landscape and artificially designed infrastructure of parks and waterways. Kolkata's blue infrastructure in the form of rivers, canals, and wetlands offers multiple ecosystem services, including flood regulation and stormwater management. A reliable drainage system is the city's lifeline catered by the canal network. The emergence of sewage-fed fisheries and peri-urban agriculture of East Kolkata Wetlands (EKW) is also nature's gift and a source of livelihood for thousands. It is possible to revive the age-old tradition of water transport as a cheap, eco-friendly, and convenient conveyance mode by linking them to the major arterial roads to develop a multi-modal transport exchange. These longitudinal water bodies can provide an excellent opportunity to create a continuous

Issues identified from literature and analysis	Pre-requisites	Long-term goals	Short-term strategies
Species diversity			
Lack of systematic data on existing species, habitats, and ecosystem services	Systematic collection, classification, management, monitoring, and review of data regularly	Make local area programs for biodiversity conservation through open space planning and management, awareness of invasive species, habitat fragmentation and loss, indirect pressures on biodiversity and how they disrupt ecological processes	Conduct a city-wide biodiversity survey Rare and endangered species Endemic species Non-existent (wiped-out) or locally extinct species. Protect and manage existing native and vulnerable species

Issues identified from literature and analysis	Pre-requisites	Long-term goals	Short-term strategies
		Make concerted efforts to protect existing native species, vulnerable species, habitats and ecosystems and to re-establish species which once existed	Encourage appropriate development that mimics pre-urbanisation ecology and biodiversity to reducing the impact on native species
Natural areas (Urban-blue-and-green-spaces)			
Lack of conservation strategies for urban biodiversity and habitats	Formulation of strategies and systematic conservation of the existing natural areas	Restore the degraded habitats and improve the quality of existing habitats	Preserve, create, and maintain optimal green spaces in each neighbourhood for proportionate distribution
Lack of natural areas and poor ecosystem services	Qualitative and quantitative upgradation of natural habitats	Prepare and implement biodiversity management plans, outlining the approach to biodiversity conservation and habitat management work in large areas of nature reserves like wetlands, urban parks, sanctuaries, national lakes, riverfront, and canal banks	Prepare holistic strategies to revive, restore and reinstate the degraded natural streams and wetlands to maximise the socio-cultural, economic, and ecological benefits they offer
Lack of connectivity between the existing natural habitats	Establishment of connectivity between valuable natural habitats	Increase ecological connectivity through biodiversity corridors within and beyond the core city to protect regionally significant natural areas.	Identify the existing and potential ecological cores and connecting corridors through a city-wide review of existing and proposed natural or landscaped (anthropogenic) Urban-Blue-and-Green-Spaces (UBGS).
Institutional capacity			
Lack of coordination between the governmental and non-governmental bodies involved in biodiversity and environmental management.	Interdepartmental collaboration for a holistic approach towards biodiversity	Take a holistic and multi-sectoral approach that will directly or indirectly contribute towards the enhancement of biodiversity and ecosystem services	Utilise, protect, and upgrade local biodiversity and ecosystem services without compromising their value to people and the environment
Lack of sufficient budget for biodiversity management	Ensure a minimum proportion of funds allocated towards biodiversity and environmental management	Recognise biodiversity and healthy ecosystems as an aspect of good economic development	Improvement in the existing policy and legislation, collaboration between the relevant stakeholders
		Mainstream the national level targets and strategies into local (city-level) biodiversity plan	Develop and implement a city-level biodiversity strategy and actions plans with possibilities of periodic upgradations

Issues identified from literature and analysis	Pre-requisites	Long-term goals	Short-term strategies
Public participation			
Lack of knowledge, awareness, and participation among the citizens	Awareness, public participation and social inclusion for all ages and sections of the society	Reinforce human connections with natural areas	Promote regular direct contact with nature through a variety of open spaces (such as allotments, private gardens or incidental green areas, school grounds, environmental education centres and city farms, urban parks, and informal wildlife areas).
		Increase public participation and community engagement to encourage positive actions that support environmental conservation.	Maximise the health benefits of green spaces
		Engage and optimise local knowledge, networks, and resources, and improve local capacities.	Schools should arrange visits to local parks, urban parks, and other natural areas more frequently
		Increase public understanding and awareness of biodiversity and ecological processes to improve social resilience, health, and well-being	Appointment of knowledgeable people to create awareness and motivate students towards nature and biodiversity

Table 1.
Biodiversity conservation and community resilience for climate change preparedness for the city of Kolkata.

network of greenery within the city with their dynamic, complex landscape and rich aqua-terrestrial interface.

4. Conclusions

Urban biodiversity is an integral component of urban ecology and a significant parameter of sustainable development. Biodiversity is a basis for establishing a resilient ecosystem, and diverse ecosystems provide essential ecological services for humans, such as food, timber, pollination, and climate mitigation. Loss of urban biodiversity is one of the most pressing concerns, a widely discussed topic in environmental research and international policies. Cities need to understand that addressing biodiversity loss and climate change helps to achieve climate, ecological, and social resilience. Conserving urban biodiversity and restoring and connecting natural areas are essential for the community's well-being. Appropriate regulations and policies can significantly reduce biodiversity loss through land-use policies while providing physical and mental health benefits for city dwellers. Restricting construction on the precious East Kolkata Wetlands (EKW) can maintain natural ecosystems, conserve biodiversity, and prevent natural risks. The EKW is a natural protection against floods and other natural calamities. Contributions from the national government, international organisations, the private sector, residents, and academicians are required to address biodiversity conservation adequately.

Author details

Souporni Paul[1*], Suchandra Bardhan[2] and Sankeerthana Ananthula[1]

1 GITAM School of Architecture, GITAM (Deemed to be University), Hyderabad, India

2 Department of Architecture, Jadavpur University, Kolkata, India

*Address all correspondence to: spaul@gitam.edu

References

[1] Crutzen P. Geology of mankind. Nature. 2002;**415**:23. DOI: 10.1038/415023a

[2] Mahbub P, Goonetilleke A, Ayoko G, Egodawatta P, Yigitcanlar T. Analysis of build-up of heavy metals and volatile organics on urban roads in Gold Coast, Australia. Water Science & Technology. 2011;**63**(9):2077-2089

[3] SCBD. Cities and biodiversity outlook - Action and policy, secretariat of the convention on biological diversity, Montréal. 2012. Available from: http://www.cbd.int/en/subnational/partners-and-initiatives/cbo/cbo-action-andpolicy-executive-summary/cbd-cbo1-book-f-web-rev2

[4] Seddon N, Chausson A, Berry P, Girardin CAJ, Smith A, Turner B. Understanding the value and limits of nature-based solutions to climate change and other global challenges. Philosophical Transactions Royal Society B. 2020;**375**:20190120. DOI: 10.1098/rstb.2019.0120

[5] UNCED. United Nations Conference on Environment and Development. 1992. Available from: http://www.un.org/geninfo/bp/enviro.html. [Accessed: 14 September 2017]

[6] UNEP. 21 Issues for the 21st Century: Result of the UNEP Foresight Process on Emerging Environmental Issues. Nairobi, Kenya: United Nations Environment Programme; 2012. pp. 16-18

[7] CBD: India - Main Details, n.d. Convention of Biological Diversity. Available from: https://www.cbd.int/countries/profile/?country=in. [Accessed: 09 December 2017]

[8] Ghosh AK. Ecology and Environment of Calcutta. Government of West Bengal: Calcutta's Urban Future; 1991

[9] Mukherjee J, Chakraborty G. Commons vs Commodity: Urban Environmentalisms and the Transforming Tale of the East Kolkata Wetlands. Urbanities. 2016;**6**(2):78-88

[10] Mukherjee S, Bebermeier W, Schütt., B. An overview of the impacts of land use land cover changes (1980-2014) on urban water security of Kolkata. Land. 2018;7(3):1-25. DOI: 10.3390/land7030091

[11] Ramachandra TV, Aithal BH, Sowmyashree MV. Urban structure in Kolkata: metrics and modelling through geo-informatics. Società Italiana di Fotogrammetria e Topografia (SIFET). 2014;**6**:229-244. DOI: 10.1007/s12518-014-0135-y

[12] Mukherjee J. Sustainable flows between Kolkata and its peri-urban interface: challenges and opportunities. In: Allen A, Lampis A, Swilling M, editors. Untamed Urbanisms. New York: Routledge; 2015. DOI: 10.4324/9781315746692

[13] Paul S, Bardhan S. Framework for quantitative assessment of urban-blue-and-green-spaces in a high-density megacity. International Review for Spatial Planning and Sustainable Development. 2022a;**10**(3):208-301. DOI: 10.14246/irspsd.10.3_280

[14] Paul S, Bardhan S. Landscapes to eco-scapes: Prescriptive transitions for urban waterways for a sustainable megacity. Urban Water Journal. 2022b. DOI: 10.1080/1573062X.2022.2086883

[15] Staff Reporter. Study bares loss of open space in city. The Telegraph. 17 February 2017. 2017. Available from: https://www.telegraphindia.com/states/west-bengal/study-bares-loss-of-open-space-in-city/cid/1397757, [Accessed 19 November 2019]

[16] Ghosh AK, Samling CL, Ghosh S, Chatterjee NN. Changes in tree species diversity in Kolkata: 1946-2010. Indian Journal of Landscape Systems and Ecological Studies. 2013;**36**(2):1-21

[17] Ghosh S. Urban Biodiversity of Calcutta: Flowering Plants, Butterflies, Birds and Mammals. Kolkata: Zoological Survey of India; 2010

[18] Paul S, Bardhan S. Biodiversity accounting of cities: A case study of Kolkata, India. WIT Transactions on Ecology and the Environment. 2017;**226**:815-826. DOI: 10.2495/SDP170711 Available from: https://www.witpress.com/elibrary/wit-transactions-on-ecology-and-the-environment/226/36344

[19] Paul S, Bardhan S. Reinstating nature in a city with the help of biodiversity indicators: A framework for a resilient Kolkata. Urban India. 2018;**38**(2):115-135 Available from: https://www.niua.org/urban-india/urban-india-2018-jul-dec

Chapter 11

Perspective Chapter: A Conceptualization of Measuring People's Activity in Sustaining Urban Life

Haider Jasim Essa Al-Saaidy

Abstract

The criterion of human activity could be one of the critical points in dealing with street life mainly and urban vitality frequently. An attempt to classify the distinct patterns of activities is recently required. The potentiality of a street is to formulate people's interaction and responses to the street edges and reactions to each other. This chapter highlights the more significant outline regarding human activities and their patterns, besides their classifications. In this regard, the chapter aims to create a conceptual framework to form a platform for analyzing and studying the effectiveness of street life at the micro level. In this chapter, a research type is a descriptive-analytical approach; regarding the method, it employs a broad spectrum of related literature reviews. Consequently, the street edge generates the three main (and their oppositional) human activities. These activities can be categorized as necessary versus optional, individual against social, and staying as opposed to moving. Increasing the calls toward the sustainability of urban livability and street life versus decreasing the motor-based street has become a more significant demand. The street is a vital milieu that offers different opportunities for those who present and use such urban space as the street.

Keywords: sustainability, urban form, street edge, activity pattern, human activity

1. Introduction

In this chapter, the street and social life will be addressed according to the three categories of activities observed along the street edge: necessary versus optional, individual versus social, and staying versus moving. These three types of activities occur in a street when people respond either to the street's edge or interact with each other. The primary objective of this research is to quantify the different responses of people who use the street. The observations could be based on the ethnographic technique that allows the observer to record people's reactions and interactions without affecting people's behavior. This chapter aims to create a conceptual framework of human activities with a clear classification related to the three categories: necessary, individual, and social, and their opposites. Furthermore, there is a lack of a hard border between these activities when experiencing street life. However, priorities play a crucial role in distinguishing between different street life activities. Urban context grants different opportunities for those who use the urban spaces to share their varied responses and ambitions and involve in such activities as a conversation, sitting, walking, staying, standing, and observing when the street links the accommodation of such activities and actions in a certain milieu [1–3].

The urban ingredients could be range from fine level to the hard scale of components and elements that give the distinct characteristics of the entire urban pattern. Also, analyzing and examining the urban parameters can be typified into different levels based on the study's primary purpose: micro level, macro level, and local and global level. In this regard, comprehending the link of the spatial qualities at a micro and macro level involves knowledge of how spatial factors in an urban context are arranged around each other and the influence of the configurational pattern on the street network and people's behavior. Bianca [4] denotes that the physical setting signifies, "... every genuine cultural tradition, architecture, and urban form" and that this "... can be seen as a natural expression of prevailing spiritual values and beliefs .... it is an outcome of tradition and daily practices which correspond to certain spiritual principles" ([4], p. 22). People who use different urban spaces can be categorized into two key activities: walkers (movement-on, movement-through) and stayers. The objectives of these two kinds differ in terms of the purpose of the destination [5].

Responding to the behavioral settings in an urban context is subject reciprocally to the parameters of the surrounding environment and its entities. Different studies and scholars addressed broadly and thoughtfully the people's response and their interaction whether with the street space or between each other as social interaction [6–9]. Other studies dealt with micro level, where people come to interact with just adjacent street edge [10–15]. Moreover, some scholars went further in order to analyze the interrelationship between the private-public edge of the street and how individuals could respond to it [1, 14–16]. The behavioral studies in relation to the environmental setting also addressed significant findings in highlighting the symbiotic mutual relationship between the incentive of the physical settings as the built environment and the individuals' response and their reactions [17, 18].

The centrality and integration in computing the street network and how a certain link(s) play a key role in shaping not only the people stream but also how forming the individuals' behavior along with the edges. These studies of centrality offered different approaches in order to make a correlation between people and urban space, such as multiple centrality assessment (MCA) and space syntax (SS) [19–21].

2. Method and materials

In this chapter, the research method is a descriptive-analytical approach, on the subject of the method, it uses a broad spectrum of an associated literature review. In order to capture the more significant vocabulary of people activities, the chapter tends to classify activity into two main approaches: urban form and human activity and street characteristics and activity pattern. The materials have been derived from previous studies regarding the urban form and peoples' activities. Moreover, identifying human activities depends formally on recent research. Besides, there is no hard border between human activities. Instead, one can conduct multiple acts or responses toward the characteristics of the built environment synchronously.

2.1 Urban form and human activity

In behavior-environment interaction, Canter [22] argues that "the environment providing perceptual stimuli ... [and] also be thought of as a filter ... we are always in the environment to carry out certain activities, and we usually carry out these activities with other individuals ... this is the fact that we actively modify, build and influence our physical surroundings." For this reason, Canter [22] alludes that, "the physical environment surrounds and supports all human activities. It is, therefore, expected that the study of human-environment should be as complex and multi-faceted as is the range of studies of human behavior." Many urban studies and research examined the nature of influences of the built environment on human activities, whether psychological, social, and functional, such as Brownson et al. [23], Caro [24], Craig et al. [25], Greenwald and Boarnet [26] Handy et al. [27], King et al. [28], Shay et al. [29], and Sun et al. [30].

A considerable number of scholars addressed the interrelationship between human activities and use patterns. The density and diversity of land use analysis determine the nature of activities that are likely to take place; these accord with the uses along a street. Jacobs [31] refers to that, "in dense, diversified city area, people still walk ... the more intensely various and close-grained the diversity in an area, the more walking [where] life attracts life." Montgomery [32] states that activity is an inevitable result that derives from two distinct but quite related concepts: vitality and diversity. The former means a successful urban space and its characterization from others. It considers the pedestrian flow where people in and around the street can benefit from the facilities that the street offers, at different times of the day. It is a place where people meet for various purposes; "... the presence of an active street life, and generally the extent to which a place feels alive or lively. Indeed, successful places appear to have their own pulse or rhythm, a life force" ([32], p. 97) (**Figure 1**).

Montgomery [32] suggests that secondary diversity relates to enterprises and services which respond to primary uses and offer different amenities in serving consumers' needs. Thus, "areas of high development density can be planned to accommodate and stimulate mixed-use and self-generating secondary diversity, [in this regard], diversity must be sufficiently complex to stimulate public contact, transactions and street life ... for this to happen, streets need to be active, to accommodate and generate diversity, and they must be permeable. They must also engender a sense of belonging, familiarity and the respect of users" ([32], p. 103, 105, 109). Canter [22] states that, to enable space operation, it is necessary to examine the relationship between several groupings of spatial interactions; people in relation to physical objects, people with regard to other people, groups in relation to physical environment, and groups with regard to other groups.

Moreover, land use, to some extent, deals with large-scale (global and local or even neighborhood), while the activities pattern tends to address uses at the micro and street scales. In a sense, the oldest built environment in the most traditional area emanates from the street scope and daily needs that might be characterized as high-frequency activities. While in a preplanned order, the activity pattern is controlled by the top-down approach that determines the activities according to land use-based action. The difference between the traditional and modern neighborhoods adds to the other characteristics that define the activity pattern, such as the street pattern and block-plot system, besides the land use itself. These play a significant role in formulating the relationship of people to the street edge. To track the activity pattern, the current study addresses three types of human activity, namely: necessary versus optional, individual as opposed to social, and staying against moving. Three pairs of activities are likely to happen in the street in different ways, with various densities.

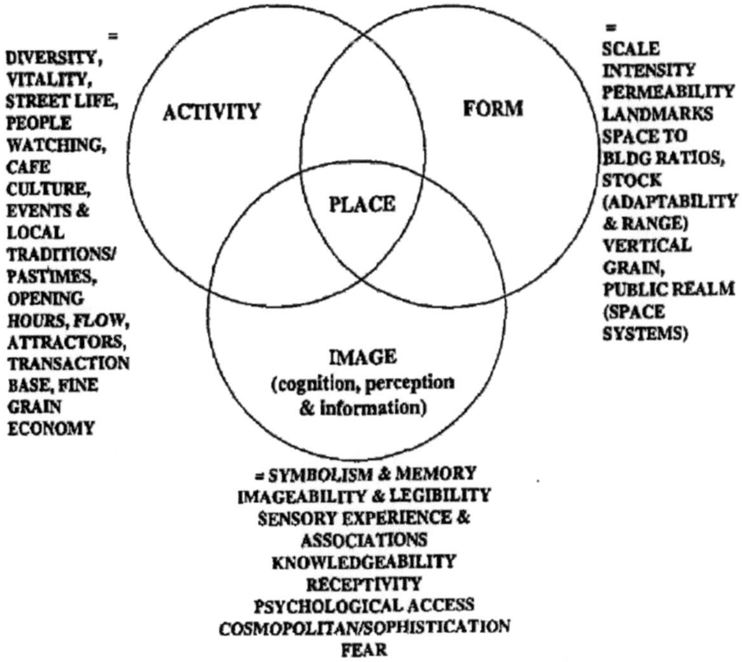

Figure 1.
Three essential entities in formulating the vitality of urban place: activity, form physical attributes, and image (meaning and conception), Montgomery [32]. Source: based on Canter [33]; Gehl [7] and Punter [34].

Thus, the street scale is the domain to examine human activities after defining each kind of activity and disclosing the ability of the street edge to formulate the interaction, whether between people and the edge (private and public) or between people themselves. This interaction with its interfaces (human-edge) is responsible for promoting street life and, in turn, social interaction.

2.2 Spatial configuration and activity pattern

Gehl [7] states that, "urban structures and planning influence human behavior and the ways in which cities operate." This attracts attention to the humanity of the street as a pulsatile path through various activities. Jacobs diagnoses different issues that cause a decrease in the social life of outdoor spaces. Jacobs [31] states that, "street in cities serve many purposes besides carrying vehicles, and city sidewalks—the pedestrian part of the street—serve many purposes besides carrying pedestrians ... a city sidewalk by itself is nothing." A community organizes its functions spatially, and this affects human behavior through the distribution pattern of activities and their locations along the street; these might also be characterized by the integrated street that has a high number of connections to other streets within a short metric distance [35].

Despite having their own private spaces, people still demand spaces to interact socially and/or economically with other members of the community. Human activities tend to take their place in a physical space, where the mechanism of the organization of activities depends on society itself (privately and officially). In turn, they impact the built environment as well. Consequently, the spatial configuration of a built environment helps to shape human behavior in the urban space based on the possibilities for social control, and opportunities for economic activity and social interaction [35].

During their study of the traditional area at South Bank of London, Penn et al. [36] diagnosed four factors, which work together to enable apparently natural relationships between various types of activities and facilities. These factors can be summarized as follows: "the network of streets and spaces in an area produces a pattern of natural pedestrian and vehicular movement. Most movement in urban areas is through movement. Once shops have set up on a street, they become destinations for to movement in their own right. And through movement alone is not enough to produce a successful urban shopping area" ([36], p. 82). Hence, the movement pattern and street network with land use play a significant role in emanating the activities in certain urban places. Hillier [37] adopts centrality as a process to study the center, such as in a city. Apparently, the concentration of mixed use of activities and functional diversity are placed in a prominent location in a city, called a center or subcenter. Also, Hillier [37] adopts the term "live centrality" to express the spectrum of activities that collectively form the centrality elements, such as retail, markets, catering, entertainment and other activities that engage people, and in turn, benefit from movement. Van Nes [38] states that the location of functions or activities depends on the configuration of the grid pattern where movement takes place.

Recently, the emphasis has been placed on the notion of the active and inactive edge in terms of theoretical or practical investments in the urban field. These refer to the capacity of the edges to attract and hold people by creating and continuing social activities, which thereby make the urban edge a socio-spatial factor in the urban fabric [39]. This meaning also is denoted by Alexander [40] who states that an effective and attractive edge that takes a scalloped shape includes various activities as pockets; moreover, "to make the space lively, the scalloped edge must surround the space completely" ([40], p. 601). Marshall [41] distinguishes between patterns of physical entities and patterns of activity or use. The former concerns roads, land use areas, built area, and open spaces, while the second addresses trip-making, commercial activities, or trajectories of movement [41] (**Figure 2**). Moreover, Marshall [42] refers to that, "the social space of streets is the single contiguous public off of which private spaces are carved. ... in this sense, the public-private filtering of the building-plot-street system enables settlements to exist—they enable large agglomeration of humans to coexist in a limited area. This is why the streets are not merely voids between blocks of buildings, but must be seen as integral to the concept and fabric of a city."

At the global and local scale, the spatial configuration of urban elements; plot, block, and street network are the essential entities that lead human activities and control the distribution pattern of activities. Furthermore, the underlying characteristics

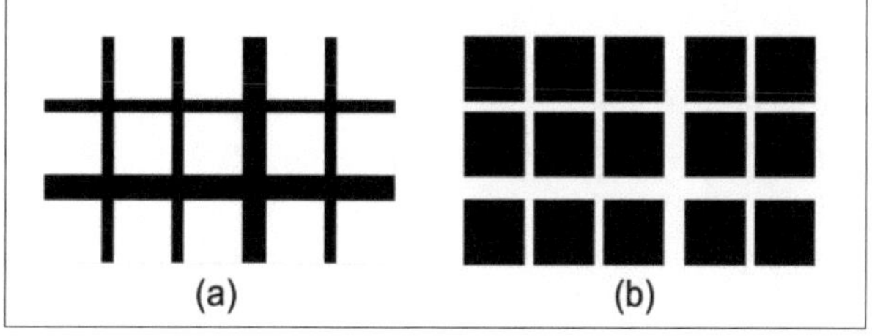

Figure 2.
The classification depends on the purpose of its application. (a) The transport modeller might see a 'cross-nodal network' of routes where they constitute the adjacent buildings, while (b) a planner or developer might see a 'square tessellation' of land parcels where the streets constitute the edge of buildings. Source: based on Marshall ([41], p. 23).

of each urban element (in terms of shape, size, number, and its interrelation to the other components in space), also determine the type of present or potential activities. Regarding the micro scale, the street as a transitional edge is constituted by adjacent buildings and their characteristics play a crucial role in controlling the interrelationship between the private and public space and the level of interaction between them.

2.3 Street characteristics and activity pattern

Computing the value of the street in terms of its proportional position to the surrounding links within the network system is another significant aspect in evaluating the level of human activity and the degree of its presence in a particular area. Can and Heath [10] suggest that, when integration and connectivity in spatial patterns are higher, social interaction will increase. Additionally, modern urban areas are likely to be more introverted than the inner city and traditional regions. Most often, activities in modernist areas tend to be located on the edge, where the primary route contrasts with inner parts in the traditional city. Carmona, Tiesdell et al. [43] state that the term street and other labels, such as boulevard and avenue, reveal design elements lacking in the term "road." The main goal of these labels is to accommodate and reconcile the demands of the movement, social life space, and urban activities. Collectively and substantially, they need to emerge in the same physical space.

Thus, the notion of the street is to define social space and to secure an effective connection with the whole network system [43]. Different terminologies and definitions are given to identify the street space itself and its relation to others. Kostof [44] argues that the history of the street so far has to be addressed, either as an urban form or as an institution. On the one hand, the street belongs to the architecture and urban study archives as it is a physical phenomenon. The street is an entity that consists of a roadway, usually a pedestrian way, and bordering buildings; however, the street as an institution is a critical theme. In this regard, the street has an economic role and social importance.

The traditional aim of the street is to enable the movement to exchange goods, conduct social exchange, and communicate. These three aspects are inseparably related to the street space, where those activities are located, and then to the whole street life ([44], p. 189). Porta et al. [45] explain the relationship between street centrality and economic activities in a certain area of Barcelona by adopting the measure of multiple centrality assessment. The aim of using MCA analysis is to quantify the centrality value and its relationship to economic activities. According to Porta et al. [45], two kinds of economic activity can be recognized: primary and secondary. In the relationship between economic activities and spatial networks lies a fundamental question about the mechanism of distribution of activities in the urban context and the role of urban structure, functions and other capabilities in shaping this relationship [45].

3. Identifying activity pattern

The aimed of urban space usage is to create an interactional relationship between individuals as occupants of space and activities within a behavioral framework. This relationship can also cover the nature of linkages between individuals and with the kind of activities present in the same space. The reinforcing factors in a particular space might be utterly different from one space to another, and between regions and countries, according to a series of considerations. In other words, what may be an enhanced factor in a given space might be a debilitating factor in another one place [46]. A clear pattern of activity relates to the classification process of compound parameters, which increases in limited areas or specified spatial dimensions. However, minimal or single settings mainly affect large-scale classifications, and this can refer to the comprehensive analysis of commercial streets. This is likely to be irrelevant in creating the distinctive urban characteristics of a whole city. Rapoport [47] argues that the environment is perceived by people, which causes a reaction before any act is taken to specifically analyze it. There are two types of meanings embodied in a built environment, latent, and manifest, since that environment apparently grants cues for behavior and represents a form of people's behavior. These two types of meanings relate to the nature of the activity of buildings within a whole urban context. This can be classified into four components for each activity: (1) the activity proper, (2) the specific way of doing it, (3) additional, adjusted, or associated activities that become part of the activity system, and (4) the meaning of the activity ([47, 48], p. 15) (**Figure 3**).

In this respect, Chapin [49] refers to an outline working schema based on human activity systems, which includes two steps: the first is behavioral constructs through the spatial structure of the city and a physical construct as the second step. According to Chapin [49], "environment is construed not only as a structure of land uses and communications channels with physical dimensions but also as a structure of institutions with significant social and economic dimensions, all influencing and begin influenced by human activity." To examine and evaluate a human activity that could occur in an environment, Chapin [49] suggests a linear scale that consists of minimum dissatisfactions and maximum satisfaction. Two factors can govern the stimulus of activities that are primarily reflected in human behavior; these are *pull* factors and *push* factors, where the first decision is mostly based on *push* factors, and the second is based on *pull* factors. Furthermore, people stand between these *push-pull* factors in response to their surrounding physical environment (**Figure 4**).

Street space must be conceived as an outdoor milieu that contributes to the sense-making of a place through a considerable number of activities and facilities.

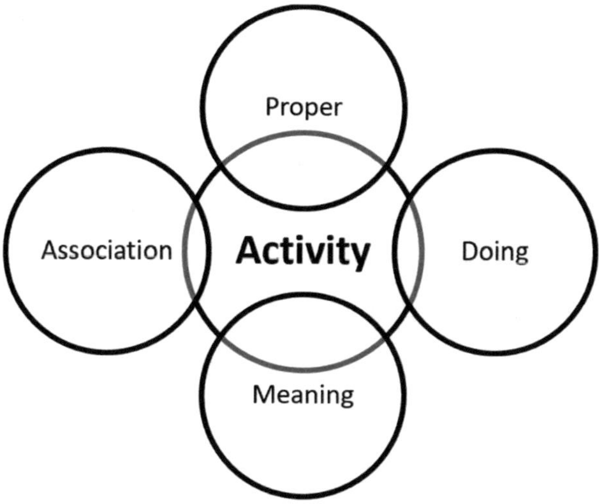

Figure 3.
A relationship among four sections of activity. Source: based on Rapoport [48].

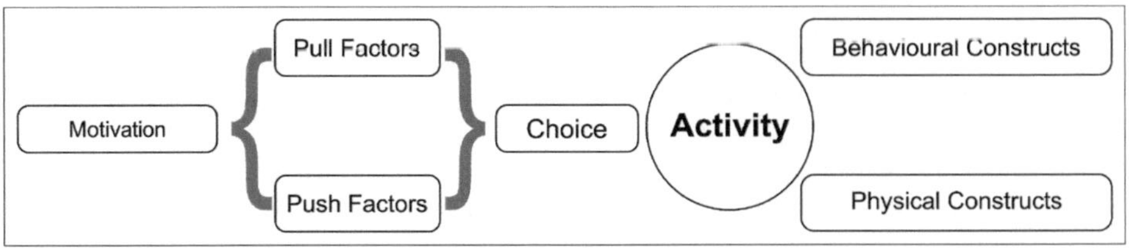

Figure 4.
A linear relationship among three components of behaviour. Source: based on Chapin [49].

This includes a place to relax and enjoy the urban experience, a venue for a range of different activities, including entertainment, sport and play areas, a site for civic or political functions, and most importantly a place for walking or sitting-out. A strict interrelationship between the public space and people who live and work around it is an essential aim in raising the public space value through its spatial performance [50]. A street is a place to meet people and is a place for people to have an opportunity to be intimate, anonymous and in some way private. It also promotes concentrated spaces for face-to-face activity [50].

The definition of activities is an essential characteristic of understanding and effectively interacting with urban space and acknowledges a symbolic and meaningful pattern. This helps to form the crucial determinants of activities and social interactions, so that they coincide with the surrounding environment. This occurs by decoding the process of communication between people and the urban context, which is mediated by the activities [47]. In this respect, physical form and activity are congruent, and this enables the adoption of the idea of the activity-based place [6]. Thus, according to Jacobs [51], "the interplay of human activity with the physical place has an enormous amount to do with the greatness of a street. It is difficult or impossible to separate the two ... streets are settings for activities that bring people together."

In his seminal work, Gehl [11] refers to three types of activity: necessary, optional, and social activity (**Figure 5**). This classification of urban space activities can cover a large range of behaviors throughout people's presence in space. It might start with a small indicator, such as talking between people and ending with a festival or an annual activity, to promote the value and quality of space. Diversity is considered one of the leading characters of three types of activities regarding the urban context [7, 11, 52]. Thus, the street space in a city is a shared milieu that embraces people and activities alike. This sharing covers the responsibility of designing the street space, where spaces are always much broader than the specialized expertise in designing the built environment. In other words, the design of a space, cannot be separated from daily life and the social activities of urban areas as well as other aspects and institutional considerations in making a decision [43].

3.1 Necessary versus optional activities

Necessary activities can be defined as a kind of compulsory act with a diverse degree of participation among others. The occurrence of these activities is not affected by the physical scope. It fluctuates due to dependence on the doer, where these activities constrain the interactions as a crucial part of their life, and regardless of the conditions of the surrounding environment [11]. The second is optional activities, which are based on the desires of participants and the extent to which they help to motivate the individuals to take part in these activities. Exterior conditions are regarded as one

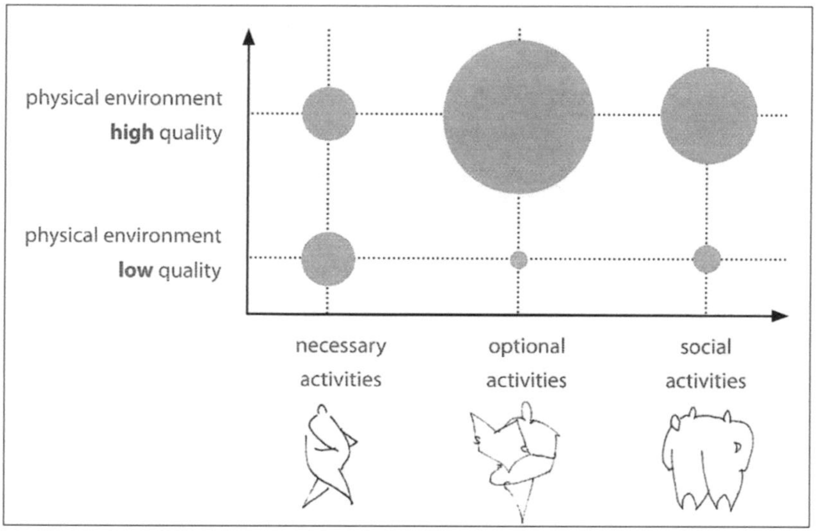

Figure 5.
Graphic representation of the connection between outdoor quality and outdoor activity. An increase in outdoor quality gives a boost to optional activities. The increase in activity level invites a substantial increase in social activities. Source: Gehl [7].

of the most critical factors in achieving this kind of activity, one of which is the built environment and natural conditions. Opportunities for occupancy of the place can increase if there are high-quality conditions within the place. Individuals are automatically attracted by places, which include what people like and desire [11].

Jacobs [31] tends to distinguish between two main activities: primary and secondary uses. The former is employed to bring people to particular places that function as an attractive point, such as offices, factories, and dwellings, as well as the main destinations of, for example, educational institutions, entertainment centers, and recreation. In addition, museums, libraries, and galleries are primary uses. The secondary uses serve the primary purposes and include a wide range of activities and events [31]. Porta et al. [45] offer a description of primary activities by stating that, "primary activities are characterized by a larger-than-local market or catchment area; they are typically highly skilled, larger, or more specialized economic activities, such as wholesale, industry, and those not related to the public or not mainly serving the end-users; and their location choice is more likely to be driven by a formal top-down decision-making process" ([45], pp. 1476–1479). The distribution pattern of economic activities has significant implications since it influences the availability of land for particular economic uses and for people who benefit from these activities [53].

Furthermore, Porta et al. [45] write descriptively about secondary activities by saying that a local market or catchment area characterize secondary activities, and these are typically retail and other services that serve the regular needs of the populace on a daily or frequent basis. Accordingly, secondary activities can be defined as the kind of economic setting that sustain and embody the sense of a lively and walkable local community at the scale of the neighborhood [45]. The primary uses are associated with, and represent, necessary activities, while the secondary uses are optional and social activities according to [7, 11]. Moirongo [54] refers to the relationship between street function and activity pattern. He states that the street is characterized by heterogeneous and mixed uses that tend to be for optional and social activities. In comparison, the street is identified by monofunctional and homogeneous uses that contain necessary activities. This, however, emphasizes the earlier statements that necessary activities are obligatory for those accustomed to attending a regular activity for a particular purpose, such as school, work, shopping, or waiting for someone. Meanwhile, the optional activities relate to a degree of desirability, besides the impulse factors that attract people's engagement with activity.

3.2 Individual versus social activities

Social activities tend to see individuals gathering in public spaces. These activities cover a wide range of people's acts while they are in a place. Spontaneity is one of the features of these activities, where people, at the same place, can meet each other and talk. It seems to be quite difficult to isolate and separate these three activities from each other within a certain space; however, the weight of activity may vary from one space to another [11]. Thwaites et al. [39] pay more attention to social activity as a key factor in social restoration. Thus, social activity is, "soft, active or engaging edges are commonly associated with a social activity, usually generated by their capacity to hold the attention of passers-by. Such edges are also associated with having transitional qualities defining an overlap of adjacent realms, their social activity related to accessibility across it and opportunities to be stationary coupled with things that can hold attention" ([39], p. 81).

The need to return to social space and urban social activities has become a persistent demand in order to promote the vitality of a city. Shared places provide a platform to create an integral relationship between various aspects of a city's elements and its people. One of the aspects of shared space is to reduce motor flow and constrain its speed. Therefore, shared urban places are pedestrian orientated and give rights to people to invest in the street as a social milieu. The role of social interaction in an urban context is witnessed through its components, namely streets, sidewalks, and parks, which represent intermediate spaces between activities and human behavior. One of the concerns about urban space is the meaning of context in terms of cross-connection between activities and space [43].

The range of opportunities that could be offered by the street edge contributes significantly to creating social life and attracting people to share their activities with others. In this regard, Jacobs [31] prioritizes the commercial, entertainment, and other aspects represented by, for example, restaurants, shopping centers, retailers, and cafes. When located on the sidewalks of a street, these activities, help to entice people to use the street. If it does not attract social life and the sharing the street edge, the street functions as a transmit channel. Moreover, the activities generated by people help to promote street life, where people can be seen as magnetic points to attract others. As such, Jacobs [31] alludes that, "the sight of people attracts still other people ... people's love of watching activity and other people is constantly evident in cities everywhere."

According to Jacobs [31], the primary mixed uses must be concurrently active with diversity. Effectiveness depends foremost on people who tend to use spaces for different purposes and reasons. However, efficiency, according to Jacobs [31] means, "that the people using the streets at different times must actually use the same street [including] among them, people who will use some of the same facilities. All kinds of people can be presented [and] the mixture of people on a street at one time of day must bear some reasonably proportionate relationship to people there at other times of day." Engwicht [5] defines the street according to aspects that have been ignored, and particularly in consideration of modern thought. Therefore, street reclaiming means, "[an] exchange space —an outdoor living room for social, cultural, and economic exchange, [and] as a place for the adult and those on the margin of society to share their street wisdom, [also] the street for adult play: people-watching, promenading, water, art celebration, festivals, eating, or just hanging out, [and] as the stage upon which those at the margin of society can make a contribution to community life. A vibrant street life is essential for any egalitarian, democratic society to a healthy political process" ([5], pp. 17–54).

3.3 Staying versus moving

People's activities regarding the street milieu can be classified into two trends, namely: *through street-based movement* and *to street-based movement*, including those who are staying and moving. The reciprocal role between staying and moving relates to the pedestrian flow and the street edge, which governs the level of stopping or walking. Movement is a crucial key to expressing a dynamic street. Carmona et al. [43] state that, "movement is fundamental to understanding how places function. Pedestrian flows through public space are both at the heart of the urban experience and important in generating life and activity ... most shops, for example, are not a sufficient magnet and have to be well-located with respect to the existing movement patterns. The land-use activity merely reinforces/multiplies the basic movement."

The opportunity for people to be able to stay or move in a space correlates to the space's characteristics one of which is intervisibility and permeability through-in space. This is what Hillier [55] calls the strategic value of isovist when he argues that this makes intuitive sense because, if the primary activity of those who stay in public spaces is people-watching, then "... strategic spaces with areas close to—but not actually lying on—the main lines of movement are optimal." For this reason, Jacobs [31] states that, "a city's collection of opportunities of all kinds, and the fluidity with which these opportunities and choices can be used, is an asset—not a detriment—for encouraging city— neighborhood stability." The street edge is responsible, to a large extent, for generating the three main (and their oppositional) human activities. The street edge grants the opportunity for those who belong to a different gender, age, and cluster to use the street for different purposes and activities. Hence, the value of the street is not limited to only vehicular movement but offers a broad spectrum of opportunities for people to advantageously use the street.

4. Conceptual framework of quantifying human activities

In this chapter, the idea of extracting conceptual layout in quantifying different patterns of activity could play a significant role in examining human behavior. Moreover, the street edge as an independent factor also contributes substantially to forming peoples' responses. Gehl [3] states that the dramatic evolution that markedly changed the atmosphere of cities, public spaces, and public life in the twentieth century, was the influx of motor cars in substantial numbers. He also states that the "city space continued to function as important social meeting place in the twentieth century, until the planning ideals of modernism prevailed and coincided with the car invasion" ([11], p. 25). However, modernism carried a new perspective of scale and proportion based on its ideal thoughts, which stand away from what people perceive in order to their desires and aspirations with meaningful and comfortable [7].

The scale can be divided into two levels: the scale of the urban context in the city and the scale of buildings in the city. The other dramatic changes in reading the city are that the city has become a milieu for cars rather than people, and the destruction of the human scale of the city in terms of high-rise buildings and wide roads with larger open spaces, which are far from a human scale and tend to be antipedestrianism [56].

In this chapter, the ethnographic technique is an investigation that can be adopted through the observation of people's behavior throughout the streets. Ethnography examines how persons behave in a space without any direct connection to them; this is achieved by conducting a direct field observation. To collect data from the field, synchronically, the need is to record pedestrian flow, non-pedestrian flow in motorized (cars, vehicles, motorcycles, etc.) spaces, and nonmotorized flow (bikes and human-powered cart). They also recorded other observed activities. The observation could be functioned for different time per day, such as morning, noon and afternoon, and during weekdays and weekends, when each recording period covered approximately an efficient interval. Furthermore, this meant coexisting with street life and people in the field study. Two fundamental types of movement occur in the street: pedestrian flow and vehicular movement as non-pedestrian flow. The first sort is classified according to kind, amount, and proximity. The kind considered the gender and age of people, and the volume, which calculated the number of pedestrians. The proximity referred to the grouping of pedestrians (**Figure 6**).

The conceptual framework is to evaluate the street edge and the extent to which the edge, with its embodied activities, attracts people. In addition, this captured the age and gender of people and how they behaved; for example, whether they walked and stayed, or were alone or in a group. The level of pedestrian flow through the street and how people respond to both the private and public edge is a crucial question for the nature of the street life. Similarly, the physiognomies of the street edge play a key role in formulating people's behavior and in controlling the interrelationship between both the human-edge interface and the human-human interface. Concerning non-pedestrian (vehicular) movement can be considered one of the leading issues in studying a street network, particularly where most of the street is designated for vehicular-based standards rather than human-based dimensions.

The conceptual framework includes two main sections: pedestrian movement and non-pedestrian stream. The former covers three categories, namely: group, volume, and kind each one deals with a specific demand of data. A group pattern represents the number of individuals to be grouped and how the different groups are scattered across the street, and the diversity of these clustering(s) in terms of members who belong to them. A volume pattern refers to the number of persons who move-to and move-through in a certain segment of the street. In this chart, a kind pattern illustrates two classes: gender and age, where the people are symbolized into male and female. Also, the kind pattern embraces another division, that is age pattern, where the individuals can be labeled into three different ages: child, teenager (young), and adult (elder) people. The second part of the classification allocates for the non-pedestrian flow, in this class, there are three lists: nonmotorized, hybrid, and motorized. The nonmotorized flow deals with two items: bike and human-powered cart, and hybrid class contains motorcycles, vehicles, and other, besides, motorized alike (**Figure 6**).

In this flowchart of classification, on the furthest right of the diagram, a column illustrates the levels of information in relation to each row of the conceptual framework. The level of interfacing is for a human-human interface meaning that the relationship between people themselves across the street life. A scope is a domain of dealing with a certain range of street segment observation such as these levels as micro,

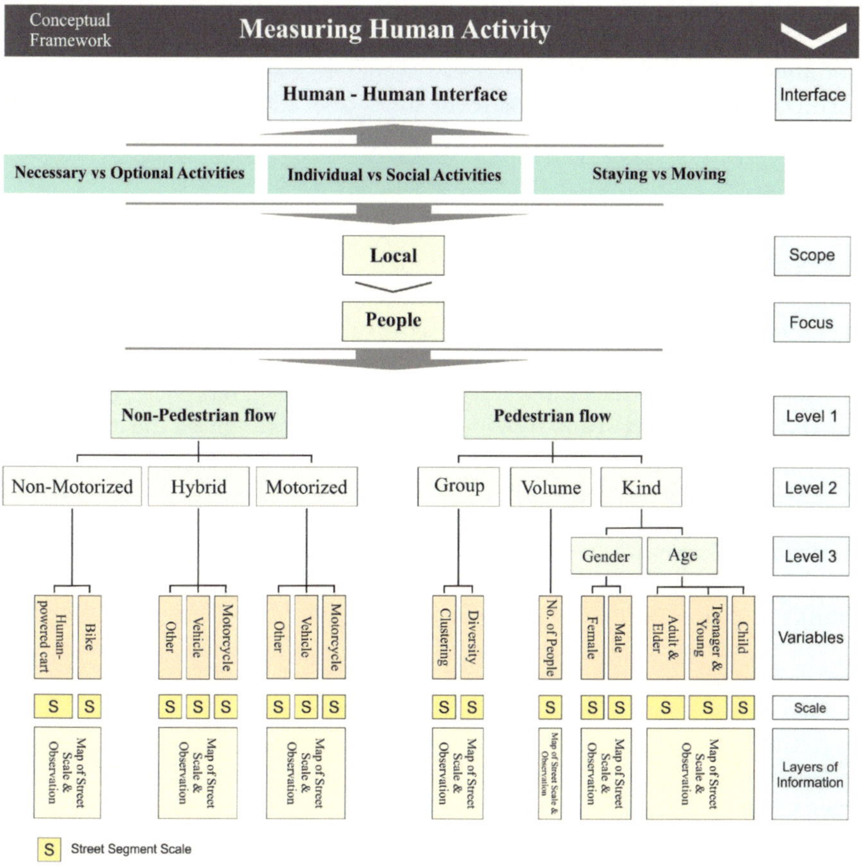

Figure 6.
A Conceptual framework of the movement in the street into two main flows: Pedestrian flow and Non-pedestrian movement. Source: Drawn by the author.

macro, local, and global scale in measuring different activities. A focus indicates the chosen sample of this framework and that is people who would be observed and experienced in street life. Following, the levels that link the different rows of the flowchart are flexible and expandable concerning the circumstance of each case separately. The variables that were listed through the conceptual framework, these variables were interpreted thoroughly before and then giving the definition for each one. A scale is an indicator to define the scope of each variable in studying street life, for instance, street segment scale, block scale, and neighborhood scale. A layer of information, in the flowchart, is an important scheme that helps to deal with specific data and input evidence and materials that would be invested and employed in a study (**Figure 6**).

Mapping individuals' behavior through the street life and how people meet each other for different purposes, such as walking, chatting, sitting, and sharing an event, this procedure entails a spectrum of plans and scenarios. For example, the conceptual framework and its variables and indicators would be a research process map in order to draw the layout of future studies regarding urban life, street life, and social interaction. Raising the appeals toward the sustainability of urban life and street knowledge against minimizing the car-based street has become a more essential need. The street is an active ambience that grants various possibilities for people who come and utilize such different urban space as the street. This conceptual framework has been designed to deal with people within the surrounding environment. Additional details of examining the physical settings of the adjacent street edges require more concentration on the anatomical characteristics of such street edges.

5. Conclusion

This chapter was an attempt to draw the main layout of different patterns of activities that could perform in an urban street setting. In this sense, the street edge promotes how individuals might respond to the surrounding environment differently. Three significant actions could be reflected in people's behavior when they move through the street: necessary, individual, and staying; moreover, their opposite actions are: optional, social, and moving. This chapter highlighted the meaning of these activities, and it provided the conceptual framework through projecting its vocabularies on the street observation. Human activities and how people behave are not independent phenomena within the urban context; however, one can recognize the level of the relationship between those who use the street and the street itself. The spatial configuration of the urban form and the street's characteristics are ranked most important in affecting the individual's response to the built environment.

However, creating an obvious border of human activity is not straightforward, and it is challenging for the observer to distinguish people's movement and their intent and purpose. Therefore, the outlines and presents findings from the ethnography technique, namely, to record human movement and behavior is a methodical technique. Nevertheless, the direct observations of, and coexistence with, people in the observed streets and the capturing of notes are more significant and helped to enrich future studies. The classification of different activities that could be observed and experienced in a street space contributes, to a large extent, to diagnose the different abilities of people's responses and reactions. The conceptual framework chart granted a new vision of a research map that would employ in order to study street life.

The main variables and indicators included in the conceptual framework would support capturing the fine characteristics of the varied reactions that could come from those who experience street life and different social interactions. The ability of resilience of the conceptual framework is not limited to what has been done, but instead, the outline of these variables is open for more progress procedures. Finally, the reliability and validity of such a conceptual framework are needed meaning that future research could seek to apply this framework in order to get the level of proof of the variables and how that could be interpreted and explained the findings, and in turn, promote and update the current and new variables and indicators. However, the chapter had been focused on the activity pattern and, theoretically, apart from the built environment as an independent factor. For this reason, there is an opportunity to give more attention to the physical setting and the different levels of urban scales, such as micro, macro, and local and global levels.

Author details

Haider Jasim Essa Al-Saaidy
Department of Architectural Engineering, University of Technology-Iraq, Baghdad, Iraq

*Address all correspondence to: haider.j.essa@uotechnology.edu.iq

References

[1] Al-Saaidy HJE, Alobaydi D. Studying street centrality and human density in different urban forms in Baghdad, Iraq. Ain Shams Engineering Journal. 2021;**12**(1):1111-1121

[2] Chen H, Jia B, Lau SSY. Sustainable urban form for Chinese compact cities: Challenges of a rapid urbanized economy. Habitat International. 2008;**32**(1):28-40

[3] Gehl J. Public spaces for a changing public life. In: Thompson CW, Travlou P, editors. Open space: People space. Abingdon, Oxon, Uk: Taylor & Francis; 2007. pp. 3-7

[4] Bianca S. Urban Form in the Arab World: Past and Present. USA: vdf Hochschulverlag AG; 2000

[5] Engwicht D. Street reclaiming: Creating livable streets and vibrant communities. British Columbia: New Society Publishers; 1999

[6] Broadbent G, Bunt R, Llorens T. Meaning and Behaviour in the Built Environment. Chichester: Wiley; 1980

[7] Gehl J. Cities for People. Washington: Island Press; 2010a

[8] Willis A et al. Human movement behaviour in urban spaces: Implications for the design and modelling of effective pedestrian environments. Environment and Planning B Planning and Design. 2004;**31**(6):805-828

[9] Widodo J. Urban Environment and Human Behaviour: Learning from History and Local Wisdom. Procedia-Social and Behavioral Sciences. 2012;**42**:6-11

[10] Can I, Heath T. In-between spaces and social interaction: A morphological analysis of Izmir using space syntax. Journal of Housing and the Built Environment. 2016;**31**:31-49

[11] Gehl J. Life Between Buildings: Using Public Space. Copenhagen: Danish Architectural Press; 2010b

[12] Holland C, Joseph Rowntree F. Social interactions in urban public places. York, York: Joseph Rowntree Foundation; 2007

[13] Nooraddin H. Al-fina', in-between spaces as an urban design concept: making public and private places along streets in Islamic cities of the Middle East. Urban Design International. 1998;**3**(1-2):65-77

[14] Al-Saaidy HJE. A conceptualisation of defining and computing the street edge: Constitutedness — Permeability – Intervisibility. Civil Engineering and Architecture. 2022a;**10**(4):1260-1279

[15] Al-Saaidy HJE. In: Urban Agglomeration SB, Baiani A, editors. The Street Edge: Micro-Morphological Analysis of the Street Characteristics of Baghdad, Iraq. London, UK: IntechOpen; 2022b

[16] Zhang A, Li W, Wu J, Lin J, Chu J, Xia C. How can the urban landscape affect urban vitality at the street block level? A case study of 15 metropolises in China. Environment and Planning B: Urban Analytics and City Science. 2021;**48**(5):1245-1262

[17] Salah AA et al. Challenges of human behavior understanding. Human Behavior Understanding. 2010:1-12

[18] Van Wormer KS. Human Behavior and the Social Environment, Macro Level: Groups, Communities, and Organizations, Oxford. New York: Oxford University Press; 2011

[19] Hillier B et al. Space syntax. Environment and Planning B: Planning and Design. 1976;**3**(2):147-185

[20] Porta S et al. The network analysis of urban streets: A primal approach. Environment and Planning B Planning and Design. 2006a;**33**(5):705-725

[21] Porta S et al. The network analysis of urban streets: A dual approach. Physica A: Statistical Mechanics and its Applications. 2006b;**369**(2):853-866

[22] Canter DV. Environmental Interaction: Psychological Approaches to Our Physical Surroundings. London, UK, London: Surrey University Press: Distributed by International Textbook Co.; 1975

[23] Brownson RC et al. Measuring the built environment for physical activity: State of the science. American Journal of Preventive Medicine. 2009;**36**(4):S99-S123 e112

[24] Caro T. Behavior and conservation: A bridge too far? Trends in Ecology & Evolution. 2007;**22**(8):394-400

[25] Craig CL et al. Exploring the effect of the environment on physical activity: A study examining walking to work. American Journal of Preventive Medicine. 2002;**23**(2):36-43

[26] Greenwald M, Boarnet M. Built environment as determinant of walking behavior: Analyzing nonwork pedestrian travel in Portland, Oregon. Transportation Research Record: Journal of the Transportation Research Board. 2001;**1780**:33-41

[27] Handy SL et al. How the built environment affects physical activity: Views from urban planning. American Journal of Preventive Medicine. 2002;**23**(2):64-73

[28] King AC et al. Theoretical approaches to the promotion of physical activity. American Journal of Preventive Medicine. 2002;**23**(2):15-25

[29] Shay E et al. Walkable environments and walking activity. Final Report for Seed Grant Submitted to Southeastern Transportation Center, University of Tennessee. 2003

[30] Sun G et al. Measuring the influence of built environment on walking behavior: An accessibility approach. International Conference on Geographic Information Science, Springer. 7478. 2012. pp. 187-197

[31] Jacobs J. The Death and Life of Great American Cities. New York: Random House; 1961

[32] Montgomery J. Making a city: Urbanity, vitality and urban design. Journal of Urban Design. 1998;**3**(1):93-116

[33] Canter D. The Psychology of Place. London: The Architectural Press; 1977

[34] Punter J. Participation in the design of urban space. Landscape Design. 1991;**200**(1):24-27

[35] Van Nes A. Measuring the degree of street vitality in excavated towns: how can macro and micro spatial analyses tools contribute to understandings on the spatial organization of urban life in Pompeii? Proceedings of the 7th International Space Syntax Symposium, Stockholm, Sweden, June 8-10, 2009. Koch D, Marcus L, Steen J. edited by Trita-ARK. Forskningspublikationer, 2009: 1. 2009

[36] Penn A et al. Configurational modelling of urban movement networks. Environment and Planning B-Planning & Design. 1998;**25**(1):59-84

[37] Hillier B. Centrality as a process: Accounting for attraction inequalities in deformed grids. Urban Design International. 1999;**4**:107-127. DOI: 10.1080/135753199350036

[38] Van Nes A. Road Building and Urban Change, A morphological and configurative explanation of how ring roads change the pattern of distribution of shops in city and town centres. Proceedings of the 3rd International Space Syntax Symposium, 2001. 2002

[39] Thwaites K et al. Socially Restorative Urbanism: The Theory, Process and Practice of Experiemics. UK: Routledge; 2013

[40] Alexander C. A Pattern Language: Towns, Buildings, Construction. New York: Oxford University Press; 1977

[41] Marshall S. Urban Pattern Specification. London: Institute of Community Studies; 2005

[42] Marshall S. Cities Design and Evolution. New York, NY: Routledge; 2007

[43] Carmona M et al. Public Places, Urban Spaces: The Dimensions of Urban Design. Amsterdam; Boston: Architectural Press/Elsevier; 2010

[44] Kostof S. The City Assembled: The Elements of Urban Form Through History. London: Thames and Hudson; 1992

[45] Porta S et al. Street centrality and the location of economic activities in Barcelona. Urban Studies. 2012;**49**(7):1471-1488

[46] Whyte WH. The Social Life of Small Urban Spaces. Washington, D.C.: Conservation Foundation; 1980

[47] Rapoport A. The Meaning of the Built Environment a Nonverbal Communication Approach, The University of Arizona Press, USA, Original published: Beverly Hills, Calif. London: Sage; 1990. p. 1982

[48] Amos R. Human aspects of urban form: Towards a man-environment approach to urban form and design. In: Rapoport A, editor. Urban and Regional Planning Series. Vol. 15. Pergamon Press; 1977

[49] Chapin FS. Activity systems and urban structure: A working schema. Journal of the American Institute of Planners. 1968;**34**(1):11

[50] Thompson CW. Urban open space in the 21st century. Landscape and Urban Planning. 2002;**60**(2):59-72

[51] Jacobs AB. Great Streets. Cambridge, Mass: MIT Press; 1993

[52] Gehl J, Svarre B. How to Study Public Life. USA: Island Press; 2013

[53] Parker DC, Meretsky V. Measuring pattern outcomes in an agent-based model of edge-effect externalities using spatial metrics. Agriculture, Ecosystems & Environment. 2004;**101**(2-3):233-250

[54] Moirongo BO. Urban public space patterns: human distribution and the design of sustainable city centres with reference to Nairobi CBD. Urban Design International. 2002;7(3):205-216

[55] Hillier B. Cities as movement economies. Urban Design International. 1996;**1**:41-60. DOI: 10.1057/udi.1996.5

[56] Wiedenhoeft R. Cities for People Practical Measures for Improving Urban Environments. New York, London: Van Nostrand Reinhold; 1981

Chapter 12

Smart-Field of the Farm, First Relevant Link of Transparent Path of Grapes in the Farm to Fork Value Chain: A Review

Constantin Vîlcu, Steve Vanlanduit, Resul Kara, Aurora Ranca and Mohamed Bourouah

Abstract

Data-based farming facilitates the zonal management (Smart-FIELD) in the smart farm (SF) vineyards. The classic SCADA for data acquisition and processing system, used especially in rapid industrial processes, can also be adapted for the slow processes (during a vegetation period) of horticultural crops (including the vineyards). The Smart-SCADA core software activities will develop through AI modules the CPS technique for integrating IoT devices with operational applicability in vineyards SF. The Web Smart-SCADA core interconnection with the European platforms in use (UTOPIA, ATLAS, DEMETER, PANTHEON, SmartAgriHubs, ...) will add value in the transparency of the traceability of the F2F value chain of grapes.

Keywords: precision agriculture, multispectral monitoring of crops, IoT in vineyard smart farming, toward zero pollution in the farming environment, smart SCADA core in vineyards

1. Introduction

A little while ago, the European Green Deal (EGD) established the Farm to Fork Strategy for Next Generation Europe (NGE) in the agricultural field (R-D & Innovation, management of farms and the smart operations in real farming environment). Long ago, much ink flowed on paper for industrial development and the effort to move agriculture to a higher level. Now we have agriculture-4.0, where the analog pen has turned into a digital tool. We can write in machine code, thanks to the ICT technology, on the virtual paper from the SSD hard drive of the PC for the variable rate actuation of the execution elements in the processes of horticultural crops (vineyard inclusive).

Data-based farming will represent the agriculture of the future. In this sense, ICT systems have a crucial role in data acquisition (metadata), analysis and decision support (DSS), action with variable rate of actuation (VRA) in crops processes, and promotion of agri-food products through the IoT technique, [1]. The hardware-software integration of these systems in any agricultural sector allows the farmer to obtain higher harvests with reduced expenses, i.e., a profitable ROI (the first point of view - that of the farmer). The second point of view is at the other end of the F2F value chain - end user, and represents the needs of the end consumer to have quality (healthy) and cheap products. Between these two extreme points of view, the TZP-UTOPIA-F2F concept seeks an optimal path and conforms to the Farm to Fork Strategy. The paper addresses the horticultural sector of the culture processes in open spaces and focuses the activities of work in the vineyards for SF digitization.

As rationale for SMFs that represent a significant part of EU horticultural landscape, the paper approaches vineyard farms regarding F2F Strategy. The digital transformation of this agricultural sector (through data-driven farming) is crucial for addressing global issues related to climate change and implicitly in the sustainability of F2F agri-food value chain. As method, the TZP-UTOPIA-F2F concept will develop a smart-SCADA core system adapted to the needs of horticultural SMFs and implement the software modules that enable a scalable, easy and economically feasible approach of the management of vineyards farms in the digital context of extreme climate change constraints.

Smart-FIELD of the farm, is the first relevant link at the potential applications of TZP-UTOPIA-F2F concept and will combine new ICT technical solutions with innovative learning systems to achieve the adoption of digital technology in vineyard farms. As impact and potential benefits, the TZP-UTOPIA-F2F works make an in-depth analysis of the socio-economic issues of precision farming in vineyards, an assessment of the return on investment and the development of sustainable business models, will provide new and innovative possibilities for vineyards SF to produce the necessary and healthy food in the context of the F2F Strategy.

2. Edge-activities in operational FIELD of vineyard Smart Farm

2.1 Why, Smart-FIELD in vineyards?

In order to obtain quality and profitable products, farmers from vineyards must respect and even permanently adapt the technological links of the culture system to climatic constraints. Their cost price must cover all the necessary products (fertilizers, pesticides, fuels, etc.) and, at the same time, be competitive on a sufficiently saturated market, such as the European one. Farmers are currently facing two major constraints, the high price of inputs and the demand of consumers to find cheap and healthy products on the market. In this context, we meet the demands of consumers by ensuring a diminishing of the inputs and a traceability system of wine products from the farm, to finished products such as grapes intended for fresh consumption, fresh must or preserved by pasteurization and wines of different types and levels of quality. The system can come to the aid of certification bodies, for example ecological certification, or those that verify and attest the origin and quality of wines. Through the various signs provided to manufacturers for application on labels (logos, holograms, QR codes), the degree of consumer information increases considerably (MUR) [2, 3].

2.2 How, do we operate in the vineyard?

Farmers are heavily investing in digital technologies to improve their productivity and to safeguard their harvest from losses due to drought, pests, etc. Multispectral imaging systems installed on mobile platforms (like mobile robots, drones) are a valuable tool to reach this goal. Unfortunately, the imaging systems and platforms that are currently available are expensive, and difficult to use (by farmers without expertise in computer programming). The InViLab research group of the University of Antwerp will develop a low-cost multispectral vision system (consisting of industrial cameras and image processing units integrated in a compact housing). State-of-the-art machine learning algorithms will be developed to monitor the growth and health of the vineyards (both at plant and at field level). The algorithms will be deployed on edge platforms to realize real-time performance. In the work the methodology developed by InViLab is applied to vineyards, even though the applicability of the framework is much wider: seaweed monitoring and monitoring of vertical greenery systems applied to buildings (UAB) [3, 4].

2.3 How do we have, F2F traceability?

Improvements in IoT devices have been going on for many years. Especially in recent years, with the use of IoT in the field of farming, healthier, more efficient and traceable products have been grown. By evaluating the data collected from the environment and products that are grown by various methods, the way to obtain more perfect products has been opened. Similarly, in the light of the data obtained by using the surrounding sensors, it is possible to intervene immediately in adverse situations that will affect the development of the product. In this study, both product-related data and data about environmental conditions will be collected quickly from various sensors located in a wide area of agricultural land. Thanks to the advanced IoT network, all data will be collected in the central communication unit and situations that require urgent intervention will be decided. Necessary actions can be taken in emergencies. The data collected from the environment will be sent to the cloud system for processing together with the actions taken and all other data. In the next phases, the feedback from the customers will also be recorded in this system, and thus the products will be tracked from the farm to the table. In addition, the farmers will have the opportunity to monitor their products grown in the field and in the F2F value chain (DU), [3, 5].

2.4 What is needed to solve the problem?

An ICT system durable solves the problem by digitizing the SF vineyard. For data-driven farming is needed the integration of a Web SCADA (data acquisition and processing system). Smart-SCADA core software activities will develop through AI modules the CPS technique for integrating IoT devices with operational applicability in SF vineyards. The interconnection with the UTOPIA platform will add value to the TZP-UTOPIA-F2F concept. The graphic transformation for mathematical modeling of the F2F process is presented in **Figure 1**, from organizational chart 1. (a) – F2F agri-food chain management for grapes products, to the graph 1. (b) – Graphic diagram of the F2F agri-food value chain: 1-Farm, 2-Fork, 3-Crop storage, 4-Food processing, 5-Distributor transporter, 6-Agri-food market, 7-Food health safety (FHS), 8-web

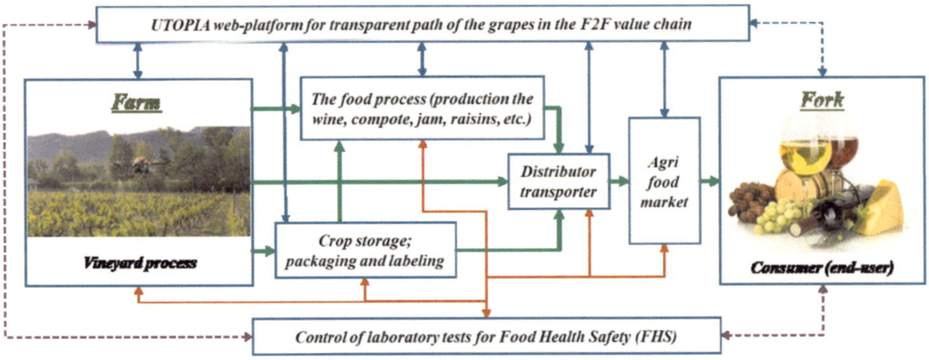

1. a)

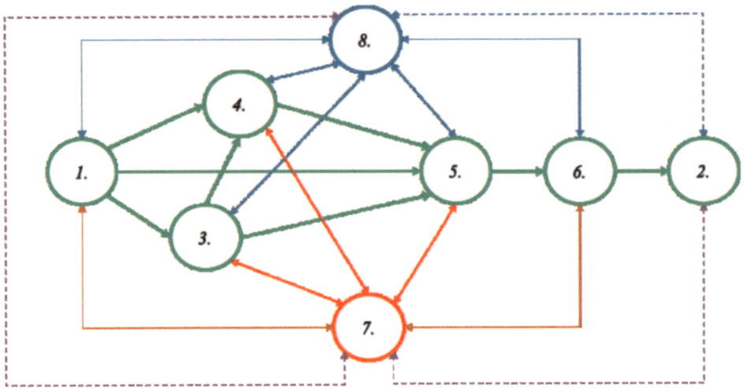

1. b)

Figure 1.
Transparent traceability of agri-food products from Farm to end-user.

platform UTOPIA, which is the basis of Fuzzy neural processing in the software module integrated of TZP-UTOPIA-F2F concept, (INMA), [6–8].

3. Materials and methods

3.1 Excellence

The World is currently facing several challenges, [9]:

- At current growth rate, by 2050 there should be over 9.5 billion people to be nourished;
- Chemical based mass-agriculture, despite higher short-term yields, negatively impacts food safety and will lead to soil impoverishment on the long-term and reduced yields;
- The overall Farm to Fork value chain produces 25% of worldwide emissions directly contributing to climate change crisis;
- 75% of F2F emissions are located in the agricultural environment.

The paper has chosen to address of TZP-UTOPIA-F2F concept for Agri-food systems enabled by interconnected digital technologies that are more transparent to consumers, farmers and other stakeholders along the agri-food value chain, as a strong collaborative engagement to impact positive for a healthy life.

From the multiple sectors of activity of Agriculture presented in the tree graph in **Figure 2**, the branch was chosen: Agriculture → Crop farming → Horticulture → Olericulture. Why horticulture? Horticulture represents, through its activities in open and closed spaces (greenhouses), the "art" of vegetable crops in agriculture. Vineyards and fruit farms cover an extensive area of Europe and the automation of process crops in these areas is experiencing a high development. The TZP-UTOPIA-F2F concept is for data-processing in vineyard crops in order to promote the grapes product in the F2F value chain.

Ultimately, TZP-UTOPIA-F2F concept has been conceived to create value for farmers and consumers through the Farm to Fork value chain, while restoring value to the Natural Capital value chain (see **Figure 3**). The farm is the main link in the F2F value chain, obtaining organic food products (e.g., grapes) must be promoted transparently so that end-users know that the consumed product is fresh and healthy.

By assessing all indicators related to the use of intrants, such as pesticides and fertilizers, as well as monitoring the state of nutrients, Web Smart-SCADA core

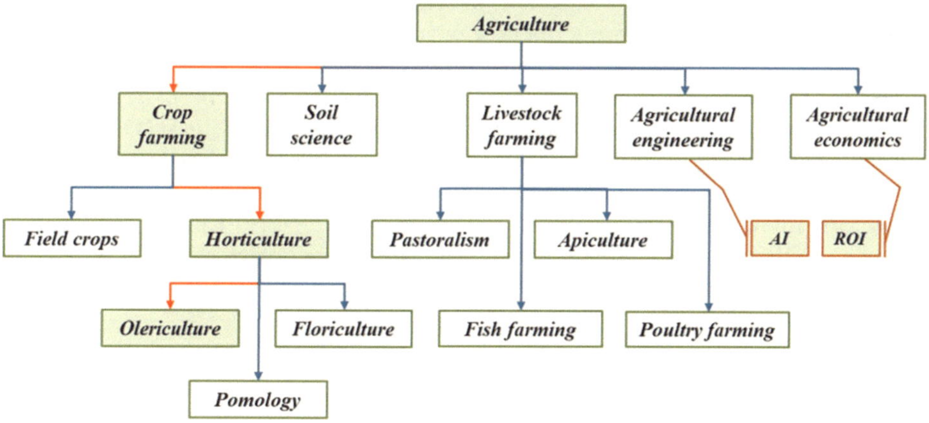

Figure 2.
Summary of branches of agriculture.

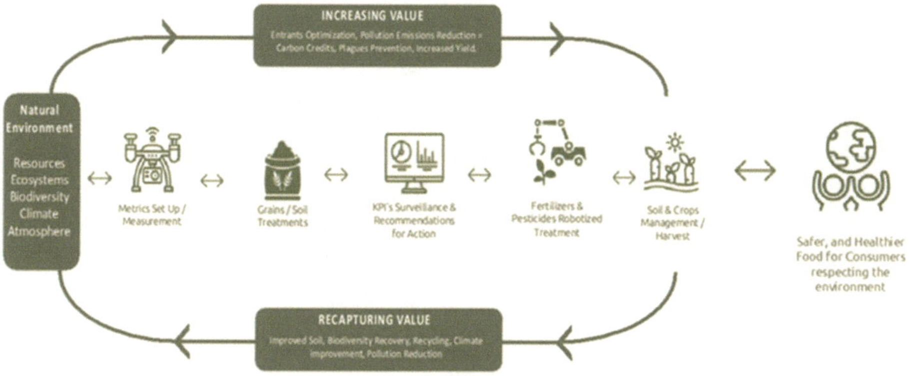

Figure 3.
Method of restoring the value to Natural Capital value chain.

through AI software modules will provide farmers with automatized call-to-action recommendations to farmers. Once farmer validates call-to-action SF will automatically provide point-specific agriculture surgical treatments (watering and other intrants) with newly developed operating plans compliant pesticides, fertilizers and other nutrients, reducing water pollution, polluting emissions and intrants waste across the cycle of farming production. This will have a positive impact on the health of the agri-food products produced (e.g., grapes).

3.2 The overall concept and progress beyond the state-of-the-art

Regardless of the field of the agricultural environment (plant or animal kingdom), data-based agriculture requires data acquisition systems, IoT sensors and devices, IT technique, or in a word CPS that allows the storage of monitored data in large databases, necessary for analysis further for an intelligent management of the agricultural process.

The aim of the work is to digitize the vineyards by implementing the ICT technology in Smart Farm, consisting in Web Smart-SCADA core integration and the scalable software packages for easy methods of learning sustainable farm management toward zero pollution. Today, precision farming or agriculture-4.0 is no longer a new concept in farming operations management using UAV drones (for measurement, aerial observation and response to crop variability) and/or GRS robots (for resource efficiency even in cases where they are limited). The use of these MAS systems in a unitary hard-software framework of TZP-UTOPIA-F2F concept, leads through the F2F strategy, to the achievement of the following targets: (i) obtaining high quality farming crops; (ii) ROI economic profit optimization; (iii) integrated environmental protection to TZP, (iv) Cloud connection with the UTOPIA platform to allow full visibility of the health status of grapes in F2F value chain for a complete and reliable traceability of agri-food products to the end user.

The TZP-UTOPIA-F2F concept has a high degree of novelty through Web smart-SCADA core that unifies software relevant preliminary results from agriculture-4.0 for advanced vineyard SFs management. This paper, in relation to the current state of the art technology, facilitates the transition to agriculture-5.0 or agriculture based on management of large databases in cloud (metadata). After the Industrial Revolution, especially since the advent of mechanization and during the Green Revolution, farmers and their machines worked together effectively to cultivate crops with a tendency to digitize agriculture.

A new approach driven by digital technology implies that growers need to act as supervisors of their crops rather than as workers, thus avoiding repetitive, demanding and tiring tasks in the field. In this modern agricultural setting, agricultural databases are key, and the information-based management cycle provides the practical approach that combines the concept of Smart-FIELD with agricultural tasks. Agriculture based on large databases on crops and the farming environment, with the help of MAS systems that incorporate state-of-the-art AI techniques, lays the foundations for sustainable farming of the future (agriculture-5.0), [3].

3.3 Ambition of works

Unmanned Aerial Vehicles (UAV) and Ground Robotic Systems GRS, as collaborative Multi-Agent Systems (MAS), make a crucial contribution to improving precision farming by controlling the TZP global index for sustainable agriculture. Monitoring the Inputs in agricultural process and the state of crops vegetation are essential in management of SF. The processing of data recorded for the monitored crops influences managerial decisions and implicitly the ROI indicator of vineyard farm.

The components of the abiotic environment (air, water, soil) are vital as inputs together with the planting material for a sustainable SF. The current trend is to make smart farming by automatically controlling inputs and outputs. Supervisory control and data acquisition of agricultural processes involves the introduction of high-performance technologies to streamline the farming process and ensure control of production in ecological conditions. All farmers try to get as much agricultural production as possible at the lowest cost.

To this end, the paper proposes the creation of a Smart SCADA core for vineyard precision farms with an integrated package of Smart-FIELD software modules for easy learning of the new generation of farmers for the management of SF. The TZP-UTOPIA-F2F model will be physically materialized through a dedicated Web SCADA core system by parameterization to any SFs that integrates it.

The human operator in the farm control center is continuously interconnected with MAS systems and on-site sensors to control the process and pollution of the agricultural environment. Thus, TZP-UTOPIA-F2F allows real-time assessment of crops and the agricultural environment without the need to move the human operator. The main target of the works is therefore the digital creation and implementation of modular software packages that allow a scalable and easy approach to the management of vineyard farms in the sustainable context of TZP. It's shown in **Figure 4** the automation process diagram for vineyard farms in the new TZP-UTOPIA-F2F concept of F2F strategy, [3, 6].

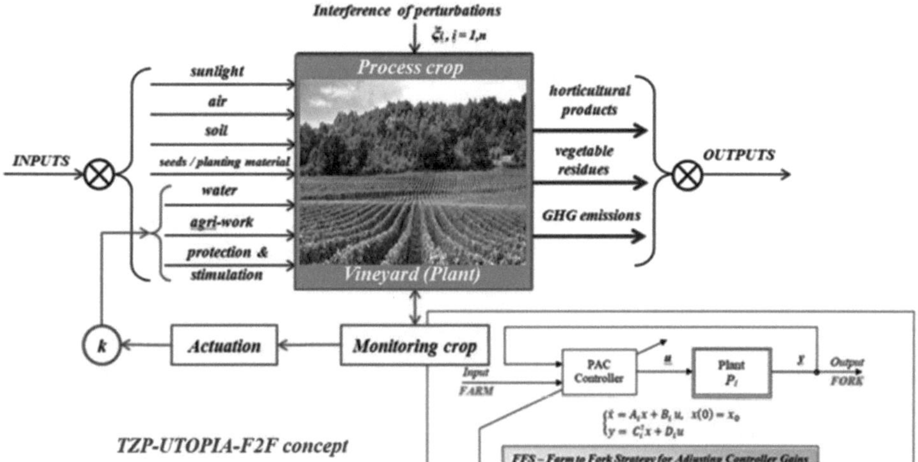

Figure 4.
Automation process for vineyards farms in the new TZP-UTOPIA-F2F concept of the F2F strategy.

3.4 List of parameters for Smart-SCADA core

The hard- and software implementation of the Web Smart-SCADA core in the SF vineyard, requires the existence of a minimum equipment's and IoT devices to ensure

Integrated equipment in smart farming (vineyards)	**#**	***List of monitored parameters for TZP-UTOPIA-F2F concept***			
		Abiotic component		***Symbol***	***Name***
 [10]: air sensors*	1	***Air***	General indicators	T_a	Air temperature
	2			H_a	Air humidity
	3			p_a	Barometric pressure
	4			Ws	Wind speed
	5			Wd	Wind direction
	6		Special indicators	TSI	Solar irradiation (Sun intensity - lighting) [W/m^2]
	7			CO	Carbon monoxide
	8			CO_2	Carbon dioxide
	9		Critical indicators	PM	Dust particles (PM – particles matter)
	10			FW	Frost warning
	11			NO_x	NOx gases
 [11]: water sensors*	12	***Water***	General indicators	T_w	Water temperature
	13			pH	pH
	14			DO	Dissolved oxygen
	15			EC_w	Electrical conductivity
	16			Wt	Water turbidity
	17		Special indicators	Ra^*	Rain amount [l/m^2]
	18			Ri^*	Rain intensity [l/m^2/h]
	19			ORP	Oxygen reduction potential (Redox) [V]
	20		Critical indicators	P_{est}	Pesticides
	21			O_{rth}	Orthophosphates
	22			NO_3	Nitrate
 [12]: soil sensors*	23	***Soil***	General indicators	T_s^*	Soil temperature
	24			VWC^*	Soil moisture (volumetric water content in soil)
	25			EC_s	Soil electrical conductivity
	26		[13]: Special indicators	N	Azote (Nitrogen)
	27			K	Potassium, (Kalium)
	28			Mg	Magnesium
	29		Critical indicators	IW^*	Soil water tension (Irrometer Watermark measures moisture available for plants)
					...

Integrated equipment in smart farming (vineyards)	#	*List of monitored parameters for TZP-UTOPIA-F2F concept*			
		Abiotic component		***Symbol***	***Name***
 [14]: MAS* - remote sensing	30	***Crops***	Remote sensing - Satellite	*EVI*	Enhanced vegetation index
	31				...
	32		Remote sensing - Drones	*NDVI*	Normalized difference vegetation index
	33				...
	34		Remote sensing - GRS	*NDVI*	Normalized difference vegetation index
	35				...
 [15]: other* sensors	36	***Other***	General indicators	LW^*	Leaf wetness (Meter Group PHYTOS 31)
	37			pH_v	pH
	38			SO_2	sulfur dioxide
	39			*Rs*	reducing sugars
	40		Special indicators	*YAN*	Yeast assimilable nitrogen
	41			*Tsc*	tartrate stability by conductivity
	42			*Bt*	bentonite
	43		Critical indicators	*Sc*	sugar content
	44				...
	45	...	...		...

**See Reference.*

Table 1.
List of the usual parameters monitored in the SF vineyards.

monitoring the list of parameters in **Table 1**. The multitude of sensors in the field must ensure a minimal list of monitored parameters of the culture through the processing of which in the DSS software module results in those VRA operations when the situation in the field requires it.

3.5 Originality of the proposed approach of article

The originality of the approach proposed in this paper is the definition and calculation of the quality global indicator TZP of the agricultural environment toward zero-pollution within the processes of agricultural crops as a basic principle of precision farming.

The TZP principle – Toward Zero Pollution is the principle of supervisory control of inputs (air-water-soil) and outputs in a horticultural process to strive for "zero-pollution" through advanced management of inputs/outputs using MAS systems. In the following, a combined global indicator function is defined, which provides an index of the degree of pollution of the three abiotic components of the air-water-soil agricultural environment. This is represented by a matrix function with the index "TZP" and (f_q) variables, randomly determined at the t time throughout the T period of a horticultural process:

$$\left[F_{TZP}\left(f_q,\ t\right)\right] \tag{1}$$

Only the monitored values of the polluting parameters of the abiotic environment will be taken into account. Carrying out the successive replacements according to the stages described in the paper [12], for a work variant adapted to the horticultural process in the vineyard, the following TZP global indicator function is obtained:

$$\left[F_{TZP}\left(f_q,\ t\right)\right] = \left\| \begin{matrix} \left(\prod_{i=1}^{l} x_i(t)\right)^{\frac{1}{l}} & 0 & 0 \\ 0 & \frac{1}{m} * \sum_{j=1}^{m} y_j(t) & 0 \\ 0 & 0 & \frac{\sum_{k=1}^{n} \omega_k \bullet z_k(t)}{\sum_{k=1}^{n} \omega_k} \end{matrix} \right\| \tag{2}$$

The attached determinant to the global indicator function $[F_{TZP}]$ take values in the [0,1] range. A subunit value in the vicinity of "0 + 0" indicates that process in the vineyard is environmentally friendly.

$$det.\left[F_{TZP}\left(f_q,\ t\right)\right] = \left|F_{TZP}\left(f_q,\ t\right)\right| \overset{t=T}{\rightarrow} 0 \tag{3}$$

where: $t \in (t_0, T)$, T being the period of a vineyard production process (grapes maturity period).

The SCADA software program in LabView code automatically creates the color graphic representation shown in **Figure 5** required for product packaging labeling and grapes promotion in the F2F value chain [3].

3.6 Minimum devices and mandatory equipment for Smart-SCADA core in vineyard SF

A complete list for equipping a pilot station as a smart vineyard farm is a complex problem because it depends on the following factors: economic-financial, geographical, climate, agrarian policy of the vineyard farm. It's shown in **Figure 6** the specification of the first device necessary for monitoring the farming environment in vineyard SF. Mobile app for remote crop monitoring with Meteobot provide, [16]:

- Current and historical data from weather stations;
- Local weather forecast for 10 days;
- Agronomic indicators such as rain sum, temperature sum, etc.;
- Weather notification for frost, intensive rain, etc.

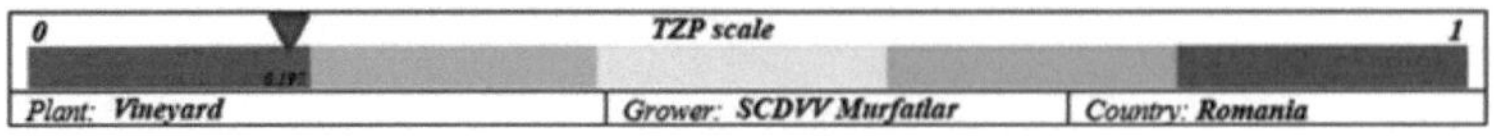

Figure 5.
Graphic representation of the TZP quality global indicator.

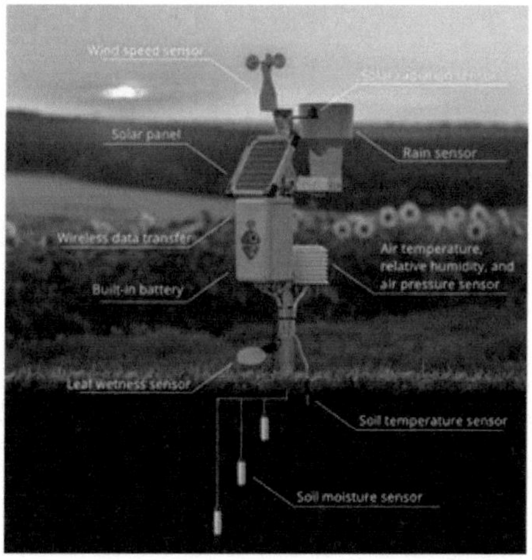

Data specification equipment.
Complete agrometeorological weather information:
- Rain;
- Wind;
- Soil moisture and soil temperature;
- Air temperature, air humidity and air pressure;
- Leaf wetness and solar radiation (options);
- Upgradable upon request;
- Weather history for each field.

Figure 6.
Agricultural farm weather station Meteobot®Pro model, [16].

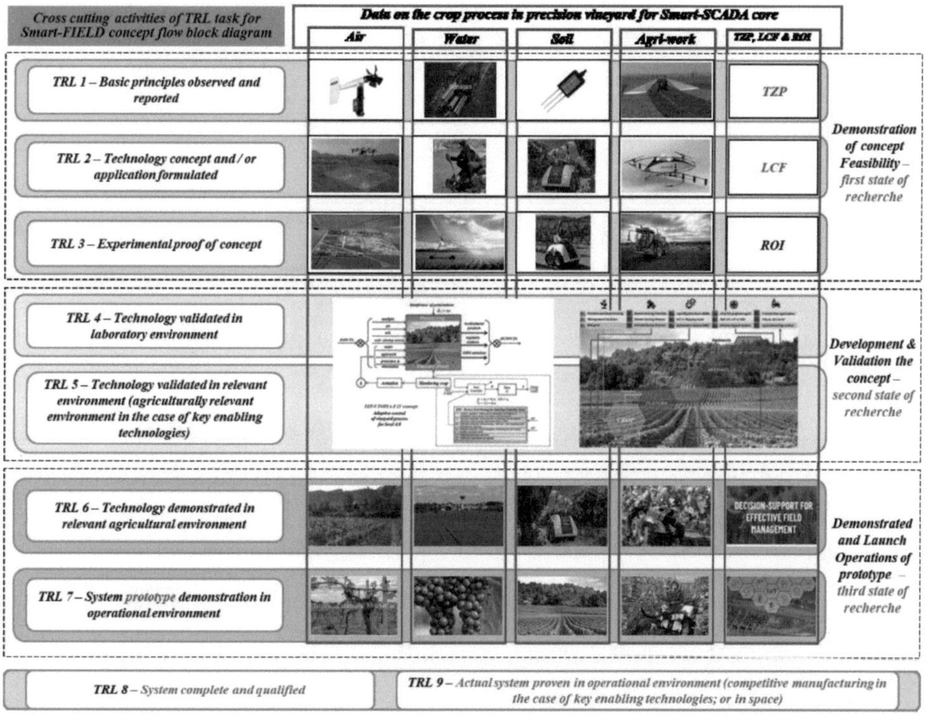

Figure 7.
Cross cutting activities of TRL tasks for the Smart-FIELD and flow to prototype.

3.7 Implementation the Smart FIELD to prototype level in vineyard SF

Realization of the TZP-UTOPIA-F2F concept brings an extra novelty through the hard- and software elements created for digitization of vineyards SF. Complete automation of farming processes in vineyard SFs through centralized digital management from the farm PC of UAVs, GRS (as MAS) and of execution elements (for plowing, arable land preparation, sowing, irrigation, protection, stimulation, harvesting, transport and storage), makes easier activity of the farm manager in making decisions regarding ROI. The cross-cutting activities for TRL tasks of Smart-FIELD within the TZP-UTOPIA-F2F concept are shown in **Figure** 7, [3].

4. Conclusion

The Smart-SCADA core model of TZP-UTOPIA-F2F concept to be developed and tested in real Pilot farm (MUR) as an easy method and **S**mart technique to **F**ostering **I**nnovativ**E** **L**earning for **D**igital agriculture-4.0 (Smart-FIELD) is built on four pillars as basic principles: (i) monitoring the NDVI & EVI vegetation index of crop; (ii) TZP - toward zero pollution of crops process; (iii) ROI - Return on Investment, (iv) Transparent traceability of agri-food products between the F2F for the end user and other stakeholders. The basic Smart-SCADA package and AI scalable software modules are based on deterministic mathematical models that define the stated principles.

Smart-FIELD – Easy techniques to promote innovative learning for digital agriculture toward Zero-Pollution, will be addressed through applied research in the MUR Case Study within the TZP-UTOPIA-F2F project, as integrated smart actions in the SF vineyard. Transparent traceability of agri-food products between Farm to Fork for the end user and other stakeholders will be guaranteed for grapes, through the veracity of data monitored and transmitted from the smart vineyard. Web Smart-SCADA core

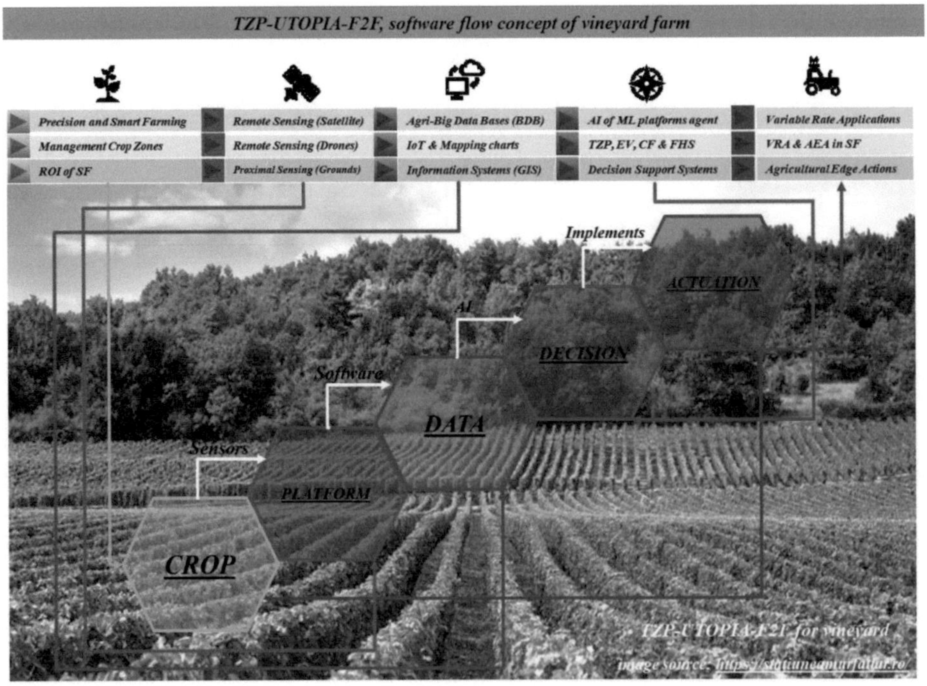

Figure 8.
Smart-SCADA core architecture of flow process from CROP to VRA actuation for data transparency in F2F chain value.

software flow of TZP-UTOPIA-F2F concept for SF vineyard as future research directions in agriculture-5.0 is shown in **Figure 8**.

Acknowledgements

We thank the fellow co-authors of this paper for this collaboration. We also thank the entities to which we are affiliated for the trust given to the idea and concept presented in the work and its development in a European project (e.g., ICT-AGRI-FOOD Call).

Conflict of interest

"The authors of this paper declare no conflict of interest."

Acronyms and abbreviations

AI	Artificial Intelligence
LCF	Low Carbon Footprint
CPS	Cyber-Physical System
MAS	Multi Agent Systems
DSS	Decision Support System
NIR	Near InfraRed
EV	Edge Vision
NDVI	Normalized Difference
EVI	Enhanced Vegetation Index
FHS	Food Health Safety
ROI	Return on Investment
F2F	Farm to Fork
SCADA	Supervisory Control and Data
GPS	Global Positioning System Acquisition
GRS	Ground Robotics System
SF	Smart Farm
IoT	Internet of Things
SMF	Small and Medium Farm
SRS	Smart Robotic Systems
UAS	Unmanned Aerial Systems
TZP	Toward Zero-Pollution
UAV	Unmanned Aerial Vehicles
TZP-UTOPIA-F2F	Toward Zero-Pollution concept in Farm to Fork
VRA	Variable Rate Application

Author details

Constantin Vîlcu[1*], Steve Vanlanduit[2], Resul Kara[3], Aurora Ranca[4] and Mohamed Bourouah[5]

1 National Institute of Research-Development for Machines and Installations Designed for Agriculture and Food Industry – INMA, Bucharest, Romania

2 University of Antwerp, Faculty of Applied Engineering, InViLab Research Group, Antwerp, Belgium

3 Düzce University, Faculty of Engineering, Department of Computer Engineering, Düzce, Turkey

4 Research and Development Station for Viticulture and Winemaking Murfatlar, Constanța, Romania

5 Hahn-Schickard-Gesellschaft für Angewandte Forschung e.V., Villingen-Schwenningen, Germany

*Address all correspondence to: constantin.vilcu@inma.ro, constantin.valcu57@yahoo.com

References

[1] Saiz-Rubio V, Rovira-Más F. From smart farming towards agriculture 5.0: A review on crop data management. Agronomy. 2020;**10**(2),207:1-21. DOI: 10.3390/agronomy10020207

[2] Ranca A, Artem V, Dina I, Parcalabu L, Iliescu M, Tabaranu G, et al. Experimentation of grapevine cultivation in organic system, on five different Romanian vineyards. Bulletin of UASVM-CN – Horticulture. 2016, 2016;**73**(2):175-181 Available from: http://journals.usamvcluj.ro/index.php/horticulture/article/view/12206/10047

[3] Vîlcu C, Vanlanduit S, Kara R, Ranca A, Bourouah M, Savu T, et al. Toward zero-pollution, an edge utopia in vineyard field. The TZP-UTOPIA-F2F concept: Digitally promotes the transparent path of grapes in the Farm to Fork value chain; Part I: Smart-FIELD of the farm, first relevant link of transparent path of grapes in the farm to fork value chain. A revue. In: Proceedings of the ISB-INMA'TEH 2022 International Symposium; 6-8 October 2022. 2022. p. 5. Available from: http://isbinmateh.inma.ro/archive.html

[4] Sels S, Ribbens B, Vanlanduit S, Penne R. Camera calibration using gray code. Sensors. 2019;**19**(2),246:1-11. DOI: 10.3390/s19020246

[5] Şentürk A, Kara R. Low-cost air pollution monitoring with wireless sensor networks. Journal of Multidisciplinary Engineering Science and Technology. 2017;**4**(10):8497-8505

[6] Caba I-L, Marin A-M, Voicu G, Maican E, Duţu I, Brăcăcescu C, et al. Precision agriculture, the role of agridrones for a sustainable farming environment. A review. ISE – Agriculture & Food, The 9th International Conference, ISE-Burgas, Bulgaria. Vol. 9. 2021. pp. 303-315. Available from: https://www.scientific-publications.net/en/article/1002170/

[7] Maican E, Vladut V, Vilcu C, Sorica C, Dorian M, Mirea D-P, et al. Hybrid renewable energy systems for isolated farms. A review. INMATEH Agricultural Engineering Journal. 2019;**59**(3). DOI: 10.35633/inmateh-59-09

[8] Vladuţ V, Caba I-L, Constantin A-M, Voicu G, Maican E, Vîlcu C, et al. Principles of integration the agri-drones in agricultural production environments. New concepts towards agriculture-5.0. Published 2021-12-29, Issue Vol. 51. No. 2 (2021), Annals of the University of Craiova - Agriculture, Montanology, Cadastre Series, Section Working Group 4 - Management, agriculture mechanization and cadastre, Craiova, Romania. 2021. Available from: https://anale.agro-craiova.ro/index.php/aamc/article/view/1269

[9] Gutiérrez F. Special Issue Application of Decision Support Systems in Agriculture. MDPI, Digital Agriculture, 31 May 2022. 2022. ISSN 2077–0472. Available from: https://www.mdpi.com/journal/agriculture/special_issues/Decision_Support_Systems_Application. [Accessed: 20221027]

[10] Air Sensors. Available from: https://www.baranidesign.com/weather-station-packages. [Accessed: 2022-10-17]

[11] Water Sensors. Available from: https://www.ysi.com/applications/stormwater. [Accessed: 2022-10-17]

[12] Soil Sensors. Available from: https://precisionagricultu.re/soil-sensors-the-new-part-of-precision-agriculture/. [Accessed: 2022-10-17]

[13] Wallor E, Bourouah M, Kersebaum KC, Gebbers R. Assessing sampling strategies and soil sensors performance in the detection of field scale variability of plant-available nitrogen. The 12th European Conference on Precision Agriculture in Montpellier, ECPA. 2019. pp. 571

[14] MAS – Remote Sensing. Available from: http://w.sensilize.com & https://www.debate.com.mx/losmochis/Drones-agricolas-revolucionan-la-agricultura-regional-20190623-0033.html. [Accessed: 2022-10-17]

[15] Vineyard Sensors. Available from: https://hannainst.ro/brands/wine-line. [Accessed: 2022-10-17]

[16] Agricultural farm weather station Meteobot®Pro model. Available from: https://meteobot.com/en/weather-stations/. [Accessed: 2022-10-27]

Chapter 13

Perception of Soundscape in Landscape

Banu Chitra Mookiah

Abstract

Landscape or nature-related design is mostly focused on the visual aspects. As a result of increased urbanization, the exploration of the possibilities of a tranquil environment concerning landscape enhances the positive effect on the quality of the urban population. This can be achieved by understanding and characterizing the sonic environment in such a setting. In connection to the visual landscape aspects, acoustical cognition in terms of perception of the environment is important. In modern scenarios; the concept of soundscape is used to discuss the quality of the environment. In line with this, the chapter reviews the state of the art of literature on various definitions, perceptions, and theories of landscape in conjunction with the theories of the soundscape and classification of soundscape elements. Also, it reviews the soundscape dimension in the landscape through the function of green spaces and its impact on quiet/tranquility in an urban context by understanding the role of natural and manmade landscape elements on sonic perception along with various methods of data collection commonly used for soundscape research.

Keywords: landscape, landscape perception, soundscape, sound mapping, urban environment

1. Introduction

The improvement of the soundscape concept evolves with the concept of human beings who perceive the world in a multisensory manner [1]. Human beings can communicate to the world through the five senses seeing, hearing, smelling, touching, and tasting. However, sounds environ us everywhere. With the rapid growth of urbanization in the 20th century, urban communities are discontented with the quality of the urban environment, especially with the urban acoustic environment. In [2] performed a pioneering field study of the urban soundscape in a sector of central Boston, including several subjects, and tested the perception of sounds and sights. His study strongly suggested a need for sonic planning and designs [3]. The [4] defines soundscape as an acoustic environment as perceived or experienced and/or understood by a person or people in a given context. The soundscape is usually referred to in terms of identifying and describing diverse sound sources in a particular place. In general, the soundscape is the combination of all sounds within a given location with an emphasis on the relationship between an individual's or society's perception [4]. Schafer explained the control of visual aesthetics in present societies, where a series of hearing exercises intended to create sonic awareness among people through field studies were carried out. These studies were carried out using sound measurements, soundscape recordings, and portraying various sound features. This interest directed him and a few of his soundscape colleagues to invent a few terms as

Keynote: Sounds that are continuously heard in the given location. This is also called background sound (the sound of the train in the railway station).

Sound Signals: Sounds that attract the attention of the people. This is also called foreground sound (e.g. announcements in the railway stations about the train timings, for people to listen)

Sound marks: The sound which serves as a landmark for a particular place.

Lo-fi: The sounds that cannot be heard properly due to the masking of other sounds, which acts as a disturbance

High- fi: Sounds can be distinguished since there is only little masking by other sounds.

In the process of understanding and defining the soundscape environment of a given place, the sensitivity of the people and their preference for a sonic environment decide the overall soundscape quality of that place. As per the studies, with the increase of age people tend to have more inclination toward sounds relating to Nature and human activities whereas the young crowd was tolerant towards mechanical sounds and loud music. The preference criteria for different sounds add to the character of a space. Further, it is also observed from the studies that these sounds have a strong connection with the people's landscape preferences, particularly in the absence or presence of desirable and undesirable sounds, more than in the acoustic environment. In most cases, landscapes of these spaces are designed whereas soundscapes are not designed. Ren et al. [5] in their study explored that there is a strong connection between the preference for soundscape and landscape elements.

2. Characteristics of sound

Sounds are caused due to the vibration created by the motion of a source. The more sounds experienced by people are transmitted through the air [6]. The vibrations created cause oscillation in air pressure which results in sound waves that a human ear can detect [7]. Sounds are considered to carry information from the environment. It acts as one of the major components to communicate with our surroundings [8]. Sound also holds the ability to awaken the emotional response of people both in positive and negative ways. There have been various studies on the various categories of sound based on sound sources [9–12]. The general three categories of sound source classification are natural, technological, and human sounds [12] as shown in **Figure 1**. Based on the perception of the people the sounds from natural sources are considered to be pleasant whereas sounds from technological sounds are considered to be unpleasant and the sounds from human beings are regarded to be tolerable [14–16]. In addition to this, the perception or preference is connected to other physical components such as a source of the sound, the context of the sound, and personal preference [17]. Constant exposure to unpleasant sounds or noise can harm the auditory system along with other nonauditory health effects such as cardiovascular disease, sleep disturbance, annoyance, hypertension, etc. It also leads to mental problems [18]. The constant noise exposure also reduces the quality of the environment [19]. In addition to negative effects on health. As per the World Health Organization [20], there are 1.0 and 1.6 million Disability Adjusted Life Years

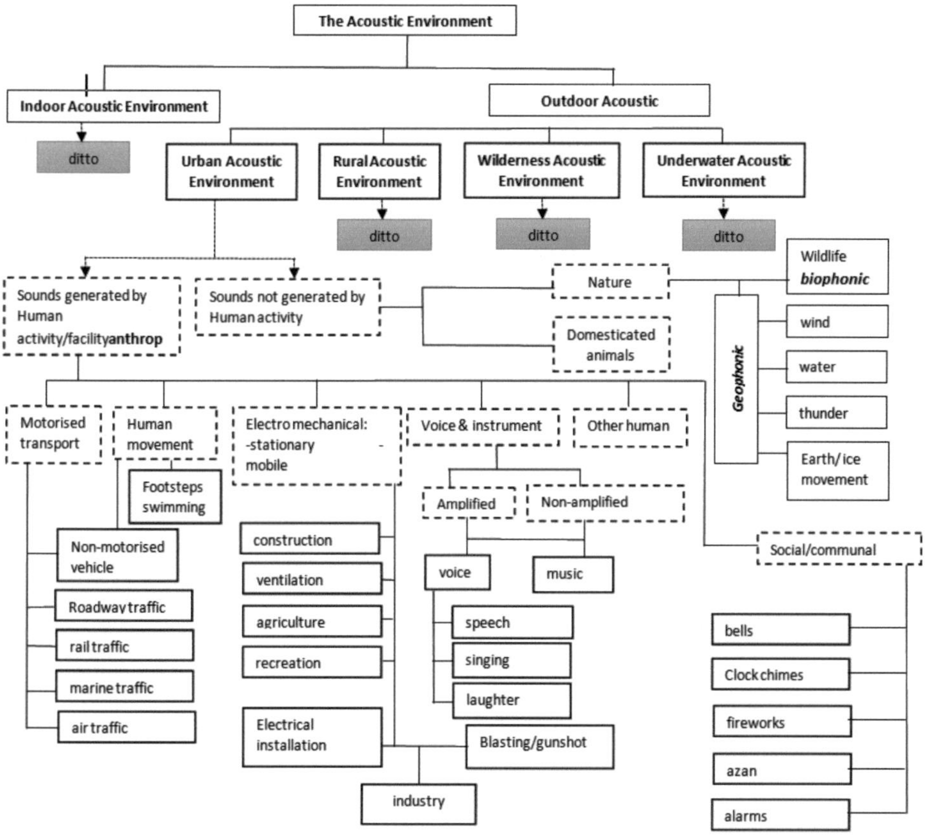

Figure 1.
Taxonomy of sound source classification (Source: [13]).

(DALYs) annually due to the harmful effects of noise exposure. However, to act as a relief there are a growing number of studies that show that natural sounds such as birds chirping, water sounds, etc, can be used to reduce the stress level [16, 21].

2.1 Theories and concepts of soundscape

Several scholars proved that nature sounds could act as a healing medium in connection to anxiety and stress [22–24]. As an essential component, to relieve the people from the negative impact of urbanization the concept of the soundscape was initially developed by the geographer J. G. Grano in the year 1929 who explained the shifting of sound from animal sound to mechanical sound in the agrarian landscape [25]. In 1969, Southworth carried on the study of the soundscape with different subjects on the perception of sounds and sights [2]. Further his study strongly recommended the need for sonic planning [26]. The Canadian musician Murray Schafer in the year 1977 coined the term soundscape in his book Tuning of the world with the summary collected in his World Soundscape Project [27]. The project by Schafer and his colleagues recommended the approach, understanding, and development of soundscape by introducing many methods and concepts which form the rationale for soundscape study. Followed by Schafer and Truax in the year 1978 study on the terminological dictionary and also introduces a series of distinctions and conceptualizations that builds the relationship between soundscape and noise research [28]. Moreover an organization was formed in the year 1993 to coordinate soundscape research is known as World Forum for Acoustic Ecology (WFAE). In the year 2014, the International Organization for Standardization (ISO) defines the soundscape as the "acoustic environment as perceived or experienced and/or understood by a person or people, in context" [3]. Around the 1990s, the focus of noise studies shifted toward soundscape perception. Generally, the human reaction to noise is related to acoustical factors such as sound pressure level. In contrast, non - acoustical factors such as expectations of the users, context, etc also play a vital role [29]. The soundscape approach to noise includes a qualitative perspective which is very constructive from the user's perspective. As mentioned earlier, the city sound experience was first experimented with in the World Soundscape Project in the 1960s. Southworth, a city planner in the year 1969 investigated the character of the city sounds with visual aspects. He suggests the four strategies to deal with city sounds based on planning as cautious positioning of noisy activities, new kind of street design, unique vehicular design and most importantly masking of unwanted noise by addition of wanted sound [2]. Later in the 1970s, the research institute CRESSON highlighted the relationship between sound, people, and the environment. The most notable publications of CRESSON are Sonic Experience – the wanted and unwanted sounds which can be seen from the sound quality perspective of open spaces. Another approach [29] proposes two ways of observing sound's sonic activities and spatial aspects. They also suggested the desirable decibel limit of 65-70 dBA above which introduction of new sounds is not appropriate unless the reduction in already existing sound pressure level is exercised. Further, there have been suggestions to introduce the map to illustrate how soundscape can be integrated into the planning process of the UK [30] which was later incorporated by De Coensel et al. integrating various activities such as sound walk and soundscape simulations [31, 32] created a framework to improve soundscape with three stages describing soundscape, factors influencing soundscape, and relating possible design interventions to user's perspectives which were evaluated through the Kano model.

2.2 Tools and techniques used for soundscape studies

There have been various tools and techniques to conceptualize the soundscape to understand the role of the soundscape in experiencing the environment which serves as an influential tool in the assessment of soundscape. One of the models developed by Hedfors in 2003 is the model of prominence which gives the distinction between the total of prominent sounds to the total of background sounds in a given environment [33]. This model characterizes the soundscape as powerful, mild, clear, and crowded to describe the sonic environment. Axelsson et al. included the two dimensions of "pleasantness" and "eventfulness" to understand the variations in the sound environment based on the listening experiment with one hundred listeners. Later a principal component analysis model was developed with the inclusion of soundscape characteristics such as exciting, monotonous, calm, and chaotic [14]. Jennings et al proposed an illustrative framework to describe soundscape perception that considers aspects such as direction, proximity, foreground, and background of the sonic environment [34]. A system based on the evaluation of soundscape in urban open spaces was suggested by Zhang & Kang based on the source, space, people, and environment [29]. Later a detailed taxonomy was proposed to understand the relationship between environment and types of sound sources [35]. Another model was introduced by

Herranz-Pascual et al. to seek the interaction between person, activity, and place in the environment [36]. The approach suggested a broad way of identifying sound sources and improving the soundscape which is based on the identification of wanted and unwanted sounds and the possibilities to control them. Based on the urban scholar and sound artist [37] a tool was established for urban sound design named 'Sonic rupture' which was centered around five approaches for designing addition, subtraction, disclosure, passion, and transformation.

2.3 Methods of data collection in soundscape research

There are various methods for the collection of soundscape data as sound walks, Listening tests, behavioral observations, and narrative interviews

2.3.1 Listening test

The method of listening tests is generally carried out to perform the sound evaluation under the controlled condition without which it may be affected by the external parameters [19]. Generally, this method is broadly used to evaluate the influence of visual aspects without the support of visual material [38]. For carrying out the study, a replica of the outdoor environment is usually created. It gives more focus to the participant's response in response to the sonic environment [39]. This data collection can experiment with a group that has no prior experience and with a varied range of participants [39, 40] who can be recruited through the mail [39, 40]. However, the interpretation of the data collection requires a significant determination of the hearing capacities of the participants.

2.3.2 Interviews

This method is commonly used in assessing the sound environment related to sound quality. This is widely used in evaluating environmental noise and other noise pollution [41]. This method is normally conducted with the interaction of users of the space from the micro-level as residents of the space to the macro- level users of the space as urban space as parks, plazas, squares, etc. This includes the investigation of sound quality along with different parameters based on the objective of data collection [42, 43]. The data collection using interviews was explored in various studies [44–48]. The various parameters used for the interview method was adopted on the basis of soundscape expectation [49, 50], soundscape description [51], soundscape preference [52, 53], soundscape memories [49], soundscape perception [51].

2.3.3 Sound walk

The sound walk method implies the method of conscious listening to the sound environment which helps in exploring the multimodal aspect of the surrounding [54]. This method is carried out through both qualitative and quantitative data collection [16, 46, 47]. As per quantitative data the data collection procedure it varies concerning the measurement (duration, measurement, seasons, etc) and varies with a collection of data based on the various objectives and for qualitative data, it differs with the size of the sample, duration of the sound walk, size of the participants, etc [55, 56]. Sound walk procedure is generally carried out individually or in groups based on the prefix trail routes using a structured protocol [27, 56]. There have been various studies that have used the sound walk as a tool for the interpretation of soundscapes [57–59]. After the initial learning, most of the scholars used the sound walk as an effective tool for investigating soundscape based on the objective. Many sound walk studies are conducted in groups [30, 58]. While conducting the study it is important to maintain a specific distance between the participants to avoid the effect of footsteps [58]. In contrast to the group sound walk, the individual sound walk can be performed at diverse times and days based on the objectives [59, 60]. In the 1970s the soundscape took place in urban, rural, and different locations. However, recent studies focus mainly on the urban contexts especially urban parks, urban squares, and urban streets [30, 57, 58, 60, 61]. The study of the sound walk is generally conducted in locations where there is a broad range of sound sources [58]. The range of context varies based on the objective of the study.

2.3.4 Focus group

Focus groups are the additional modality of data collection in which the aim is to facilitate discussion based on the specific issue put forth by the scholar. A certain topic of discussion comes out with an unbiased opinion. This method engages the reflective state of mind of the participants about their previous experience of soundscape based on the discussion. It also helps the participants to express their ideas with the agreed response from other participants [46, 62].

2.4 Methods of mapping in soundscape research

There have been various methods of mapping soundscape based on the data collected on the measurements and these maps are found to be an effective tool in the assessment [63]. The noise maps that presently exist and that are recommended by EU Directive are in 2D. For the creation of urban sound maps, the model which is based on numerical methods is broadly used which provides quality inaccuracy [64]. However, these maps have considerable limitations in terms of both sound sources and the dynamics of a sound environment. Sound maps created based on sound measurements helps to improve the mapping of the soundscape [65–67]. Various applications allow participatory sensing to increase the potential of mapping the sound environments in the city based on measurements [68–71]. To know the time and spaced relationship concerning such measurements, certain knowledge of the interpolation methods can be created to produce sound maps that show the spatial and temporal aspects of the sound [72–74]. Studies show that shortening the time of recording proved the 15 min sampling period as relevant [74]. The shortening recording time of 5 min was also found in the literature; however, it should be compensated by a large number of measurements [69]. The dynamics in the urban environment can be explained with the help of spatial characteristics of the environment. He also mentioned that the representativeness in the space based on spatial interpolation of the sound environment is a very important factor. Several types of research on describing the methods have been explored using urban sound level interpolation as Kriging methods, multi quadratic interpolation [67, 75], and (IDW) methods [17, 67, 75]. There have been various studies that created the interpolation of sound maps through fixed sound measurement stations which provides a useful insight for the city level [75, 76]. However, this is not possible as the distance of the measurement station is more and these can be explored by model-based methods. Various studies have been suggested for model-based sound maps for measurements suggesting that spatial interpolation methods can also be based on perceptual assessments [65, 77, 78].

3. Theories and concepts of landscape

The origin of the landscape was traced back to the 5th century A.D by J.B, Jackson in his book "Discovering the Vernacular Landscape" [79]. In the fifteenth century, the reasonable depiction of the landscape which emphasizes the visual character and symbolic meaning was found in Renaissance paintings [80]. There are significant references that define the landscape as a picture representing natural inland scenery [81]. According to the landscape convention of Europe, Landscape is defined as an area perceived by the people, whose character is the result of the action and interaction of natural and/or human factors' [82]. This landscape became an expression of human thoughts, human beliefs, and human intuition or feelings. Later it was incorporated into different parts of the world. The original intent of the word was to define a specific parcel of land, and later a particular bounded scene as an object for painting. As per Jackson the word 'land', means a bordered territory and also refers to soil and territory, and as 'organized land' based on the characteristic of the people who made it. Landscape expresses the (visual) expression of territorial identity. It also refers to subjective observation and understanding [83]. In general, the varied definitions imply that landscapes are the portrayal of surroundings and aesthetics. Moreover, there is various understanding through definitions mainly to understand the process of human interaction and to understand the spatial dimension of the surroundings.

3.1 Landscape – A cross-disciplinary overview

The term landscape is used by a varied range of scientific disciples retaining different definitions. In broad terminology, the landscape is defined as the "total character of an area of the Earth" [84]. However, there are dynamic perspectives for the term based on different disciplinary contexts as shown in **Table 1**. According to the perspective of art, landscape images are the representations of landscape, e.g., as drawings, paintings, or photographs, which are the impressions or illusions which evolve in the mind of the observer as two-dimensional objects with added colored stains [86]. More importantly, landscapes are defined as the perception as demonstrated within the images, however as per archeology landscape was perceived as a backdrop or a setting that is characterized by the interpretation of the artist [87]. In general, as per the observations from the historical perspectives, the landscape is considered in terms of cultural aspects and is interpreted as a cultural landscape that claims landscape as a result of human actions over time which gained its importance in the 1990s. Landscapes that are engaged by people through which the identities are created. Hence as per ecology landscape is defined as the investigation of the human population which induces the changes in the heterogeneity of ecological and landscape components. Geographers investigate landscape in terms of the region with an integrated spatial view followed by region and zone and also different factors such as topography, land use, etc. While the landscape is a representation of a relatively smaller unit as per geography, it serves as the representation of the earth's surface which makes it a more powerful concept for geographers [88]. In explaining the "Principles of Geology" Charles Lyell describes the landscape as the result of external and internal factors acting upon the structure of the earth's surface. Similar to Geographers, Geologists who describe the landscape as a geomorphologic process stumbled upon in a certain environment. The human action of the landscape where the natural landscape was transformed

S. No	Various disciples	Given the meaning of landscape
1	Geographer	Landscape as features in an area
2	Historian	Landscape as a record of history
3	Architect	Landscape as townscape
4	Academic	Landscape as the analysis of meaning in the environment
5	Ideological	Landscape as an expression of property ownership
6	Landscape architect	Landscape as an object
(Source: Edward Relph, [85])		

Table 1.
Functional classification of 'landscape' meaning.

into anthropized space as a result of human's fight against space which leads to the exploitation of environment. To conclude according to historian's landscape is an analysis of complex interaction between humans and the natural environment and the investigation of the same.

3.2 Landscape perception

"Landscape is composed of not only of what lies before our eyes but what lies within our heads." [89]. Generally, perception is the process of deriving information through the senses which are an active process between organisms and the environment [90]. The perception of the environment helps us to understand the environment diversely. People usually interact with their environment for a purpose. As a result, we select spatial information related to our purpose [91]. There are two basic modes of perception auto-centric and allocentric as the first one deals with subject-oriented and the second one deals with object-centered [92]. He explains that the components of auto-centric involve the sensory quality whereas the allocentric deals with attention and directionality. The terms of perceiving the physical environment involve not only physiological phenomena by both social and cultural factors as the perception changes based on the individual experience [93]. Hence the perception of our surroundings is dynamic based on the individual [55]. In terms of the perception of the environment landscape architects must play a crucial role in comprehending this relationship.

3.2.1 Visual landscape perception

Although spatial information is received through various senses such as a sense of smell, touch, auditory, etc the sense of sight is assumed to be the most valued sense. Almost 80% of our sensory inputs are perceived visually [92]. Hence in the environmental assessment studies, the evaluation of visual landscape character is mostly taken into consideration [94]. Bourassa 1990 states the two principles of landscape aesthetics as biological and cultural in which biological aesthetics deals with the aesthetic contentment obtained from refuge or prospect whereas cultural aesthetics deals with the aesthetic contentment obtained from the landscape that contributes to cultural identity. The concept of aesthetics has evolved from history based on philosophical context.

3.2.1.1 A brief history of aesthetics in the philosophical context

The picturesque beauty of the landscape serves as a concern for the assessment of the visual quality of the landscape and its preferences. The concept of aesthetics in environmental psychology gained its importance in ancient times. The term aesthetics was coined by a German philosopher [95]. The word aesthetics is derived from the Greek word "aisthanesthai" which represents "to perceive" and aisthet which means "perceptible objects". Various philosophers connected beauty to truth and symmetry [96]. It was believed that beauty is linked to good and ethics [97]. Aristotle on the other hand, argued beauty in perspective to mathematics [98]. He insisted that beauty was associated with size and order which is carried out in Rome and Greece where the approaches towards landscapes are connected to order, symmetry, proportion, and balance which rose to a concept of classism. Apart from this, the concept of modern aesthetics defined beauty as a certain composition of colors and Figures giving happiness to the beholder which was subjective [99]. Beauty beyond the expression of pleasure or joy was perceived as the perfection of sensitive cognition based on intellectual directions [100]. Landscape beauty on the other hand is related to our social and natural structure [101]. In the 19th century, beauty was perceived in terms of romanticizing Nature. During this period landscape was viewed as an object which has some intrinsic qualities. However, the perception of objectifying landscape changed during the 20th century as the landscape is perceived as relevant to its subject which is connected to the people's experience [102]. Mainly in landscape two approaches of visual landscape assessment were observed as subjective and objective. Subjective assessment is the assumption of visual quality in relevance to the observer whereas the objective approach deals with the characteristics and physical characteristics of the environment. Four various prototypes involve the perception of visual assessment of landscape as experts, psychophysical model, experiential model, and cognitive model

3.2.1.1.1 The expert's prototype

This is a paradigm based on the judgment of the experts based on the visual quality of landscapes. These are based on the characteristics of landscapes as vegetation, texture, color, landform, etc, and based on this model the natural ecosystems gain their aesthetic value. The disadvantage of this prototype is its inefficiency to consider the users' perceptions [103]. Moreover, the model was criticized due to its consideration of only experts based on the perception of the visual environment; however, it is mandatory to incorporate the people's opinions about the landscapes which the model failed to do so.

3.2.1.1.2 The psychophysical prototype

In contrast to the expert's prototype, the visual quality of the landscape is evaluated in terms of public opinion in the psychophysical prototype. The techniques used for evaluating this prototype are ranking and categorizing for evaluating the visual quality of the landscape [104]. The main objective of this model is to perceive the landscape as an external factor without conscious thinking based on its incentive nature.

3.2.1.1.3 The cognitive prototype

The cognitive paradigm centers on the reason to understand why people prefer specific landscapes which are theoretical approaches. This prototype differs from the psychophysical paradigm as it influences aesthetic judgment based on the visual quality of the landscape. Mostly tools such as semantic differential analysis and a checklist of adjectives are considered to evaluate the preferences and meanings of such landscapes. Mostly it is based on the evolutionary theories on environmental perception however it neglects the physical environment and focuses on meanings associated with landscapes

3.2.1.1.4 The experiential prototype

The experiential prototype approach is commonly explored by geographers who practice this prototype in "sense of place" studies. This focus on human-environment interaction is based on their experience. The experiential approach focuses more on

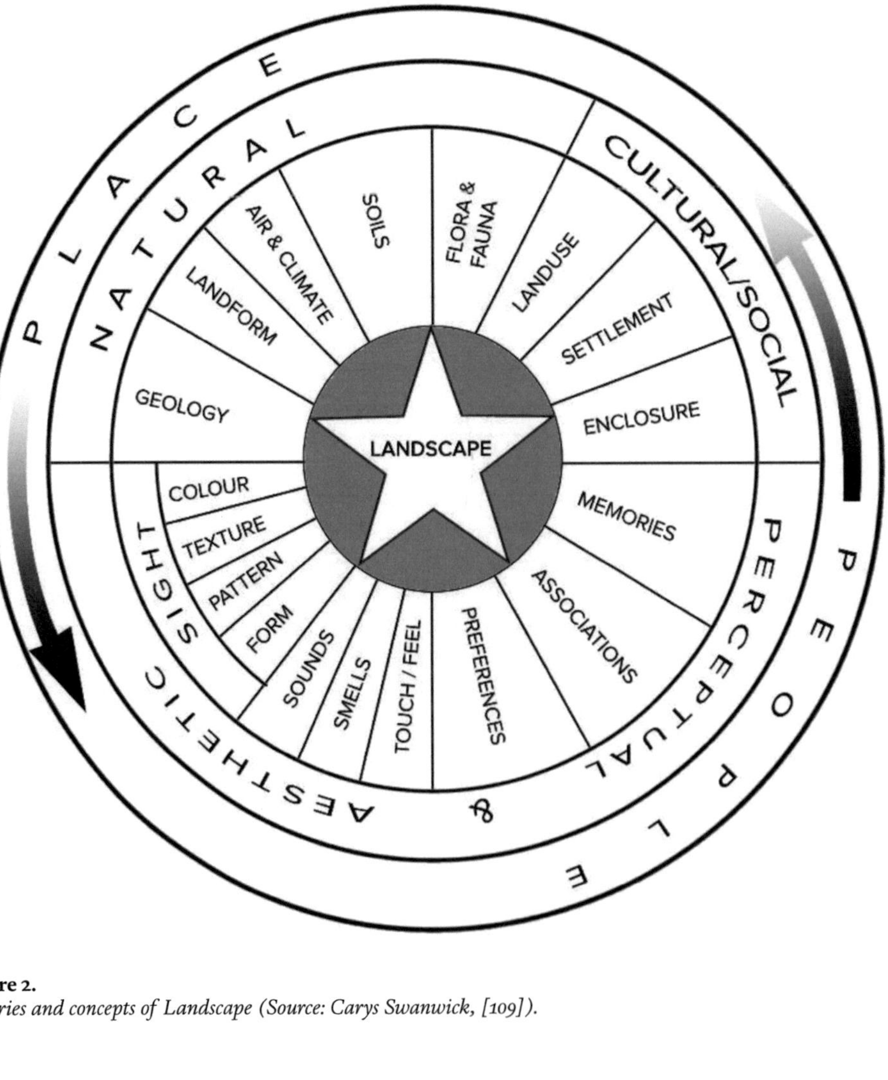

Figure 2.
Theories and concepts of Landscape (Source: Carys Swanwick, [109]).

subjective than cognitive and psychophysical paradigms. Hence the measurement of the dependability and validity of the results is tough. The prototypes classified contribute to the overall comprehension of environmental perception. However, making use of the public as the center of the prototype is essential [105] assessed that landscape architects and geographers intend to use a theoretical and a psychological approach in landscape assessment while psychologists use mainly theoretically derived psychometric methods. Hence it is very much essential to relate different disciples to resolve the conflicts.

3.3 Significance of landscape

Landscapes have a significant role in enhancing the human experience which has a wide range of theoretical perspectives [106]. This landscape serves as a cultural image in which it represents the pictorial way of representing human surroundings [107]. As a result of new concepts and approaches, the landscape ensures a remarkable change [90]. Landscape integrates the dynamic and functional relationship between the components which serves as an expression of an ecological, economic, and social organization [108] as shown in **Figure 2**.

Landscape helps us to perceive and depict the importance of its resources through conservation and enhancement of the same [110]. In general, the landscape originated from the concept of Nature's philosophy in which human beings are considered as a part of nature which is described as ecological humanism.The landscape can be seen as physical geography which is perceived as a concept of space rather than trying to create new boundaries [111]. The authors insisted that landscape can also be perceived as archeology and human geography in which the allocation, distribution, and people were compared and incorporated with the background information of the natural environment. Landscapes are argued as "symbolic environments" which are created by human acts that confer meaning to the environment [112]. Landscape can also act as social constructions that are mediated through the collective human experience as it emerges the people with the social setting with the engagement of people with the material world.

4. Role of green spaces in an urban context

As a result of growing urbanization, the configuration of recreational space in connection to environmental problems such as noise has become a challenge since roads and traffic sounds are inevitable in the urban context. Green spaces have an important role in addressing these issues as they can increase the health and well-being of the citizens in the urban context. They also had the potential to provide habitat for biodiversity, especially in urban parks. Moreover, these spaces bring tranquility to the urban context which is evident in many studies [113] made a specific questionnaire in seeking an answer for quietness in the urban environment. In response to that, a large group of participants responded that they visit green spaces to relax. Therefore, visiting quiet places or going outside actually meets a need for quietness in the urban context. Quiet is not a core requirement for such acoustic preference in the outdoor acoustic environment. Core requirements include congruent soundscape and landscape, and dominant wanted sounds in a place over, and not masked by, unwanted sounds [114]. There have been various efforts taken to protect such places. E.g. a decade ago the Environmental noise Directive 2002/49/EC suggested the competent authority implement a noise action plan intended to protect the quiet areas as urban parks. There have been various definitions of quiet areas as quiet areas are those that contribute to the society's culture and have a specific character and it does not represent the absence of sound [13]. In addition to that, the END defined the specification of the quiet area as 55 dB L_{day}. It is observed that the possibilities for quiet recreation in urban environments can have a positive effect on the quality of life in the urban setting [13, 115] worked on the first-ever evaluation of the tranquility of open spaces for the characteristics of both acoustic and visual stimuli. They found that the maximum sound pressure level and the percentage of landscape features present at a location were the key factors influencing the tranquility of the urban context. With this, it is evident that green spaces especially urban parks help in exploring the tranquility of the citizens in an urban context.

5. Soundscape dimension in Landscape

As a result of urbanization, more concerns related to sustainability aspects of the landscape are taken into consideration by various scholars. In the early 1960s and 1970s the concept of sustainability under landscape evolve from the ecological point of view which is stated with various concepts such as urban metabolism [116], landscape urbanism [117], and "design with nature" [118]. In general, the aspects of sustainability include aspects of social, environmental, and economic concerns where it could be agreed upon sound, especially in connection to social sustainability [33, 119]. Due to the increased urbanization nature, related places act as a pause for stress and have been important places for restoration [120, 121]. Especially in an urban scenario, these spaces act as a key element in providing tranquility [12]. The extent of landscape architecture is a diversified one. It ranges from the micro level as the design of individual plant beds to the macro level as planning at the city level. In 1847, Sir Joseph Paxton designed Birkenhead Park in England by reclaiming the existing marshland. Landscape architecture as a profession gained its value in the year 1858, after the winning proposal of Calvert Vaux and Fredrick Law Olmsted for the design of the central park, New York [122]. However, the field had its significance much earlier when Geoffrey and Susan Jellicoe portrayed their first recognized examples of landscape designs in cave paintings in France and northern Spain, which dated back between 30,000 and 10,000 BC [123]. The field gained its significance in the direction of environmental-related approaches with the idea of a few works of notable landscape architects as "design with nature" [118]. McHarg gives an ecological concept that is chained to the environmental services which laid the foundation for environmental sciences. Because of the regional environment, the emergence of multifunctional landscape design has become an important paradigm that addresses varied societal pressures such as population growth, and degradation of the environment [124]. Mc Harg's design-with-nature concept is a precursor for the multifunctional landscape design in which he focuses on social, natural, and cultural processes and establishes the relation between man and nature. Various systems address the integration of nature for multifunctional benefits which is the result of Mc Harg Design with nature concept [118]. The main feature of this system is the integration and facilitation of several ecosystems along with consideration of human interventions which becomes an integral part of the ecosystem [118]. With this evolved the concept of landscape urbanism which is considered to be an aspect of sustainable consideration [117]. As an inspiration from McHargian landscape planning, landscape urbanism focuses on natural science with the conception of ecology as the main theme. It attempts to engage the natural and cultural system which is on par with the McHargian concept which explains the best fit between culture and nature. In his book, design with nature he defined landscape architecture as the process of superimposing and synthesis land information. He also incorporated sound data as a source of noise [33]. It has been proved that the landscape architects are well-thought-out of sound in their experiments. For instance, on the measurement of Lascaux and other caves, [125] found that the image representing animals was placed in such a way where the aural qualities of such caves would reverberate sound which makes it feasible for the animals to come alive. Moreover, the conscious embellishment of garden design features such as water features in Renaissance Italy, suikinkutsu in Edo-era Japan, and mechanical singing birds of oriental gardens were a few interesting examples of how sound was incorporated into the landscape. In line with other aspects of landscape architecture, perception, understanding, and function of sound-space associations have undeniably diverse among practitioners. Soundscape planning a resemblance to landscape planning involves design or management to influence the acoustic environment of a place to improve the human perception of such an environment. These environments are continuously shaped by both social and cultural characteristics of society. In general, the urban environment is an ecological entity that comprises three factors landscapes, soundscapes, and people [57]. Hence, the diverse human senses and the physical environment should be considered together in interpreting urban environments [126].

5.1 Influence of natural and manmade landscape elements on soundscape perception

Landscape factors have been well-thought-out in numerous studies about sound and soundscape perception [2, 115, 127]. People's landscape preferences play a vital role in a significant correlation with the sounds, especially in the presence or absence of wanted and unwanted sounds more than acoustic characters [128]. The classification of sound sources according to [129] is shown in **Figure 3**. Biophony refers to the sounds which are produced by biological organisms such as birds and insects [10]. Geophony refers to running streams, rain, and waves whereas anthrophony refers to sounds produced by humans. Among all these sound sources, anthrophonic sounds are considered the most dominating sounds because of their capacity to produce high energy [129]. It is also considered to mask biophony and geophony sound sources due to its high energy. In broad terms, sound sources mingle and interact with one another.

Biophony involves the sounds which are produced by living organism as birds and insects that are considered to be the most frequent biophonic sound producers [10]. The pitch and frequency of their songs depending on whether they are habitat to the natural or urban environment. In general, some birds in an urban setting have been observed to sing more at the night due to the less intrusive anthropogenic sounds [129]. Correspondingly geophony may have an impact on biophony as wind or rain which suppresses the bird's sound or biophony. In many cases, anthropogenic sounds are considered to be the dominant as they mask the biophony and Geophony due to the strong energy produced. The high influence of these sounds can be mitigated by camouflaging them with the addition of positive sounds because relatively enhancement of pleasant sounds was found to reduce the perceived loudness and botheration of the receivers. In urban scenarios sounds from nature are typically taken as a means

	Biophony	Geophony	Anthrophony
Sounds produced by	biological organisms	geophysics	humans
Examples	Birds, insects, amphibians mammals	Wind, running streams, rain movements of earth, thunder, waves	Machines, vehicles, sirens bells, traffic, music, language
Annotations	signals are complex because they carry information	driven mostly by climate	more common during the daylight hours

Figure 3.
Classification of sound sources (Source: [129]).

to enhance the acoustic quality [130, 131]. Factors such as the presence of trees and natural features contribute to psychological perception by users, even in areas with higher sound limits than those that define as quiet areas. Natural sounds are those which are produced by organisms (biophony) or by the physical environment (geophony) [43]. It is proved from various studies that good soundscape quality areas are those with highlighted natural sounds whereas low soundscape quality areas are those with highlighted technological sounds [132, 133]. These factors cannot be neglected in the soundscape assessment of urban spaces according to [52]. Through interviews, [134] showed that natural sounds such as birds twittering and the wind are the most expected sounds than sounds from road traffic and aircraft fly-over which are considered to be most annoying. Natural sounds and landscape features have a significant part to play role in improving the perception of the soundscape [16, 135].

5.1.1 Role of water body in soundscape perception

There have been various categories of water bodies that have a significant contribution to the enhancement or masking of sounds. In an ecological waterscape, acoustic comfort is an important element associated with landscape experience. The water sounds such as streams and waves of lake sounds were selected as effective natural sounds to mask urban noises. In general ponds, lakes, pools, and puddles fall into the still-water category whereas waterfalls, rivers, brooks, fountain jets, and cascades fall into the moving water category [136]. The level of the water sounds should be similar to or not less than 3 dB below the level of urban noises [46]. It is observed from various literatures that water elements such as jets, fountains, running water, etc. have some influence on the soundscape of an urban environment, especially in urban parks. It has been noted that sounds from water features improve the urban soundscape also mask the unwanted background noise in parks [13]. They can also be used to effectively mask other irritating sounds [134]. Brown in 2003 suggested that rushing water can be used as an acoustic camouflage of traffic sounds [137, 138] experimented with fountains and proved that water sounds may have an indirect impact on soundscape quality by camouflaging the cap urban of hearing the road-traffic noise.

They conducted a field experiment to explore whether water sounds from a fountain had a positive impact on soundscape quality in a downtown park. In total, 405 visitors were recruited to answer a questionnaire on how they perceived the park, including its acoustic environment. Meanwhile, the fountain was turned on or off, at irregular hours. Water sounds from the fountain were not directly associated with ratings of soundscape quality. Rather, the predictors of soundscape quality were the variables "Road traffic noise" and "Other natural sounds". The former had a negative and the latter had a positive impact. However, water sounds may have had an indirect impact on soundscape quality by affecting the audibility of road traffic and natural sounds. Traffic noise reduction can be achieved by the introduction of water-oriented sound in urban open spaces. Coensel et al. [31] suggested that adding fountain sound with low temporal variability reduces the loudness of road traffic whereas, the addition of bird sound enhanced soundscape pleasantness and eventfulness. The addition of the latter was more effective in curbing traffic noise. They also agree with the results from laboratory studies that water sounds may mask road traffic sounds, but that this is not straightforward. Thus sound should be brought into the design scheme when introducing water features in urban open spaces especially since flowing water can improve the restoration experience [139].

5.1.2 Role of greenery in soundscape perception

As water features influence the soundscape, the greeneries also help to influence the soundscape [140] suggested that tree belts help with noise reduction in open spaces and urban parks which decreases the stress on environmental noise for people. Besides, [141] proved through experimental research that the effects of hedges, as a result, of a combination of physical noise reduction, influence people's perception. The acoustical effect produced by the hedges can be understood as a combination of physical noise reduction and its influences on perception. In a study that involved the measurement of light vehicle noise reduction by hedges, thick dense hedges were found to provide only a small noise reduction at low speeds. Whereas, higher noise reductions were found to be associated with an increased ground effect. It is also proved that even the trunks could disperse sound [29]. Acoustically soft material, for instance, as found in vegetated soil can be used to reduce unwanted sounds while at the same time producing other ecosystem services [133]. New physical structures of densified urban spaces, if planned appropriately, can be used as screens that separate city spaces acoustically [142, 143] proved that viewing natural landscapes as vegetation and other natural elements generally creates a stronger positive health effect than viewing urban landscapes as concrete, buildings, and other man-made structures. This is proved by quantitative (EEG evaluation) and qualitative (the questionnaire survey) methods which indicate that landscape plants can cause inflated levels of noise reduction and psychological noise reduction. There have been various researches that show the impact of vegetation on soundscape [128, 144, 145] with the help of different methods of data collection. This shows that acoustic dynamics can be linked to vegetation structure, even on a micro-scale.

6. Conclusion

The paper reviews the state of the art of literature on various definitions, perceptions, and theories of landscape in conjunction with the theories of the soundscape and classification of soundscape elements. Also, it reviews the soundscape dimension in the landscape through the function of green spaces. In the modern scenario, most conservation aspects are focused on the preservation of natural biodiversity, however, diminutive thought is provided to the acoustic heritage which greatly contributes to the sense of the place. The same can enhance the social character of the place. Moreover, the tranquil ambiance is an essential aspect to appraise the concerned environment, especially in urban areas. In terms of landscape, the auditory sense provides visual awareness in terms of activities. A better consciousness of sound and other senses generates a more pleasant understanding of the surrounding landscape. Hence the integration of landscape and soundscape considerations has to be incorporated at the planning level in terms of the design of urban spaces. In this paper, the landscape element of vegetation and water body has been discussed. However, further research can be extended to understand the impact of other landscape elements as different footpath materials, soil, seating, hoardings, etc. in relation to soundscape.

Conflict of interest

There is no conflict of interest.

Author details

Banu Chitra Mookiah
School of Planning and Architecture, Vijayawada, India

*Address all correspondence to: banuvipin@gmail.com

References

[1] Lindquist M. Lange. Sensory aspects of simulation and representation in landscape and environmental planning: A soundscape perspective. In: Innovative Technologies in Urban Mapping. Champions: Springer; 2014. pp. 93-106

[2] Southworth M. The sonic environment of cities. Environment and Behavior. **1**:49-70

[3] ISO I. 12913-1: 2014. Acoustics—Soundscape Part 1: Definition and Conceptual Framework. ISO: Geneva, Switzerland; 2014

[4] Truax B. Soundscape composition as global music: Electroacoustic music as a soundscape. Organised Sound. 2008;**13**(2):103-109

[5] Ren X, Kang J, Zhu P, Wang S. Effects of soundscape on rural landscape evaluations. Environmental Impact Assessment Review. 2018;**70**:45-56

[6] Long M. Architectural Acoustics. London, UK: Elsevier; 2005

[7] Moore BC. An Introduction to the Psychology of Hearing. London, UK: Brill; 2012

[8] Bechtel RB, Marans RW, Michelson WE. Methods in Environmental and Behavioral Research. Michigan, US: Van Nostrand Reinhold Co; 1987

[9] Aiello LM, Schifanella R, Quercia D, Aletta F. Chatty maps: Constructing sound maps of urban areas from social media data. Royal Society Open Science. 2016;**3**(3):150690

[10] Krause B. The Great Animal Orchestra: Finding the Origins of Music in the World's Wild Places. London, UK: Hachette; 2012

[11] Matsinos YG, Mazaris AD, Papadimitriou KD, Mniestris A, Hatzigiannidis G, Maioglou D, et al. Spatio-temporal variability in human and natural sounds in a rural landscape. Landscape Ecology. 2008;**23**(8):945-959

[12] Nilsson ME, Berglund B. Soundscape quality in suburban green areas and city parks. Acta Acustica united with Acustica. 2006;**92**(6):903-911

[13] Brown AL, Muhar A. An approach to the acoustic design of outdoor space. Journal of Environmental Planning and Management. 2004;**47**(6):827-842

[14] Axelsson Ö, Nilsson ME, Berglund B. A principal components model of soundscape perception. The Journal of the Acoustical Society of America. 2010;**128**(5):2836-2846

[15] Van Renterghem T, Vanhecke K, Filipan K, Sun K, De Pessemier T, De Coensel B, et al. Interactive soundscape augmentation by natural sounds in a noise polluted urban park. Landscape and Urban Planning. 2020;**194**:103705

[16] Chitra B, Jain M, Chundelli FA. Understanding the soundscape environment of an urban park through landscape elements. Environmental Technology and Innovation. 2020;**19**:100998

[17] Hong JY, Jeon JY. Influence of urban contexts on soundscape perceptions: A structural equation modeling approach. Landscape and Urban Planning. 2015;**141**:78-87

[18] Beutel ME, Jünger C, Klein EM, Wild P, Lackner K, Blettner M, et al. Noise annoyance is associated with depression and anxiety in the general population-the contribution of aircraft noise. PLoS One. 2016;**11**(5):e0155357

[19] Yang W, Kang J. Acoustic comfort evaluation in urban open public spaces. Applied Acoustics. 2005;**66**(2):211-229

[20] World Health Organization. Burden of Disease from Environmental Noise: Quantification of Healthy Life Years Lost in Europe. World Health Organization, Regional Office for Europe; 2011

[21] Hägerhäll C, Taylor R, Cerwén G, Watts G, Van den Bosch M, Press D, et al. Biological mechanisms and neurophysiological responses to sensory impact from nature. In: van de Boch M, editor. Oxford Textbook of Nature and Public Health. Oxford, UK. 2018. pp. 79-88

[22] Annerstedt M, Jönsson P, Wallergård M, Johansson G, Karlson B, Grahn P, et al. Inducing physiological stress recovery with sounds of nature in a virtual reality forest—Results from a pilot study. Physiology & Behavior. 2013;**118**:240-250

[23] Medvedev O, Shepherd D, Hautus MJ. The restorative potential of soundscapes: A physiological investigation. Applied Acoustics. 2015;**96**:20-26

[24] Saadatmand V, Rejeh N, Heravi-Karimooi M, Tadrisi SD, Zayeri F, Vaismoradi M, et al. Effect of nature-based sounds' intervention on agitation, anxiety, and stress in patients under mechanical ventilator support: A randomised controlled trial. International Journal of Nursing Studies. 2013;**50**(7):895-904

[25] Granö JG, Granv JG. Pure Geography. Baltimore, US: Johns Hopkins University Press; 1997

[26] Porteous JD, Mastin JF. Soundscape. Journal of Architectural and Planning Research. 1985;**1**:169-186

[27] Schafer RM. Notes for an Experimental Music Course. Ontario: Berandol Music Limited; 1967

[28] World Soundscape Project, Truax B. The World Soundscape Project's Handbook for Acoustic Ecology. Arc Publications, Simon Fraser University; 1978

[29] Zhang M, Kang J. Towards the evaluation, description, and creation of soundscapes in urban open spaces. Environment and Planning. B, Planning & Design. 2007;**34**(1):68-86

[30] Adams MD, Bruce NS, Davies WJ, Cain R, Jennings P, Carlyle A, Cusack P, Hume K, Soundwalking as a methodology for understanding soundscapes

[31] Coensel BD, Vanwetswinkel S, Botteldooren D. Effects of natural sounds on the perception of road traffic noise. The Journal of the Acoustical Society of America. 2011;**129**(4):EL148-EL153

[32] Cain R, Jennings P, Poxon J. The development and application of the emotional dimensions of a soundscape. Applied Acoustics. 2013;**74**(2):232-239

[33] Hedfors P, Berg PG. The sounds of two landscape settings: Auditory concepts for physical planning and design. Landscape Research. 2003;**28**(3):245-263

[34] Cain R, Jennings P, Adams M, Bruce N, Carlyle A, Cusack P, et al. SOUND-SCAPE: A framework for characterising positive urban soundscapes. The Journal of the Acoustical Society of America. 2008;**123**(5):3394-3392

[35] Brown AL, Kang J, Gjestland T. Towards standardization in soundscape preference assessment. Applied Acoustics. 2011;**72**(6):387-392

[36] Herranz-Pascual K, Iraurgi I, Aspuru I, Garcia-Pérez I, Santander A, Eguiguren JL. Integrating soundscape

criteria in urban sustainable regeneration processes: An example of comfort and health improvement. Sustainability. 2022;**14**(6):3143

[37] Lacey J. Sonic Rupture: A Practice-led Approach to Urban Soundscape Design. New York, USA: Bloomsbury Publishing; 2016

[38] Bjerre LC, Larsen TM, Sørensen AJ, Santurette S, Jeong CH. On-site and laboratory evaluations of soundscape quality in recreational urban spaces. Noise & Health. 2017;**19**(89):183

[39] Guastavino C, Katz BF, Polack JD, Levitin DJ, Dubois D. Ecological validity of soundscape reproduction. Acta Acustica united with Acustica. 2005;**91**(2):333-341

[40] Sudarsono AS, Lam YW, Davies WJ. The effect of sound level on perception of reproduced soundscapes. Applied Acoustics. 2016;**110**:53-60

[41] Bassarab R, Sharp B, Robinette B. An updated catalog of 628 social surveys of residents' reaction to environmental noise (1943-2008). Novum. 2009;**2009**:9-18

[42] Kogan P, Turra B, Arenas JP, Hinalaf M. A comprehensive methodology for the multidimensional and synchronic data collecting in soundscape. Science of The Total Environment. 2017;**580**:1068-1077

[43] Rehan RM. The phonic identity of the city urban soundscape for sustainable spaces. HBRC Journal. 2016;**12**(3):337-349

[44] Kang J, Zhang M. Semantic differential analysis of the soundscape in urban open public spaces. Building and Environment. 2010;**45**(1):150-157

[45] Kogan P, Arenas JP, Bermejo F, Hinalaf M, Turra B. A Green Soundscape Index (GSI): The potential of assessing the perceived balance between natural sound and traffic noise. The Science of the Total Environment. 2018;**642**:463-472

[46] Jeon JY, Lee PJ, You J, Kang J. Acoustical characteristics of water sounds for soundscape enhancement in urban open spaces. The Journal of the Acoustical Society of America. 2012;**131**(3):2101-2109

[47] Fiebig A. Acoustic environments and their perception measured by the soundwalk method. In: Internoise 2015 Proceedings. San Francisco, USA; 2015

[48] Marry S, Defrance J. Analysis of the perception and representation of sonic public spaces through on site survey, acoustic indicators and in-depth interviews. Applied Acoustics. 2013;**74**(2):282-292

[49] Patelli P. A field is to play. Enacting mental images of the soundscape. Emotion, Space and Society. 2017;**25**:44-53

[50] De Coensel B, Botteldooren D. The quiet rural soundscape and how to characterize it. Acta Acustica united with Acustica. 2006;**92**(6):887-897

[51] Axelsson Ö, Nilsson ME, Hellström B, Lundén P. A field experiment on the impact of sounds from a jet-and-basin fountain on soundscape quality in an urban park. Landscape and Urban Planning. 2014;**123**:49-60

[52] Brambilla G, Gallo V, Asdrubali F, D'Alessandro F. The perceived quality of soundscape in three urban parks in Rome. The Journal of the Acoustical Society of America. 2013;**134**(1):832-839

[53] Liu J, Kang J, Behm H, Luo T. Landscape spatial pattern indices and

soundscape perception in a multi-functional urban area, Germany. Journal of Environmental Engineering and Landscape Management. 2014;**22**(3):208-218

[54] Staśko-Mazur K. Soundwalk as a multifaceted practice. Argument: Biannual Philosophical Journal. 2015;**2**:439-455

[55] Lee SA. Environmental perception, preferences and the designer. In: Proceedings of the Lund Conference Architectural Psychology. Boston, USA. 1973. pp. 26-29

[56] Fiebig A, Herweg A. The measurement of soundscapes-a study of methods and their implications. In: INTER-NOISE and NOISE-CON Congress and Conference Proceedings 2017 Dec 7. Hong Kong: Institute of Noise Control Engineering; 2017

[57] Jeon JY, Jo HI. Soundscape assessment with classification of human perception to sound environment. In: INTER-NOISE and NOISE-CON Congress and Conference Proceedings 2020 Oct 12. Seoul, South Korea: Institute of Noise Control Engineering; 2020

[58] Schulte-Fortkamp B, Fiebig A. Soundscape analysis in a residential area: An evaluation of noise and people's mind. Acta Acustica United with Acustica. 2006;**92**(6):875-880

[59] Venot F, Sémidor C. The soundwalk as an operational component for urban design. In: PLEA. Genève; 2006

[60] Semidor C. Listening to a city with the soundwalk method. Acta Acustica United with Acustica. 2006;**92**(6):959-964

[61] Berglund B, Nilsson ME. On a tool for measuring soundscape quality in urban residential areas. Acta Acustica United with Acustica. 2006;**92**(6):938-944

[62] Bruce NS, Davies WJ. The effects of expectation on the perception of soundscapes. Applied Acoustics. 2014;**85**:1-1

[63] Liu Q, Liu Z, Jiang J, Qi J. A new soundscape analysis tool: Soundscape Analysis and Mapping System (SAMS). Applied Acoustics. 2020;**169**:107454

[64] Kephalopoulos S, Paviotti M, Anfosso-Lédée F, Van Maercke D, Shilton S, Jones N. Advances in the development of common noise assessment methods in Europe: The CNOSSOS-EU framework for strategic environmental noise mapping. The Science of the Total Environment. 2014;**482**:400-410

[65] Zambon G, Roman HE, Smiraglia M, Benocci R. Monitoring and prediction of traffic noise in large urban areas. Applied Sciences. 2018;**8**(2):251

[66] Asensio C. Acoustics in smart cities. Applied Acoustics. 2017;**117**:191-192

[67] Harman A, Dimitrov H, Ma R, Whitehouse S, Li Y, Worgan P, et al. Notifall: Ambient sonification system using water. In: Proceedings of the 2016 CHI Conference Extended Abstracts on Human Factors in Computing Systems. Bristol, England. 2016. pp. 2667-2672

[68] Aspuru I, García I, Herranz K, Santander A. CITI-SENSE: Methods and tools for empowering citizens to observe acoustic comfort in outdoor public spaces. Noise Mapping. 2016;**3**(1)

[69] Guillaume G, Can A, Petit G, Fortin N, Palominos S, Gauvreau B, et al. Noise mapping based on participative measurements. Noise Mapping. 2016;**3**(1)

[70] Maisonneuve N, Stevens M, Niessen ME, Steels L. NoiseTube: Measuring and mapping noise pollution with mobile phones. In: Information technologies in environmental engineering. Berlin, Heidelberg: Springer; 2009. pp. 215-228

[71] Issarny V, Mallet V, Nguyen K, Raverdy PG, Rebhi F, Ventura R. Dos and don'ts in mobile phone sensing middleware: Learning from a large-scale experiment. In: Proceedings of the 17th International Middleware Conference. New York, USA. 2016. pp. 1-13

[72] Geraghty D, O'Mahony M. Urban noise analysis using multinomial logistic regression. Journal of Transportation Engineering. 2016;**142**(6):04016020

[73] Morillas JM, Gozalo GR, González DM, Moraga PA, Vílchez-Gómez R. Noise pollution and urban planning. Current Pollution Reports. 2018;**4**(3):208-219

[74] Morillas JM, González DM, Gozalo GR. A review of the measurement procedure of the ISO 1996 standard. Relationship with the European Noise Directive. The Science of the Total Environment. 2016;**565**:595-606

[75] Segura-Garcia J, Calero JM, Pastor-Aparicio A, Marco-Alaez R, Felici-Castell S, Wang Q. 5G IoT system for real-time psycho-acoustic soundscape monitoring in smart cities with dynamic computational offloading to the edge. IEEE Internet of Things Journal. 2021;**8**(15):12467-12475

[76] Liu J, Kang J, Luo T, Behm H. Landscape effects on soundscape experience in city parks. The Science of the Total Environment. 2013;**454**:474-481

[77] Wei-Ming C, Liang RH. Listening is believing: Exploring the value of sounds in an audio drama board game for shaping technology futures. International Journal of Design. 2021;**15**(2):55

[78] Aletta F, Kang J. Soundscape approach integrating noise mapping techniques: A case study in Brighton, UK. Noise Mapping. 2015;**2**(1)

[79] Jackson JB. Discovering the Vernacular Landscape. Connecticut, USA: Yale University Press; 1984

[80] Vos J, Buchta E. Quantification of context effects in the determination of source-specific annoyance ratings for shooting and road-traffic sounds. Proceedings Internoise. 2000;**3**:1597

[81] Cosgrove D. Prospect, perspective and the evolution of the landscape idea. Transactions of the Institute of British Geographers. 1985;**4**:45-62

[82] Council of Europe Publishing. European Spatial Planning and Landscape Series. Piestany, Slovak Republic: Council of Europe Publishing; 2000. p. 72

[83] Lowenthal D. Past time, present place: Landscape and memory. Geographical Review. 1975;**1**:1-36

[84] Humboldt AV. Ansichten der Natur

[85] Relph E. Rational Landscapes and Humanistic Geography. London: Croom Helm; 1981

[86] Gombrich EH. The Earliest Description of Bosch's Garden of Delight. Journal of the Warburg and Courtauld Institutes. 1967;**30**(1):403-406

[87] Kant I. Prolegomena to Any Future Metaphysics that Will be Able to Present Itself as a Science. Indianapolis, USA: Manchester University Press; 1966

[88] Fairbridge RW. Encyclopedia of Geomorphology. London, UK. 2006

[89] Meinig DW. The Interpretation of Ordinary Landscapes: Geographical Essays. USA: Oxford University Press; 1979

[90] Kaplan A. Landscape architecture's commitment to landscape concept: A missing link? Journal of Landscape Architecture. 2009;**4**(1):56-65

[91] Forster PM. A brief introduction to environmental psychology. Date of Access. 2010;**22**(01):2012

[92] Porteous JD. Environmental Aesthetics: Ideas, Politics and Planning. London, UK: Routledge; 2013

[93] Knox P, Marston SA. Places and Regions in Global Context Human Geography, 3. Baskı

[94] Palmer J. Research agenda for landscape perception. In: Trends in Landscape Modelling, Proceedings at Anhalt University of Applied Sciences. Virginia, USA. 2003

[95] Baumgarten A, Chapter I. The history of philosophy as conceptual archaeology. Types of reading of the Anselmian argument. The Infinite Argument. 2012;**1**:13-33

[96] Sapir G, Barak-Erez D, Barak A, editors. Israeli Constitutional Law in the Making. New York, USA: Bloomsbury Publishing; 2013

[97] Lothian A. Landscape and the philosophy of aesthetics: Is landscape quality inherent in the landscape or in the eye of the beholder? Landscape and Urban Planning. 1999;**44**(4):177-198

[98] Zagris N. Aristotle (384-322 BC): The beginnings of Embryology. The International Journal of Developmental Biology. 2022;**66**(1):5-8

[99] Carson E. Locke's account of certain and instructive knowledge. British Journal for the History of Philosophy. 2002;**10**(3):359-378

[100] Gross SW. The neglected programme of aesthetics. The British Journal of Aesthetics. 2002;**42**(4):403-414

[101] Vandenabeele B. Burke and Kant on the social nature of aesthetic experience. In: The Science of Sensibility: Reading Burke's Philosophical Enquiry. Dordrecht: Springer; 2012. pp. 177-192

[102] Zube EH, Sell JL, Taylor JG. Landscape perception: Research, application and theory. Landscape Planning. 1982;**9**(1):1-33

[103] Lekagul A. Toward Preservation of the Traditional Marketplace: A Preference Study of Traditional and Modern Shopping Environments in Bangkok, Thailand. Virginia, USA: Virginia Polytechnic Institute and State University; 2002

[104] Isling B. A typology for the parks of Stockholm. Garden History. 2004;**1**:248-260

[105] Fenton DM, Reser J. The assessment of landscape quality: An integrative approach. In: Nasar J, editor. Environmental Aesthetics: Theory, Research and Applications. Cambridge, UK

[106] Swaffield SR. Theory and critique in landscape architecture: Making connections. Journal of Landscape Architecture. 2006;**1**(1):22-29

[107] Palka EJ. Coming to grips with the concept of landscape. Landscape Journal. 1995;**14**(1):63-73

[108] Brunetta G, Voghera A. Evaluating landscape for shared values: Tools, principles, and methods. Landscape Research. 2008;**33**(1):71-87

[109] Swanwick C. Landscape Character Assessment. Guidance for England and

Scotland; The Countryside Agency and Scottish Natural Heritage: Inverness, UK, 2002

[110] Stiles R. Landscape theory: A missing link between landscape planning and landscape design? Landscape and Urban Planning. 1994;**30**(3):139-149

[111] Braun R, Knitter D. Landscape as concept of space. eTopoi. Journal for Ancient Studies. 2021;**9**:37-53

[112] Greider T, Garkovich L. Landscapes: The social construction of nature and the environment. Rural Sociology. 1994 Mar;**59**(1):1-24

[113] Booi H, Van den Berg F. Quiet areas and the need for quietness in Amsterdam. International Journal of Environmental Research and Public Health. 2012;**9**(4):1030-1050

[114] Brown AL. A review of progress in soundscapes and an approach to soundscape planning. International Journal of Acoustics and Vibrations. 2012;**17**(2):73-81

[115] Pheasant R, Horoshenkov K, Watts G, Barrett B. The acoustic and visual factors influencing the construction of tranquil space in urban and rural environments tranquil spaces-quiet places? The Journal of the Acoustical Society of America. 2008;**123**(3):1446-1457

[116] Wolman A. The metabolism of cities. Scientific American. 1965;**213**(3):178-193

[117] Waldheim C. The Landscape Urbanism Reader. New York, USA: Princeton Architectural Press; 2006

[118] McHarg IL. Design with Nature. New York: American Museum of Natural History; 1969

[119] WCED SW. World commission on environment and development. Our Common Future. 1987;**17**(1):1-91

[120] Hartig T, Mitchell R, De Vries S, Frumkin H. Nature and health. Annual Review of Public Health. 2014;**35**:207-228

[121] Stigsdotter UK, Palsdottir AM, Burls A, Chermaz A, Ferrini F, Grahn P. Nature-based therapeutic interventions. In: Forests, Trees and Human Health. Dordrecht: Springer; 2011. pp. 309-342

[122] Turner T. City as Landscape: A Post Post-Modern View of Design and Planning. London, UK: Taylor & Francis; 2014

[123] Jellicoe GA, Jellicoe S. The landscape of man: Shaping the environment from prehistory to the present day. expanded and updated, Rev. and enl. 1995

[124] Hill K. Urban Ecologies: Biodiversity and Urban Design. Munich, Germany: Prestel;

[125] Waller SJ. Sound and rock art. Nature. 1993;**363**(6429):501

[126] Herranz-Pascual K, Aspuru I, García I. Proposed Conceptual Model of Environmental Experience as Framework to Study the Soundscape. Lisbon: InterNoise; 2010

[127] Viollon S, Lavandier C, Drake C. Influence of visual setting on sound ratings in an urban environment. Applied Acoustics. 2002;**63**(5):493-511

[128] Ren X, Kang J. Interactions between landscape elements and tranquility evaluation based on eye tracking experiments. The Journal of the Acoustical Society of America. 2015;**138**(5):3019-3022

[129] Pijanowski BC, Farina A, Gage SH, Dumyahn SL, Krause BL. What is

soundscape ecology? An introduction and overview of an emerging new science. Landscape Ecology. 2011;**26**(9):1213-1232

[130] Yang W, Kang J. Soundscape and sound preferences in urban squares: A case study in Sheffield. Journal of Urban Design. 2005;**10**(1):61-80

[131] Lam KC, Brown AL, Marafa L, Chau KC. Human preference for countryside soundscapes. Acta Acustica united with Acustica. 2010;**96**(3):463-471

[132] Curcuruto S, Asdrubali F, Brambilla G, Silvaggio R, D'Alessandro F, Gallo V. Socio-acoustic survey and soundscape analysis in urban parks in Rome. Proceedings of the Institute of Acoustics. 2011;**33**(2):830-837

[133] Nilsson M, Bengtsson J, Klæboe R. Environmental methods for transport noise

[134] Maffei L. Urban and quiet areas soundscape preservation. In: InVICongresoIberoamericano de Acústica. 2008

[135] Chitra B, Jain M, Chundeli FA. Understanding city park soundcape through ISO/TS 129132: 2018 sound walk (Method B) in association with landscape elements. In: INTER-NOISE and NOISE-CON Congress and Conference Proceedings. Seoul, South Korea: Institute of Noise Control Engineering.; 2020

[136] POUYA S. The role of landscape architecture on soundscape experience. Turkish Journal of Forest Science. 2017;**1**(2):183-193

[137] Brown AL. Using the sound of water in the city. Landscape Australia. 1994;**2**:103-107

[138] Nilsson ME, Alvarsson J, Rådsten-Ekman M, Bolin K. Auditory masking of wanted and unwanted sounds in a city park. Noise Control Engineering Journal. 2010;**58**(5):524-531

[139] Zhao J, Xu W, Ye L. Effects of auditory-visual combinations on perceived restorative potential of urban green space. Applied Acoustics. 2018;**141**:169-177

[140] Fang CF, Ling DL. Guidance for noise reduction provided by tree belts. Landscape and Urban Planning. 2005;**71**(1):29-34

[141] Van Renterghem T, Attenborough K, Maennel M, Defrance J, Horoshenkov K, Kang J, et al. Measured light vehicle noise reduction by hedges. Applied Acoustics. 2014;**78**:19-27

[142] Hellström B, Nilsson ME, Axelsson Ö, Lundén P. Acoustic design artifacts and methods for urban soundscapes: A case study on the qualitative dimensions of sounds. Journal of Architectural and Planning Research. 2014;**1**:57-71

[143] Yang F, Bao ZY, Zhu ZJ. An assessment of psychological noise reduction by landscape plants. International Journal of Environmental Research and Public Health. 2011;**8**(4):1032-1048

[144] Chojnacki B, Piechowicz J, Matys M, Piotrowski K. Noise control in urban parks in Cracow

[145] Song X, Lv X, Yu D, Wu Q. Spatial-temporal change analysis of plant soundscapes and their design methods. Urban Forestry & Urban Greening. 2018;**29**:96-105

Chapter 14

Securing Smart Grids to Address Environmental Issues in Regional Planning

Vicent Mbonye

Abstract

This chapter examines regional planning and development in relation to sustainability and highlights sustainability challenges in various regional planning case studies. Creating smart cities addresses the problems that arise from rapid urbanisation and growth of the urban population. This chapter provides an overview of smart cities and discusses several global smart city efforts. It introduces the idea of smart energy highlighting the smart grid components and how it tackles environmental challenges in regional planning. Additionally, it analyses several threats to the smart grid that may hinder its efficient operation and makes suggestions on how to deal with them so that sustainable energy may be delivered to smart cities.

Keywords: security, regional development, smartgrids, smart cities, sustainability

1. Introduction

The number of people residing in cities is expected to reach approximately 5 billion by the year 2030. A significant portion of this urbanisation will take place in Africa and Asia, resulting in profound social, economic, and environmental changes [1]. However, the fast shift to a highly urbanised population poses significant difficulties for the design, growth, and management of cities [2]. Examples include challenges with waste management, a lack of resources like electricity and water, air pollution, concerns about human health and traffic congestion [3].

A number of nations have embraced the idea of smart cities as a means of reducing the issues brought on by the rapid urbanisation and growth of the urban population in their regional planning efforts [3, 4]. For instance, Bristol city and Miltonkeynes in England, Chicago city in United States of America and Shenzhen in China. More smart cities have been built in Australia, Japan and Singapore.

In Africa, various smart city initiatives are being developed. These cities include Vision City in Kigali, Rwanda, HOPE City in Ghana, Waterfall City in South Africa, and Senegal's Akon City. It also includes, Eko Atlantic City in Nigeria and Konza City in Kenya [5].

Smart city elements include smart infrastructure, smart transportation, smart energy and smart healthcare. Smart citizens, smart governance, smart buildings, smart technology, and smart education are also included [4]. In these cities, solutions to more sustainable transport, energy, health and education have been developed and some are in development. For instance, MiltonKeyness has set up an innovative solutions to support efficient transportation through a smart transport application called MotionMap [6]. In Japan, Yokohama is characterised by smart city features within its energy, waste, transportation, and water services [7].

Dealing with the difficulties of dependable electricity becomes essential when populations move to cities and increase their energy consumption [8]. This is because reliable electricity fuels economies, governments, health care, education, and leads to poverty reduction [8, 9]. Hence, the lack of adequate and reliable power is a major constraints in economic growth [10].

The core of a smart energy system in smart cities is the smart grid [11]. Smartcities implement the smart grid to integrate decentralised sustainable energy sources, effective distribution, and optimised power usage [4]. Through the smart grid utility companies can add renewable energy sources like wind, solar, and biomass, which are more ecologically friendly than the fossil fuels like coal used in many facilities for generating large amounts of electricity [12]. For example, Japan has developed the smart grid system in Yokohama with distributed power sources, self-managed power plant that uses renewables [7]. In South Africa, smart grids electric systems are being piloted [13]. Already projects in various municipalities have enabled efficiencies and effectiveness not seen before in the municipal environment [14].

Despite the fact that the smart grid solves the numerous energy concerns, the vast interconnectedness of the systems creates a significant attack surface for intruders [15]. For instance, intruders can compromise smartgrid infrastructure like the smart meter through a denial-of-service attack that prevents critical alerts, such as alarms, from being relayed to the head-end systems [15]. In addition, they can hack the smart meter and be able to see everyday life patterns of consumers and occupancy of households [15]. Hence, these assaults put human lives in danger in addition to destroying infrastructure [16].

Safety is crucial to create environments where residents may fully benefit from economic and cultural opportunities [17]. Ensuring that a city is safe is important in order to satisfy the urban safety sustainability indicator [18]. The urban safety indicator focusses on urban security and protection against crime, as well as defence against its impact [18]. This chapter argues that threats to the smartgrid reduce the possibility of having safe liveable smart cities that are sustainable.

Safeguarding the security of residents calls for a thorough comprehension of best practices and solutions backed by adequate resources [17]. Regional planners, must be knowledgeable about the smart grid and its role in sustainable development. They should also know smartgrid domains and the threats to the smart grid. Furthermore they need to be aware of the proposed remedies these threats. This can aid in the planning and development of more sustainable cities.

2. Sustainability challenges in regional planning

Sustainability involves ensuring that people can satisfy their immediate demands without compromising the ability of future generations to satisfy their own [19]. Sustainability development goals include eradicating poverty, quality education, zero hunger, good health and well-being, sustainable energy, clean water, and sanitisation [19].

Regional planners must make sure that initiatives for regional development are centered on sustainability. However, a variety of barriers that could prevent sustainable regional planning have been recognised in the literature.

The management of suburban sprawl is one of the major problems confronting planners in many nations [20]. Despite several studies and attempts to control suburban sprawl, it still poses a serious problem since it encourages unsustainable urban development. As an illustration, in India, the pressure from the growing population and the saturation of urban regions within municipal boundaries led to the densification of the central urban centres of Dhanbad and Jamshedpur. In addition, the development of built-up land at the expense of agricultural land in Ranchi Urban is documented [21]. In a similar study conducted in India, land use in Kolkata UA was altered by urbanisation and development. However, this resulted in unparalleled urban sprawl, which greatly harmed the environment [20]. Additionally, the majority of the region was converted to urban territory at the expense of agricultural land. This also applied to swamps and aquatic vegetation [20].

Furthermore, in Ethiopia, it's reported that the town's Ehud-Gabia market was converted into a bus stop, residential area and kindergarten [22].

In China, the development of new development zones and the transformation of urbanised communities were the primary reasons of fast urban sprawl [23]. However, although real estate development was a main driving factor for rapid increase in Gross domestic product (GDP) in China, it led social problems like traffic congestion in Beijing [23]. In addition, urban sprawl in Shanghai resulted in the loss of a significant amount of farmland and agricultural operations and endangering the city's traditional architectural forms [23]. Moreover, many individuals in Guangzhou who relocated to suburbs found it challenging to adjust to new life transitions and living situations [23].

Sanitation and waste disposal are also growing concerns in regional planning. The soil, water, and air can get contaminated as a result of improper waste management. This fosters unhygienic circumstances that could cause illness, damage, or death. More than 60% of the world's population lacks access to either adequate or even safe sanitation facilities or services. Additionally, the majority of wastewater produced globally is discharged without being properly treated [24].

Waste management is crucial in the construction of sustainable and habitable communities, yet it is still difficult in many developing nations and cities. Global waste production rates are growing. Annual trash generation is anticipated to increase by 73% from 2020 levels to 3.88 billion tonnes in 2050 due to high population expansion and urbanisation [25]. In low-income countries, waste is usually burned outdoors or thrown in unregulated dumps. The environment, public safety, and health are all adversely affected by these actions. Moreover, methane produced by improper waste management contributes to climate change and may even fuel urban violence [25]. In addition, effective waste management is expensive, frequently taking up 20–50% of municipal budgets. Moreover, unsustainable waste management has a greater detrimental impact on residents of developing countries than on people of industrialised countries, especially the urban poor [25]. Additionally, urbanisation also leads to an increase in waste production. High-income nations and economies are more urbanised and produce more garbage overall and per person [26]. Hence, locations with a high proportion of growing low-income and lower-middle-income countries are anticipated to have the biggest increases in waste production. This is because it is projected that waste production increases with economic development and population expansion [26]. Waste levels in the Sub-Saharan Africa and South Asia regions are expected to nearly quadruple and double, respectively, during the course of the next three decades as a result of urbanisation and economic growth [26].

Waste management challenges are documented in literature. For instance, poor collection of waste affects the effective management of waste. As an illustration, it is reported that Senegal produces more than 2.4 million tonnes of waste per year. However, about 1.08 million tonnes remains uncollected causing harm to the environment [26]. Another issue in waste management is the absence of funding for the creation of solid waste management systems, which makes it even more important to factor in continuing operating expenses up front [26]. These factors make sustainable waste management a difficult task on the path to economic growth, and the majority of low- and middle-income countries and their cities struggle to manage them [26].

Integrated systems that are effective, sustainable, and socially supported are needed to operate this crucial municipal services [25]. Smartcity initiatives have facilitated addressing waste and sanitisation issues. For instance the city of Yokohama has implemented a garbage collection and management system based on IT, the 3R (reuse, reduce, recycle) philosophy, and bioenergy generated from solid waste [7].

Additional planning challenges for a sustainable regions arises from maintaining clean air and water in urban areas. Air is one of the most important environmental factors when examining the relationship between environmental quality in populated areas and population health and wellbeing [27]. However, the safety and well-being of society are endangered when pure air is polluted [27]. This may result in disease, injury, or even death.

In addition, the quality of the world's water is impacted by the growth of metropolitan areas. Runoff from urban surfaces, for instance, may contain a variety of pollutants, such as heavy metals, sodium, nitrate, phosphorus, trash, and rubber residue [28]. Additionally, contaminants from urban surfaces have the potential to significantly worsen the water quality of urban streams and other receiving waterways [28]. Furthermore, the way rainfall is caught, stored, and released in hydrological systems is also affected by changes to slopes, elevations, soils, and plant covering [28]. These developments result in the creation of extensive flat, impervious material regions and drainage networks, significantly changing the main hydrological routes. For instance, significant wetlands in Tianjin have gradually been replaced by extensive urban development [29].

Overpopulation is also acknowledged as a significant problem in regional planning. When there are more individuals than resources available at an area. In cities, overpopulation presents a number of challenges. Firstly, it prevents a nation from using its resources effectively [30]. Secondly, it also leads to production methods that are inefficient and out of touch with consumer demand. For example, most people who work in agriculture regularly try to support their families on small pieces of land even if doing so does not ensure the family's long-term survival [30]. In India, Overpopulation has resulted in inadequate health care facilities, inadequate education, and early marriages [30].

Industrialisation is another major challenge in regional development. Pollution and environmental degradation rise as a result of industrialisation and increased capital intensity. With a contribution rate of 61%, industrial activities have been shown to be a significant contributor to the global threat of environmental pollution [31].

Furthermore, energy utilisation is a key challenge in regional planning. The majority of the world's energy is now produced by fossil fuels (coal, oil, and natural gas), which cause climate change to worsen and air quality to decline. Cities currently use 75% of the world's resources and energy, which results in the production of 80% of greenhouse gases. Therefore, there might be a significant detrimental influence on the ecosystem in the next decades [4]. City planners hence need to plan for sustainable energy solutions.

Other challenges in regional development include the lack of involvement of stake holders in the realisation of regional development goals and insufficient funds [22].

In this section, the challenges to regional planning were discussed. These challenges are all unique in their own right. For instance, it is important to decrease industrial pollution since it can harm the quality of the air and water. Additionally, inefficient waste management practices could jeopardise the health of residents. Finally, the larger issue caused by urban growth must be addressed and combated via sustainable urban development. A variety of urban sustainable development solutions may be used to address urban sustainability concerns. Regional planning and the need for sustainable energy are covered in the next section.

3. Regional planning and sustainability energy

The International Energy Agency (IEA) estimates that cities account for 71% of CO2 emissions and use around 67% of all energy produced globally [32]. Urbanisation will account for 66% of the global population by 2050, which would increase energy use and CO2 emissions by up to 80–90%.

A totally decarbonised electricity sector must serve as the foundation for a net zero energy system [33]. However, the energy sector is responsible for around 60% of all greenhouse gas emissions worldwide. This is because, more than 60% of the power used in the globe today is still produced from fossil fuels. That percentage must fall to 26% by 2030 in order to be compatible with the scenario of net zero emissions by 2050. To reach this goal, the implementation of low and zero-emission sources must accelerate rapidly [32]. Regional approaches to renewable deployment in Africa can lower overall system costs and enhance operation. Regional strategies can increase the flexibility and stability of national power networks and act as a complement to enable nations to accelerate the development of single sources while also lowering the risks associated with climate change [34].

By supporting renewable energy, which is more ecologically friendly than the fossil fuels utilised in many electric power producing facilities, the smart grid offers a solution [12]. Hence, measures that ensure the efficient operation of the smartgrid in smart cities are key in its success. This ensures the delivery of safe reliable and sustainable services to individuals.

4. Urban safety and sustainability

A smart city goes beyond the use of digital technologies for better resource use and less emissions [35]. It comprises better water supply and waste disposal systems, stronger urban transit networks, more efficient building lighting and heating systems. It also implies safer public spaces, a more active and responsive local government, and addressing the needs of an older population [35].

Safety is critical for building environments in which residents may fully use economic and cultural potential [17]. To ensure safety, coordinated policy approaches to urban safety and security are required. Also a robust knowledge foundation of good practices and effective interventions backed by suitable resources are essential [17]. The goal of urban planning and implementation is to create economically productive, environmentally sustainable, socially viable, and livable urban areas that are physically connected and interconnected [22].

One of the main goals of the 2030 Agenda for Sustainable Development's Goal 11 is to make cities and human settlements secure [17]. Through this goal member states make a commitment to promote secure and safe environments in cities and human settlements that allow everyone to live, work, and participate in urban life without fear. In this context, member states are required to adopt measures to reduce violence and promote urban safety in their nations [17].

The best methods for ensuring urban safety and security involve addressing the many factors that contribute to crime and other acts of violence, such as sexual harassment and gender-based violence. This includes strengthening local ties, fostering civic engagement and collaboration, and improving local governance of safety and security as a public good through civic engagement, place-making, vernacular arts and cultural activities, behavioural change strategies, and community development initiatives [17].

5. The role of smart cities in sustainable regional development

A smart city is allocation that allows for services to be enhanced by information, digital, and communications technologies to make them more adaptive, effective, and sustainable for the benefit of its citizens [4].

Smart cities include intelligent transportation systems (ITS), which encompass the rail, water, and air transportation systems as well as their interconnections [4]. The smart transportation system makes it simple for users to choose from a variety of modes of transportation for the cheapest, shortest, or quickest routes.

Smart healthcare in smart cities refers to a concept that combines traditional healthcare with smart biosensors, wearable technology, information and communication technology (ICT), and smart ambulance systems [36, 37]. Patient information may be accessed in real-time at various offices inside a smart hospital, or even at many smart hospitals in the same or separate cities [4, 38].

Additionally, smart cities incorporates "smart governance," which relies on concepts of good governance including transparency, accountability, collaboration, and citizen involvement. Smart governance is the skillful application of ICT to enhance decision-making through greater collaboration between many stakeholders, including the government and the general public [39].

Smart cities include the use of smart technology for the design, implementation, and operation of smart cities. The ICT infrastructure is the crucial smart component of the smart city, holding all the other elements together and serving as the hub for its residents [4]. Infrastructure for services is built on physical infrastructure and may include certain ICT elements. Smartgrids and mass rapid transit systems are two examples of service components.

Everything that is physically, electrically, or digitally connected to a smart city forms its infrastructure. Smart infrastructure includes smart buildings that may have a variety of hardware, software, sensors, and smart appliances for various automated activities, such as data network, access control, power management, and lighting control [4].

Additionally, the smart city incorporates smart energy which is includes smartgrids that generate, store and transport sustainable energy using information and communication technology (ICT) [4].

5.1 Case studies of smart cities around the world

The concept of a "smart city" refers to an area where existing networks and services have been enhanced via the use of digital technologies for the benefit of both its citizens and businesses. The following are some of a smart city's components: smart infrastructure, smart buildings, smart technologies, smart government, smart education, and smart citizens [4].

To address the problems and challenges associated with urbanisation, several nations and towns have started their own smart city initiatives. For instance, Bristol is renowned for its smart city and sustainability projects; in 2017, the city was named top in the UK Smart City Index [40] and was European Green Capital in 2015.

Milton Keynes is one of the smart cities that have developed their smart city strategy and put important initiatives into effect [6]. This project's initiatives include cutting-edge methods to assist effective mobility, such as the creation of the MotionMap smart transportation app [6].

The city of Curitiba is well known for its innovative efforts in the field of sustainable urban development. Brazil's Curitiba is regarded as a premier example of sustainable urban design. It has received numerous accolades, including the 2012 Global Green City Award and the World Habitat Award for Urban Management [41].

Australia's top three smart cities are Sydney, Melbourne, and Brisbane. The most widely used technologies in these cities are those involving the internet of things, artificial intelligence, and driverless vehicles [42].

In Japan, Yokohama has developed the smart grid system with distributed power sources to provide sustainable services to individuals. In terms of waste management, it has built an IT-based collection and management system that provides efficient collection and management of waste in the city [7].

In China, Shenzhen was designated as an experimental city to spearhead China's modernisation. It has two giant telecom firms, Huawei and Tencent which are playing pivotal roles in advancing the smart city movement in China [43]. Shezen has implemented smart technologies through carefully including innovative technologies like artificial intelligence, the internet of things.

Similar methods have been employed by Singapore to promote economic development through smart city initiatives. Singapore, for instance, has earned the title of leading smart city. Singapore was ranked first among 102 worldwide smart cities by the IMD World Competitiveness Center, with the highest rating of AAA; Singapore is ranked as the seventh-smartest city in the world by the IESE Cities in Motion Index 2019 [44].

Additionally, there are various smart city initiatives being created or planned throughout Africa. For instance, Akon City in Senegal, HOPE City in Ghana, Vision City in Kigali, Rwanda, Konza City in Kenya, and Eko Atlantic City in Nigeria are a few examples [5].

The core of the "smart city" concept is the notion that information and communication technology may promote intelligent behaviour and logical, optimum decisions in urban settings (ICTs). By utilising ICTs to process information in the form of data, urban planners, investors, and developers would be able to redesign the experience of urban life using computer power, networked technologies, and scientific discoveries. Infrastructure, energy, health, safety, and other problems might be addressed by bringing objects and people continuously online and integrating sensors and the Internet into physical-virtual interactions.

5.2 Smartgrid domains

Cities currently use 75% of the world's resources and energy, which results in the production of 80% of greenhouse gases. As a result, they may suffer grave environmental consequences in the following few decades. This necessitates the idea of smart cities. The development of smart cities is a logical response to the issues brought on by the fast urbanisation and population increase of cities. Smart cities, once implemented can reduce energy consumption, water consumption, carbon emissions, transportation requirements, and city waste [4]. The smart grid is the heart of a smart energy system. The information and infrastructure that make up the smart grid are in charge of gathering data on energy usage and disseminating information about provider rates. Smart appliances like dishwashers and water heaters may be operated with an acceptable degree of energy consumption using ICT. The smart grid facilitates efficient energy storage, smart metering, and effective energy management [45].

As depicted in **Figure 1**, the smart grid can be broken down into seven domains: Operations, Distribution, Service Provider Markets, Generation, Customer and Transmission [47].

In the customer domain, Consumers can control how they use and produce energy through the Home are network [48]. The consumer is the last stakeholder, the smart grid seeks to assist the customer.

In the Service Provider/Utility Domain, Service providers manage customer accounts, billing, and energy usage. The service provider links to the customer and market domains as well as the operation domain for situation awareness and system control [46].

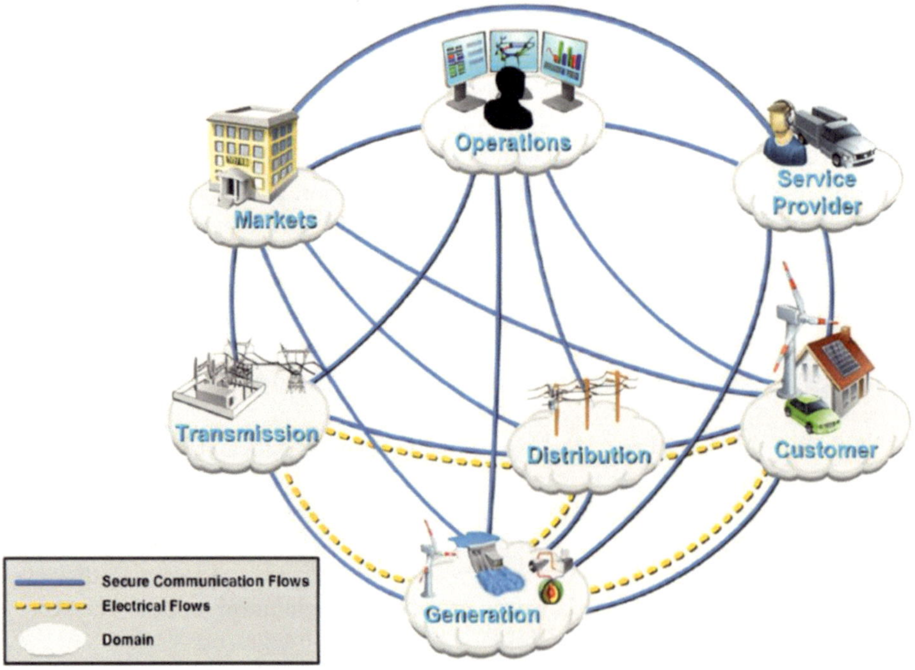

Figure 1.
Interaction of actors in different smart grid domains through secure communication flows. Source: [46].

The process of supplying electricity to the end consumer begins with energy generation. Through an interface, the bulk generating domain is electrically connected to the transmission domain and interacts with the market, transmission, and operations domains [49].

Communication with the transmission domain is crucial because without transmission, consumers cannot be served [48]. The transmission domain is in charge of the extensive transfer of electrical power via multiple substations from a generator to a distribution system [48]. The transmission domain uses several substations to move electrical energy across long distances from the generating domain to the distribution domain [46].

The transmission domain is in charge of transferring huge amounts of electrical power via a number of substations from a generator to a distribution system [48]. Electrical power production is transferred from the generation domain to the distribution domain through a large number of substations in the transmission domain [49].

Actors in the operations domain are accountable for the power system's smooth operation. The actors in charge of transporting electricity are the players in the operations domain. This department guarantees that transmission and distribution activities run smoothly [46].

Furthermore, Electrical grid resources are bought and sold via markets. Within the power system, market participants exchange prices and maintain supply and demand balance [46].

5.2.1 The role of the smart grid in ensuring sustainable energy delivery

The smartgrid provides more efficient ways to generate and distribute electricity, as well as consumers' ability to utilise electricity more effectively [50]. In electricity generation, smart grids provides numerous environmental advantages to nations by enabling the integration of renewable energy [51]. These include renewable energy sources, which are more ecologically friendly than the fossil fuels employed in many large-scale electric power producing facilities, such as wind, solar, and biomass [12]. As a result, greenhouse gas emissions that contribute to global warming may be reduced [52].

In addition, Utilities may use the smart grid to remotely monitor, troubleshoot, and manage network infrastructure. This enables faster response times to interruptions and disturbances, as well as the potential of proactive maintenance [52]. For example, if there is an unusually high voltage, the smart grid may detect it and instruct one of the network's devices to decrease the voltage [53]. This allows utilities to minimise the likelihood and frequency, as well as the cost connected to blackouts and outages [52].

Smartgrids, provide improved management and monitoring systems which allow customers to be better aware of their energy usage [52]. Customers have access to real-time production and consumption data, the ability to manage their power usage, and the ability to react to changes in the price of electricity by changing their use [51]. This not only reduces energy loss, but also saves costs incurred by the customers [11, 12].

Taking into consideration both technological and political concerns, the smart grid ensures supply security with a minimal risk of interruption. This guarantees energy independence by lowering external threats to energy supply, such as political and economic risks [51]. As regional planners plan for smart cities they need to be aware of the challenges behind powering the energy dependent infrastructure. If the smartgrid is to thrive and provide efficient service to its settlers. The planners to be prepared to deal with the many threats that threaten the smart grid. The next section highlights the smart grid domain threats to the domains.

5.2.2 Threats to the smart grid domains

There are several risks to the smart grid. For instance, smart meters (SMs) enable utility providers to gather and communicate data on residential energy use [54, 55]. As a result, consumers' sensitive information may be revealed hence putting their privacy at risk [56].

In addition smart meters may be installed both inside and outside of a house or vacation rental. As a result, they are unattended and vulnerable to an attacker who can introduce malware into the devices, causing them to send false data to the concentrator [15].

Additionally, attackers might be able to determine whether there are people on the property by analysing the consumption data from smart meters, enabling burglary [57].

Additionally, if smart meters are deployed in consumers' homes and businesses without hardware security protections, they might be vulnerable to physical attacks such tampering with metering data to cause inconsistencies in billing [58, 59]. Consumers may replace the smart meter with one that has been tampered with and shows lower usage than the actual one [60]. Additionally, the unauthorised insertion, alteration, or deletion of data or control orders in communication network traffic leads smart grid equipment to behave incorrectly [61, 62].

The retail energy provider may face the threat of intentional errors from disgruntled employees. Disgruntled employees could access the smart grid systems and inflict severe damage to the infrastructure by attacking critical smart grid nodes [63, 64].

Protection relays are prone to internal technical threats from organisational practices. A key challenge in protecting relays is that the security of many commercial relays is only guaranteed by requiring each relay to have a password. Unfortunately, because many operators do not change the default password for the sake of convenience, improper password practices have always been detected in substation-level networks [65].

Additionally, attackers have the ability to alter the readings from a variety of sensors and PMUs [66]. For instance, attackers are able to conduct a False Data Injection attack by altering the readings of several sensors and Phasor Measurement Units (PMUs) in order to introduce erroneous measurements and finally inject arbitrary mistakes into the state estimations without being noticed [66].

In addition, attackers are able to launch a jamming attack near the GPS antenna of PMUs. This enables the attacker to disconnect parts at a specific location. The region that is impacted depends on the jamming device's emission power (i.e., signal transmitter) [67].

Components in the distribution domain face external technical threats. For instance, threats to IEDs have been identified in the previous section. This also covers spoofing attacks. By performing a spoofing attack, the attacker may trick a monitoring IED into sending false close/open signals to switches, causing the protection system to malfunction and perhaps shutting off electricity to customers [68].

In addition, attacks that cause a denial of service (DoS) have been recognised as a major threat to SCADA systems. By injecting their own packets into the smart grid system's packet traffic, the attacker may overwhelm the smart grid system's packet flow until it can no longer handle or accept any more packets and fails. In addition to isolating a power substation or preventing operators from accessing SCADA systems, a distributed denial of service (DDoS) assault may purposefully delay the transmission of a time-critical message in order to breach its timing requirement. This may possibly interrupt the electricity supply or, worse, cause electrical equipment to be destroyed [45]. An attacker may cause partial or full equipment failure, making vital data inaccessible until the utility provider fixes or replaces it [59].

5.2.3 Solutions to threats in the smartgrid

To protect against customer privacy violations, customer data should be processed and kept properly since maintaining consumer privacy is a difficult task in the customer sector [69]. In addition, utility providers can employ data aggregation technologies to protect user privacy. Instead of providing their power use information directly to the utility supplier, a group of customers use this method to transmit them to a reliable third party. The utility company receives aggregated data on all users' power usage from a dependable third party [55].

Malware injection in the smart meter hardware can be addressed by implementing protection measures when installing smart meters. Stake holders should shut down unneeded physical ports or making such ports unavailable to unauthorised users [15].

Additional measures to secure the premises and control rooms have been suggested to keep adversaries or anyone intending to harm the grid out [46]. To protect smart appliances in the customer houses holds from burglary, customers can implement physical protection measures [46].

Stakeholders can place a smart meter in a safe space, like a house, or in a public spot to solve physical meter tampering and swapping, but it must be confined in a box with a lock [15].

Furthermore, an efficient defence against jamming attacks is to send arbitrary, unauthenticated packets to each wireless station in the network [70]. When packets are filtered, they are not kept on a network device if they are determined to be invalid. This method can be applied to stop jamming assaults [71].

Regarding the threat of misconfiguration, it is recommended that operators adopt good password practices that include changing the default password in substation-level networks [65].

To solve the problem of fake data injection attacks that change sensor readings. [72] recommend using a multiplicative, limited scaling factor. [67] suggest using reliable encryption and authentication methods [67].

In order to increase the resilience of smart-grid protection systems to intentional errors and unintentional errors, Fadul, et al. [73] propose the adoption of a trust-management toolkit that makes use of network-flow techniques and reputation-based trust. This toolkit can address spoofing attacks on Intelligent Electronic Devices (IEDs).

6. Conclusion

Countries are undergoing significant social, economic, and environmental changes as a result of urbanisation. This poses several difficulties for city planning, growth, and management. This chapter discusses sustainability challenges in regional planning. These include challenges in waste management, energy and water shortages, air pollution, health issues, and traffic congestion.

Regional planners are responsible for ensuring that regional development projects are oriented on sustainability. The focus of recent regional advancements has been on creating smart cities to address various issues. This chapter presents the role of Smart cities in sustainable regional development. It discusses some smart city case studies around the world and presents some solutions developed to address sustainability challenges.

Among sustainability challenges is the issue of providing sustainable energy. This is because the energy sector is responsible for around 60% of all greenhouse gas emissions worldwide. In order to reduce green gas emission levels countries must implement low and zero-emission sources. The implementation of the smartgrid is seen as a possible solution to the energy challenges in regions. The smart grid is essential to how smart cities work. It not only ensures the delivery of sustainable energy, but also addresses many sustainability issues including pollution. It offers a way to create cities that are more habitable. This chapter discusses the smartgrid domains and the role of the smart grid in ensuring sustainable energy delivery.

Safety is critical for building environments in which residents may fully use economic and cultural potential. It allows residents to fully use economic and cultural potential of regions. Urban environments that are economically dynamic, ecologically sustainable, socially viable and habitable, and safe are the aims of good urban planning. This chapter hence, also examines a number of threats to the smart grid that may prevent its effective operation. It goes ahead to offer solutions to the highlighted threats so that sustainable energy may be supplied to smart cities.

Conflict of interest

The authors declare no conflict of interest.

Author details

Vicent Mbonye
University of KwaZulu-Natal, Durban, South Africa

*Address all correspondence to: 213571072@stu.ukzn.ac.za

References

[1] United Nations Population Fund. Urbanisation. 2018. Available from: https://www.unfpa.org/urbanization

[2] Harrison C, Donnelly IA. A theory of smart cities. In: Proceedings of the 55th Annual Meeting of the ISSS-2011. UK: Hull; 2011

[3] Chourabi H, Nam T, Walker S, Gil-Garcia JR, Mellouli S, Nahon K, et al. Understanding smart cities: An integrative framework. In: System Science (HICSS), 2012 45th Hawaii International Conference. Maui, Hawaii, USA; 2012. pp. 2289-2297

[4] Mohanty SP, Choppali U, Kougianos E. Everything you wanted to know about smart cities: The internet of things is the backbone. IEEE Consumer Electronics Magazine. 2016;**5**:60-70

[5] Africansmartcities. African Cities—On the path to become Globally Competitive. 2023. Available from: https://africansmartcities.info/

[6] Cook M, Valdez A-M. Exploring smart city atmospheres: The case of Milton Keynes. Geoforum. 2021;**127**:180-188

[7] Kono N, Suwa A, Ahmad S. Smart cities in Japan and their application in developing countries. In: Low Carbon Urban Infrastructure Investment in Asian Cities. London: Springer; 2016. pp. 95-122

[8] Brummitt CD, Hines PDH, Dobson I, Moore C, D'Souza RM. Transdisciplinary electric power grid science. Proceedings of the National Academy of Sciences of the United States of America. 2013;**110**:12159-12159

[9] Williams JB. The electric century. In: The Electric Century. London: Springer; 2018. pp. 205-208

[10] Taulo JL, Gondwe KJ, Sebitosi AB. Energy supply in Malawi: Options and issues. Journal of energy in Southern Africa. 2015;**26**:19-32

[11] Alaqeel TA, Suryanarayanan S. A comprehensive cost-benefit analysis of the penetration of smart grid technologies in the Saudi Arabian electricity infrastructure. Utilities Policy. 2019;**60**:100933

[12] Bari A, Jiang J, Saad W, Jaekel A. Challenges in the smart grid applications: An overview. International Journal of Distributed Sensor Networks. 2014;**10**:974682

[13] Sustainable Energy Africa. Sustainable Energy Solutions for South African Local Government. 2019. Available from: https://www.sustainable.org.za/userfiles/smart%20grids(1).pdf

[14] South African National Energy Development Institute. Smart Grid. South Africa: SASGI; 2020. Available from: https://www.sanedi.org.za/Smart-Grid.html

[15] Shrestha M, Johansen C, Noll J, Roverso D. A methodology for security classification applied to smart grid infrastructures. International Journal of Critical Infrastructure Protection. 2020;**28**:100342

[16] Majhi SK, Patra G, Dhal SK. Cyber physical systems & public utility in India: State of art. Procedia Computer Science. 2016;**78**:777-781

[17] The United Nations Human Settlements Programme, Global Urban Safety Indicators and Monitoring Tool, 2020

[18] Rosales N. Towards the modeling of sustainability into urban planning: Using

indicators to build sustainable cities. Procedia Engineering. 2011;**21**:641-647

[19] United nations. Ensure Access to Affordable, Reliable, Sustainable and Modern Energy. 2020. Available from: https://www.un.org/sustainabledevelopment/energy/

[20] Sahana M, Hong H, Sajjad H. Analyzing urban spatial patterns and trend of urban growth using urban sprawl matrix: A study on Kolkata urban agglomeration, India. Science of the Total Environment. 2018;**628**:1557-1566

[21] Kumar A, Pandey AC, Hoda N, Jeyaseelan A. Evaluation of urban sprawl pattern in the tribal-dominated cities of Jharkhand state, India. International Journal of Remote Sensing. 2011;**32**:7651-7675

[22] Gashu K, Feyisa G. Challenges of Urban Plan Implementation in Small Towns of West Showa Zone, Oromia Regional State, Ethiopia. London, UK: Intech Open; 2022

[23] Wang X, Shi R, Zhou Y. Dynamics of urban sprawl and sustainable development in China. Socio-Economic Planning Sciences. 2020;**70**:100736

[24] Andersson K, Otoo M, Nolasco M. Innovative sanitation approaches could address multiple development challenges. Water Science and Technology. 2018;**77**:855-858

[25] World Bank. Solid Waste Management. Washington, D.C., USA; 2023. Available from: https://www.worldbank.org/en/topic/urbandevelopment/brief/solid-waste-management

[26] Kaza S, Yao L, Bhada-Tata P, Van Woerden F. What a Waste 2.0: A Global Snapshot of Solid Waste Management to 2050. Washington, D.C., USA: World Bank Publications; 2018

[27] Panagopoulos T, Duque JAG, Dan MB. Urban planning with respect to environmental quality and human well-being. Environmental Pollution. 2016;**208**:137-144

[28] McGrane SJ. Impacts of urbanisation on hydrological and water quality dynamics, and urban water management: A review. Hydrological Sciences Journal. 2016;**61**:2295-2311

[29] Li X, Zhao Y, Shi C, Sha J, Wang Z-L, Wang Y. Application of water evaluation and planning (WEAP) model for water resources management strategy estimation in coastal Binhai new area, China. Ocean & Coastal Management. 2015;**106**:97-109

[30] Sandu Z, Sukiasyan N. Overpopulation of India: Factors, implications and recommendations. International Journal of Humanities, Art and Social Studies (IJHAS). 2018;**3**(2):1-8

[31] Fayomi G, Mini S, Fayomi O, Oyeleke O, Omole D, Akinwumi I. Overview of industrial pollution activities and its curbing mechanisms. In: IOP Conference Series: Materials Science and Engineering. Bristol, England: IOP Publishing; 2019. p. 012097

[32] International Energy Agency. Latest news, commentaries and reports. Global Energy Review: CO2 Emissions in 2021. 2023. Retrieved 24/1/2023, 2023, from: https://www.iea.org/

[33] United nations. Achieving Universal Access to Electricity Developed. 2020. Available from: https://sustainabledevelopment.un.org/content/documents/17462PB1.pdf

[34] Russo D, Miketa A. Benefits, challenges, and analytical approaches to

scaling up renewables through regional planning and coordination of power systems in Africa. Current Sustainable/ Renewable Energy Reports. 2019;**6**:5-12

[35] European Commission. What are Smart Cities? 2023. Available from: https://commission.europa.eu/eu-regional-and-urban-development/topics/cities-and-urban-development/city-initiatives/smart-cities_en

[36] Ohtsuki T. A smart city based on ambient intelligence. IEICE Transactions on Communications. 2017;**100**:1547-1553

[37] Hung K, Zhang Y-T, Tai B. Wearable medical devices for tele-home healthcare. In: Engineering in Medicine and Biology Society. IEMBS'04. 26th Annual International Conference of the IEEE. San Francisco, CA, USA; 2004. pp. 5384-5387

[38] Catarinucci L, De Donno D, Mainetti L, Palano L, Patrono L, Stefanizzi ML, et al. An IoT-aware architecture for smart healthcare systems. IEEE Internet of Things Journal. 2015;**2**:515-526

[39] Pereira GV, Parycek P, Falco E, Kleinhans R. Smart governance in the context of smart cities: A literature review. Information Polity. 2018;**23**:143-162

[40] Huawei. Smart Cities Index 2017. 2023. Available from: https://e.huawei.com/uk/special_topic/solution/smart_cities_index_2017

[41] Larbi M, Kellett J, Palazzo E, Mehdipour A. Urban sustainability transitions in two frontrunner cities: Insights from the multi-level perspective. Planning Practice & Research. 2021;**36**:494-513

[42] Yigitcanlar T, Kankanamge N, Vella K. How are smart city concepts and technologies perceived and utilized? A systematic geo-twitter analysis of smart cities in Australia. Journal of Urban Technology. 2021;**28**:135-154

[43] Hu R. The state of smart cities in China: The case of Shenzhen. Energies. 2019;**12**:4375

[44] Johnston K. A comparison of two smart cities: Singapore & Atlanta. Journal of Comparative Urban Law and Policy. 2019;**3**:191

[45] Wang W, Lu Z. Cyber security in the smart grid: Survey and challenges. Computer Networks. 2013;**57**:1344-1371

[46] National Institute of Standards and Technology. Introduction to NISTIR 7628 Guidelines for Smart Grid Cyber Security. United states of America: National Institute of Standards and Technology; 2010

[47] National Institute of Standards and Technology. Guidelines for Smart Grid Cybersecurity. United States of America: National Institute of Standards and Technology; 2014

[48] Dileep G. A survey on smart grid technologies and applications. Renewable Energy. 2020;**146**:2589-2625

[49] National Institute of Standards and Technology. NIST Framework and Roadmap for Smart Grid Interoperability Standards, Release 2.0. United States of America: NIST; 2012

[50] Antunes CH, Dias L, Dantas G, Mathias J, Zamboni L. An application of soft systems methodology in the evaluation of policies and incentive actions to promote technological innovations in the electricity sector. Energy Procedia. 2016;**106**:258-278

[51] Shaukat N, Ali SM, Mehmood CA, Khan B, Jawad M, Farid U, et al. A

survey on consumers empowerment, communication technologies, and renewable generation penetration within smart grid. Renewable and Sustainable Energy Reviews. 2018;**81**:1453-1475

[52] Sanchez-Hidalgo M-A, Cano M-D. A survey on visual data representation for smart grids control and monitoring. Sustainable Energy, Grids and Networks. 2018;**16**:351-369

[53] Camachi BE, Ichim L, Popescu D. Cyber security of smart grid infrastructure. In: 2018 IEEE 12th International Symposium on Applied Computational Intelligence and Informatics (SACI). Timisoara, Romania: IEEE; 2018. pp. 000303-000308

[54] Menon DM, Radhika N. Design of a secure architecture for last mile communication in smart grid systems. Procedia Technology. 2015;**21**:125-131

[55] Guo C, Jiang X, Choo K-KR, Tang X, Zhang J. Lightweight privacy preserving data aggregation with batch verification for smart grid. Future Generation Computer Systems. 2020;**112**:512-523

[56] He D, Kumar N, Lee J-H. Privacy-preserving data aggregation scheme against internal attackers in smart grids. Wireless Networks. 2016;**22**:491-502

[57] Molina-Markham A, Shenoy P, Fu K, Cecchet E, Irwin D. Private memoirs of a smart meter. In: Proceedings of the 2nd ACM Workshop on Embedded Sensing Systems for Energy-Efficiency in Building. New York, United States: Association for Computing Machinery; 2010. pp. 61-66

[58] Gope P. PMAKE: Privacy-aware multi-factor authenticated key establishment scheme for advance metering infrastructure in smart grid. Computer Communications. 2020;**152**:338-344

[59] de Jesus Martins R, Knob LAD, da Silva EG, Wickboldt JA, Schaeffer-Filho A, Granville LZ. Specialized CSIRT for incident response Management in Smart Grids. Journal of Network and Systems Management. 2019;**27**:269-285

[60] Abdelaziz AK, Tamani N, Ghamri-Doudane Y, Karabadji NEI. Anomaly-based framework for detecting power overloading cyberattacks in smart grid AMI. Computers & Security. 2020;**96**:101896

[61] Mishra S, Li X, Kuhnle A, Thai MT, Seo J. Rate alteration attacks in smart grid. In: 2015 IEEE Conference on Computer Communications (INFOCOM). Hong Kong, China: IEEE; 2015. pp. 2353-2361

[62] Chen P-Y, Yang S, McCann JA, Lin J, Yang XJICM. Detection of false data injection attacks in smart-grid systems. IEEE Communications Magazine. 2015;**53**:206-213

[63] Sgouras KI, Kyriakidis AN, Labridis DP. Short-term risk assessment of botnet attacks on advanced metering infrastructure. IET Cyber-Physical Systems: Theory & Applications. 2017;**2**:143-151

[64] Kimani K, Oduol V, Langat K. Cyber security challenges for IoT-based smart grid networks. International Journal of Critical Infrastructure Protection. 2019;**25**:36-49

[65] Jin D, Nicol DM, Yan G. An event buffer flooding attack in DNP3 controlled SCADA systems. In: Proceedings of the 2011 Winter Simulation Conference (WSC). Phoenix, USA: IEEE; 2011. pp. 2614-2626

[66] Deng R, Xiao G, Lu R. Defending against false data injection attacks on power system state estimation. IEEE Transactions on Industrial Informatics. 2015;**13**:198-207

[67] Sun C-C, Hahn A, Liu C-C. Cyber security of a power grid: State-of-the-art. International Journal of Electrical Power & Energy Systems. 2018;**99**:45-56

[68] Akyol BA, Kirkham H, Clements SL, Hadley MD. A Survey of Wireless Communications for the Electric Power System. Richland, WA (United States): Pacific Northwest National Lab (PNNL); 2010

[69] Baloglu UB, Demir Y. Lightweight privacy-preserving data aggregation scheme for smart grid metering infrastructure protection. International Journal of Critical Infrastructure Protection. 2018;**22**:16-24

[70] Gunduz MZ, Das R. Analysis of cyber-attacks on smart grid applications. In: 2018 International Conference on Artificial Intelligence and Data Processing (IDAP). Malatya, Turkey: IEEE; 2018. pp. 1-5

[71] Huseinović A, Mrdović S, Bicakci K, Uludag S. A survey of denial-of-service attacks and solutions in the smart grid. IEEE Access. 2020;**8**:177447-177470

[72] Ma M, Teixeira AMH, van den Berg J, Palensky P. Voltage control in distributed generation under measurement falsification attacks**this work is sponsored by Chinese scholarship council (CSC). IFAC-PapersOnLine. 2017;**50**:8379-8384

[73] Fadul JE, Hopkinson KM, Andel TR, Sheffield CA. A trust-management toolkit for smart-grid protection systems. IEEE Transactions on Power Delivery. 2014;**29**:1768-1779